Thermoelectric Materials and Devices

RSC Energy and Environment Series

Titles in the Series:

1: Thermochemical Conversion of Biomass to Liquid Fuels and Chemicals
2: Innovations in Fuel Cell Technologies
3: Energy Crops
4: Chemical and Biochemical Catalysis for Next Generation Biofuels
5: Molecular Solar Fuels
6: Catalysts for Alcohol-Fuelled Direct Oxidation Fuel Cells
7: Solid Oxide Fuel Cells: From Materials to System Modeling
8: Solar Energy Conversion: Dynamics of Interfacial Electron and Excitation Transfer
9: Photoelectrochemical Water Splitting: Materials, Processes and Architectures
10: Biological Conversion of Biomass for Fuels and Chemicals: Explorations from Natural Utilization Systems
11: Advanced Concepts in Photovoltaics
12: Materials Challenges: Inorganic Photovoltaic Solar Energy
13: Catalytic Hydrogenation for Biomass Valorization
14: Photocatalysis: Fundamentals and Perspectives
15: Photocatalysis: Applications
16: Unconventional Thin Film Photovoltaics
17: Thermoelectric Materials and Devices

How to obtain future titles on publication:
A standing order plan is available for this series. A standing order will bring delivery of each new volume immediately on publication.

For further information please contact:
Book Sales Department, Royal Society of Chemistry, Thomas Graham House, Science Park, Milton Road, Cambridge, CB4 0WF, UK
Telephone: +44 (0)1223 420066, Fax: +44 (0)1223 420247
Email: booksales@rsc.org
Visit our website at www.rsc.org/books

Thermoelectric Materials and Devices

Edited by

Iris Nandhakumar
University of Southampton, UK
Email: iris@soton.ac.uk

Neil M. White
University of Southampton, UK
Email: nmw@ecs.soton.ac.uk

Stephen Beeby
University of Southampton, UK
Email: spb@ecs.soton.ac.uk

ROYAL SOCIETY OF CHEMISTRY

THE QUEEN'S AWARDS
FOR ENTERPRISE:
INTERNATIONAL TRADE
2013

RSC Energy and Environment Series No. 17

Print ISBN: 978-1-78262-323-6
PDF eISBN: 978-1-78262-404-2
EPUB eISBN: 978-1-78262-902-3
ISSN: 2044-0774

A catalogue record for this book is available from the British Library

Published by The Royal Society of Chemistry,
Thomas Graham House, Science Park, Milton Road,
Cambridge CB4 0WF, UK

Registered Charity Number 207890

For further information see our web site at www.rsc.org

Printed in the United Kingdom by CPI Group (UK) Ltd, Croydon, CR0 4YY, UK

Preface

The publication of this book is motivated by a number of recent and ground breaking advances in the field of thermoelectrics (TE), which has seen a renaissance in recent years. Thermoelectric devices have the ability to convert waste heat into useful electricity, which potentially makes TE a transformative technology for renewable energy generation. Despite this enormous potential, historically thermoelectrics have only been utilised in niche market applications, such as the powering of satellites and space probes due to the low efficiency of current TE generators. However, nanotechnologies in particular have caused a paradigm shift in thermoelectric performance through nanostructuring/nanoengineering of materials, enabling thermoelectrics to be catapulted from niche product to mainstream. Indeed, the future is extremely bright for TE technology and thermal energy harvesting is predicted to become a global multibillion dollar market by 2020 in a range of application scenarios. The current focus on fuel efficiency in vehicles and stringent legislations on carbon emissions are the main drivers for the market growth of TE technology by automobile manufacturers to extract electricity from the hot exhaust stream of cars.

This book aims to provide an up-to-date topical perspective on thermoelectric materials, fabrication and characterization, devices and applications to capture the seismic shifts that have recently occurred in this field within the UK and beyond. Whilst thermoelectrics are well-established within the US, Japan and China, they are still in their infancy within the UK and this book is intended as a reflection that many of the recent advances and innovations within this field have resulted from activities within the UK thermoelectrics community, whilst not trying to exclusively focus on them. In fact, the intention is to provide a balanced view of topics such as thermoelectric materials, devices and applications.

RSC Energy and Environment Series No. 17
Thermoelectric Materials and Devices
Edited by Iris Nandhakumar, Neil M. White and Stephen Beeby

Published by the Royal Society of Chemistry, www.rsc.org

Bismuth telluride and lead telluride have been the core materials used for commercial TE generators targeted at low and medium power generation and refrigeration since the 1950s and new materials are clearly needed to disrupt the sustainability issues associated with tellurium as well as to address the toxicity of lead telluride. The first three chapters provide an excellent overview of recent developments in materials synthesis that include zintl phases (Kauzlarich *et al.*), thermoelectric oxides (Freer *et al.*) and novel metal chalcogenides (Powell *et al.*) that hold great promise for future thermoelectric applications on the basis of their enhanced thermoelectric performance and being composed of abundant and sustainable elements. A fundamental and detailed theory of how nanostructures affect the electric conductivities, thermal conductivities and Seebeck coefficients in nanostructures is presented in Chapter 4 (Paul) that also focuses on how enhanced performance can be achieved through low dimensional structures. The importance of the complexity of the phenomenon and the uncertainties associated with existing thermoelectric measurement methods is highlighted in Chapter 5 (Cuenat *et al.*), which is followed by a description of novel high-throughput thermoelectric measurement techniques in Chapter 6 (Gao *et al.*) that enable rapid characterization of thermoelectric materials and devices. Stobart *et al.* provide a comprehensive assessment of all aspects associated with the practical design considerations of thermoelectric generators, starting from models based on the physical parameters and empirical correlations for the heat exchange processes to provide a reliable basis for design choices that lead on to numerical models supported by experimental validation to provide the next level of refinement in the formulation of design guidelines. Naylor *et al.* illustrate how electrochemical deposition methods can be employed as a low temperature and low-cost route without the need for a vacuum or extensive equipment for the fabrication of thermoelectric materials, in contrast to commonly employed high-cost approaches such as MBE and MOCVDE. The final chapter by Simpson *et al.* discusses the tremendous opportunities for thermoelectric devices in automotive power-harvesting applications with a particular emphasis on systems for electrical energy generation from automobile exhaust gases.

I would like to thank the authors of this book and my co-editors Professors Neil White and Stephen Beeby who have contributed to the making of this book, which was an elaborate and time-consuming process. I would wish for this book to become an important landmark in the thermoelectrics field that goes well beyond a 'laboratory handbook' but provides a fresh perspective of this important research field that deserves to be mainstream.

Iris Nandhakumar
Southampton

Contents

RSC Energy and Environment Series No. 17
Thermoelectric Materials and Devices
Edited by Iris Nandhakumar, Neil M. White and Stephen Beeby

Published by the Royal Society of Chemistry, www.rsc.org

CHAPTER 1

Zintl Phases: Recent Developments in Thermoelectrics and Future Outlook

SUSAN M. KAUZLARICH,*[a] ALEX ZEVALKINK,[b] ERIC TOBERER[c] AND G. JEFF SNYDER[d]

[a] Department of Chemistry, University of California, One Shields Ave, Davis, CA 95616, USA; [b] Max-Planck-Institut für Chemische Physik fester Stoffe, Nöthnitzer Straße 40, 01187 Dresden, Germany; [c] Department of Physics, Colorado School of Mines, 1500 Illinois St., Golden, Colorado 80401, USA; [d] Materials Science and Engineering, Northwestern University, Evanston, IL 60208, USA
*Email: smkauzlarich@ucdavis.edu

1.1 Introduction

1.1.1 Definition of Zintl Phases

The term Zintl phase was first used by F. Laves[1] to indicate a subset of compounds within the general class of intermetallics, named after Eduard Zintl, a German scientist who was the first to systematically prepare and structurally characterize these phases.[2] Zintl's interest was in determining what combination of elements would form salt-like structures, focusing on the heavier elements of group 13, 14 and 15. During this time,

RSC Energy and Environment Series No. 17
Thermoelectric Materials and Devices
Edited by Iris Nandhakumar, Neil M. White and Stephen Beeby

Published by the Royal Society of Chemistry, www.rsc.org

Hume-Rothery and Westgren had demonstrated the correlation of structure and valence electron concentration for intermetallics, and the compounds that Zintl studied did not fall within those structure–electron count rules. As more compounds were discovered and the transition between salt-like and metallic structures became less clear, this definition proved to be too limiting. Schäfer, Eisenmann, and Müller[3] proposed a more general definition where electron transfer is essentially complete between the alkali or alkaline earth cation and the electronegative elements that utilize the electrons such that they achieve a filled valence either by covalent bonding or by the formation of lone pairs of electrons. Therefore, these phases exhibit salt-like characteristics from the ionic bonding between the cation and the anionic unit. The anionic unit can be isolated anions; if there are not enough electrons for a filled octet, then they form covalent bonds and polyanionic units.[4] The Zintl concept provides a simple idea concerning ionic and covalent bonding within intermetallic phases, allowing for a simple description of bonding that provides insight into the structure and properties of intermetallic phases.[3,5,6] One simple way to define a Zintl phase was articulated by Nesper[7] and Miller[8] as the following: there exists a well-defined relationship between chemical and electronic structures in a Zintl phase and a chemist can understand the structure by using simple electron counting rules.

1.1.2 Charge Counting/Formal Valence Rules

Since Zintl phases fall between insulators and metals, it can be difficult to devise a consistent set of rules governing their classification. The salt-like nature of these phases often results in high melting points, high heats of formation, poor conductivities and greater brittleness than many intermetallics. There is the requirement for a well-defined relationship between their chemical and electronic structures, but that can also be a difficult criterion to implement since it requires a detailed knowledge of the structure and bonding. This particular criterion implies that Zintl phases are line compounds with narrow homogeneity widths, a restriction that would make this classification of compounds difficult to dope or manipulate electronically, and therefore uninteresting to those pursuing optimization of thermoelectric properties. In order to understand how to think about these phases, let's start by restating the obvious: all Zintl phases are composed of an electropositive atom, which is treated like a fully ionized element that provides its electrons to the more electronegative elements in the structure. These electronegative elements either use those electrons to form a closed-shell ion or, if there are not enough electrons for this, to form bonds in order to achieve a full octet of electrons.[9] These compounds are distinguished from insulators by the size of the band gap[10,11] and are typically considered to be semiconductors.

In binary Zintl phases, A_aX_x (A = electropositive metal, X = electronegative element) $8x$ electrons are required in order to achieve an octet (or closed shell) for the x X atoms, e(A) and e(X) are the number of valence electrons of A and X.

$$a \cdot e(\mathrm{A}) + x \cdot e(\mathrm{X}) = 8x \tag{1.1}$$

Typically, it is assumed that there are no bonds between A atoms and that X can have X–X bonds that are considered to be two-centre, two-electron bonds, and that the octet rule is satisfied for both elements. If this is the case, then the number of the valence electron count per formula unit of A_aX_x (VEC) is:

$$(\mathrm{VEC}) = a \cdot e(\mathrm{A}) = x \cdot e(\mathrm{X}) = x \cdot \left(\frac{a}{x} e(\mathrm{A}) + e(\mathrm{X})\right). \tag{1.2}$$

While this equation provides the VEC for the compound formula, the term in the parentheses represents the average number of valence electrons per anion, N_x. In general, this results in the classical valence rule for insulators—the $8 - N$ rule proposed by Mooser and Pearson—which provides the number of covalent bonds required to satisfy the anion valence.[12] W. Klemm proposed an additional nuance where the more electronegative partner is described as an element with the same number of electrons: a pseudoatom concept.[3] Consider the charged $X^{(ae_A/x)-}$ unit: if N_x is non-integral, then a set of pseudoatoms is required to describe the observed coordination environments. For example, heteroanions with tetrahedral units can be described as the analogous orthooxosilicate or germanate anions: $[SiP_4]^{8-}$, $[SiAs_4]^{8-}$ or $[GeP_4]^{8-}$ and $[GeAs_4]^{8-}$ where the large formal charge is balanced by means of the alkaline earth metal cation. The combination of Zintl's original proposal and Klemm's pseudoatom description is now called the Zintl–Klemm concept. Based on this electron counting model, these compounds should all be semiconductors. However, the difference between insulators and semiconductors is somewhat arbitrarily based on the bandgap and there are suggestions in the literature of either 2.5 eV[10] or 2.0 eV.[7] Because of the simple electron counting scheme, the Zintl–Klemm concept is a powerful tool for the assessment of complex main group solids and there are a number of groups working to put this on firm theoretical grounds.[13–18]

The incorporation of transition metals into these structures adds complexity and has expanded the original criteria.[9,19–23] Some of the first research in this area focused on transition metal containing compounds that are isostructural to known main group Zintl compounds.[9,20,24] Analogous to main group compounds, these compounds contained anionic units that showed isoelectronic relationships with transition metal halides and chalcogenides. The Zintl–Klemm idea of bonding has been successfully used to probe changes in electronics and bonding within the $ThCr_2Si_2$ structure type.[19,25–28] Using the Zintl concept, totally new compounds have been prepared and novel properties obtained.[9,29–32] The addition of both transition metals and rare earth ions have expanded this area considerably.[33–37]

1.1.3 Thermoelectric Zintl Compounds

Good thermoelectric materials have low electrical resistivity, low thermal conductivity and a large Seebeck coefficient. Typically, small band-gap semiconductors with carrier concentrations within 10^{19}–10^{21} cm^{-3} work better than metals or insulators. Also, a large unit cell, heavy atoms, and structural complexity generally result in good thermoelectric (TE) efficiency.

Many Zintl materials fulfill these qualifications; however, relatively few were investigated for their thermoelectric properties until the last few years.[38–40] Since these compounds are valence precise and possess the requisite small band gap as well as having complex structures, it is expected that new thermoelectrics with high efficiency may be discovered. The potential of this area for further research is demonstrated so far with a few structure types that will be described below.

1.1.3.1 *$Yb_{14}MnSb_{11}$ Structure Type*

$Yb_{14}MnSb_{11}$ is one of the first recognized phases with excellent thermoelectric performance ($zT \sim 1$ at 1200 K).[38] This compound is of the $Ca_{14}AlSb_{11}$ structure type (shown in Figure 1.1).[41] A variety of compounds of this structure type have been reported for the heavier alkaline earth cations (Ca, Sr, Ba) with Al, Ga, In, Zn, Cd, Mn, and Nb and the pnictides from P to Bi.[9,20,22,24,41–56] The rare earths, Eu and Yb, have also been prepared, along with isovalent solid solutions of a variety of compounds.[21,34–36,38,57–77] This structure type has also been prepared with a small amount of Ln^{3+} replacing some of the Yb^{2+} and a small amount of Te replacing some Sb in $Yb_{14}MnSb_{11}$.[78–80] In general, all of the compounds can be synthesized by reacting the elements in sealed Nb or Ta tubes. The tubes are sealed in fused silica tubes either under vacuum or 0.25 atmosphere of Ar and heated to

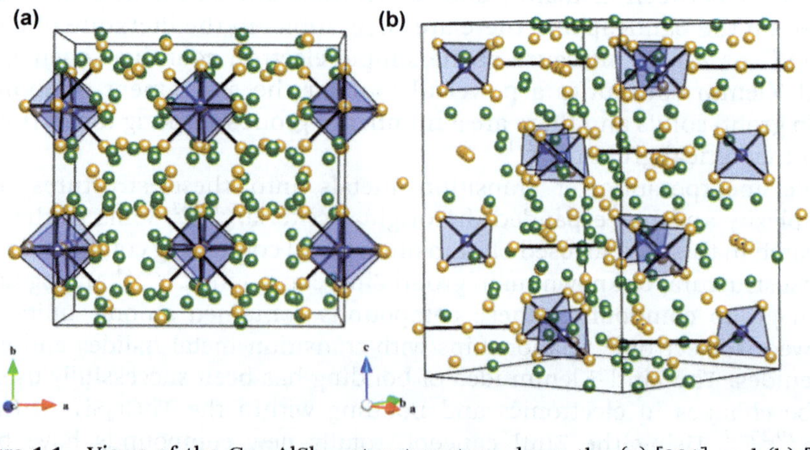

Figure 1.1 Views of the $Ca_{14}AlSb_{11}$ structure type down the (a) [001] and (b) [101] direction. Ca, Al, and Sb are indicated by the green, blue and gold spheres, respectively, and the tetrahedron is shaded.

temperatures of up to 1250 °C for periods of 24–300 h. There is evidence for reaction of the Nb tubes with some of the elements,[53] so Ta is considered the more ideal container. Quantitative yields can be obtained at low temperatures for some of the phases, but X-ray quality single crystals are formed only at higher temperatures. $Yb_{14}MnSb_{11}$ can be prepared by high temperature inductive heating.[81] In addition to direct reaction of the elements, large crystals can be prepared from Sn flux; this work provided the initial steps forward for preparing large amounts of phase pure material for detailed property measurements, including thermoelectric properties.[58,82]

The compounds crystallize in the tetragonal system, $I4_1/acd$ ($Z=8$). In the framework of valence rules, the structure of these compounds can be understood according to one formula unit ($A_{14}MPn_{11}$) corresponding to $14A^{2+}$ cations + 4 Pn^{3-} anions + MPn_4^{9-} tetrahedron + Pn_3^{7-} linear unit. The tetrahedron has $\bar{4}$ point symmetry and is translated by half along the *c* axis, while the Pn_3 units have 222 point symmetry and are staggered by 90° along the *c* axis with respect to each other. The isolated Pn atoms are 6-coordinated by cations and are located between the tetrahedral and the Pn_3 units, forming a spiral along a screw axis coincident with the *c* axis. The linear anion, Pn_3^{7-} unit, can be either symmetric or asymmetric, depending on both the identity of the cation and anion. The lighter pnictides with large cations tend to be asymmetric. The structure can also be related to the Cu_2O structure type, but forming two interpenetrating networks. The oxygen atoms are substituted by the tetrahedron and the Cu atoms are substituted by the pnictide octahedron. The central Pn atoms of the linear polyanions connect the nets with the remaining cations.[53,63,64]

1.1.3.2 $CaAl_2Si_2$ Structure Type

The $CaAl_2Si_2$ structure type (space group $P\bar{3}m1$, $Z=1$) is prevalent for compounds with the AB_2X_2 composition.[83] There has been extensive research into compounds of this structure type.[84–103] An apparent requirement for this structure type is that the B atoms are either main group or transition metals with d^0, d^5, or d^{10} electronic configuration.[104,105] Several of the compounds with the $CaAl_2Si_2$ structure type have been shown to have zT values near unity.[96,106–110] In particular, the solid solution of $Ca_{1-x}Yb_xZn_2Sb_2$ was first shown to have promise.[106] The layered structure can be described as consisting of two adjacent puckered hexagonal nets of alternating Al and Si. The A^{n+} cations alternate between the $[B_2X_x]^{n-}$ double layers (Figure 1.2). Structural relationships to other structure types, such as $BaAl_3$ and $CaSi_2$, have been described.[104]

1.1.3.3 Sr_3GaSb_3 Structure Type

Compounds with the A_3MPn_3 stoichiometry can crystallize in four different structure types. Sr_3GaSb_3 crystallizes in the monoclinic crystal system (space group $P2_1/n$)[111] and the three other structure types include orthorhombic

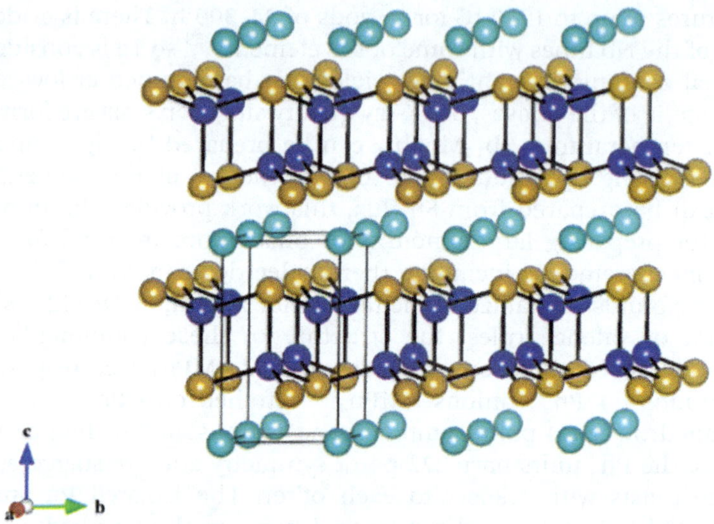

Figure 1.2 A view of the structure of $YbZn_2Sb_2$ ($CaAl_2Si_2$ structure type) showing the unit cell. Yb, Zn, and Sb are indicated by sky blue, blue and gold spheres, respectively.

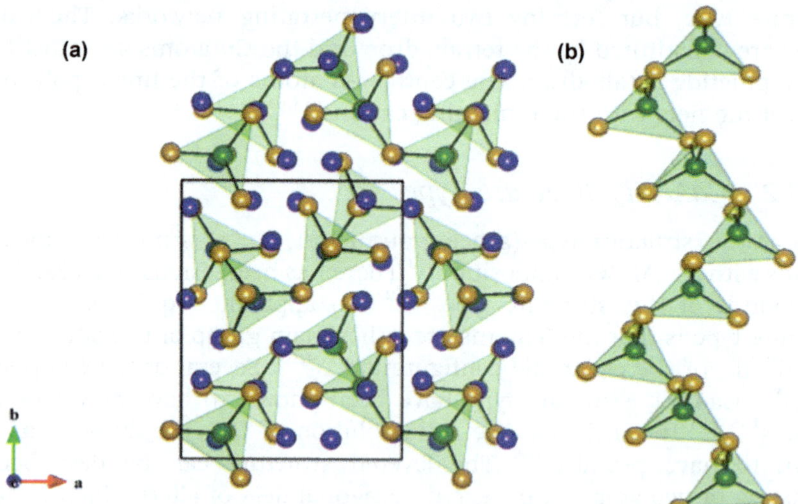

Figure 1.3 Views of Sr_3GaSb_3, showing (a) the unit cell and (b) the connectivity of the chain. The Sr, Ga, and Sb atoms are represented by blue, green and gold spheres, respectively.

Ba_3GaSb_3 ($Z=8$) and Ca_3InP_3 ($Z=4$) both in space group *Pnma*[112] and Ba_3AlSb_3, which crystallizes in the space group *Cmca*, $Z=8$.[113] All structures of A_3MPn_3 contain MPn_4 tetrahedra and the structures differ depending on how these tetrahedra are connected. Sr_3GaSb_3, shown in Figure 1.3, contains

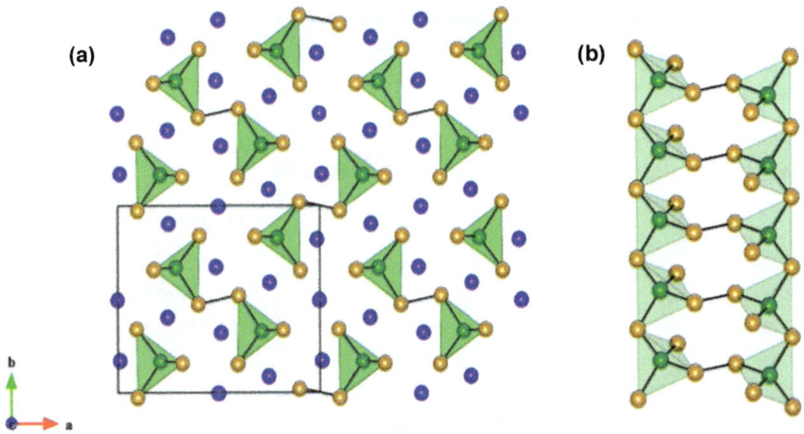

Figure 1.4 Views of $Ca_5Al_2Sb_6$ showing (a) the unit cell and (b) the connectivity of the chains. Ca, Al, and Sb atoms are represented by blue, green and gold spheres, respectively. The tetrahedral polyhedra are shaded.

infinite non-linear chains of corner sharing tetrahedra.[111,114] Various combinations of A = Ca, Sr, Ba, Eu; M = Al, Ga, In; Pn = P, As, Sb have been reported; however, a systematic study has not been performed. For example, A_3GaSb_3 (A = Sr, Ba)[111,112] have been reported, but Ca_3GaSb_3 or Eu_3GaSb_3 have not. The only Eu containing phase reported is Eu_3InP_3, which is described as very air-sensitive with high electrical resistivity.[115] To date, the thermoelectric properties of Ca_3AlSb_3, Sr_3GaSb_3, and Sr_3AlSb_3 have been reported.[116–118]

1.1.3.4 $Ca_5Ga_2As_6$ Structure Type

The Zintl compound, $Ca_5Al_2Sb_6$,[119] is well known and crystallizes in the $Ca_5Ga_2As_6$ structure type (space group *Pbam*, $Z = 2$)[120] Compounds of $Ca_5M_2Sb_6$ formula (with M = Al, Ga, In) have been prepared and all belong to this structure type.[119,121] Similar to the Sr_3GaSb_3 compound, this structure type is another example of MSb_4 tetrahedra that are connected in a unique fashion. The structure type is composed of chains of corner-linked MSb_4 tetrahedra, connected through Sb–Sb bonds to form ladder-like moieties (Figure 1.4).

1.2 Thermal Properties

The thermal conductivity of Zintl compounds is found to be extremely low, particularly at high temperature. This low thermal conductivity generally overcomes the lower electronic mobility found in these materials and provides the foundation for high zT.[122,123] Additionally, the complex structures enable phonon behaviour not observed in more simple lattices.[124,125]

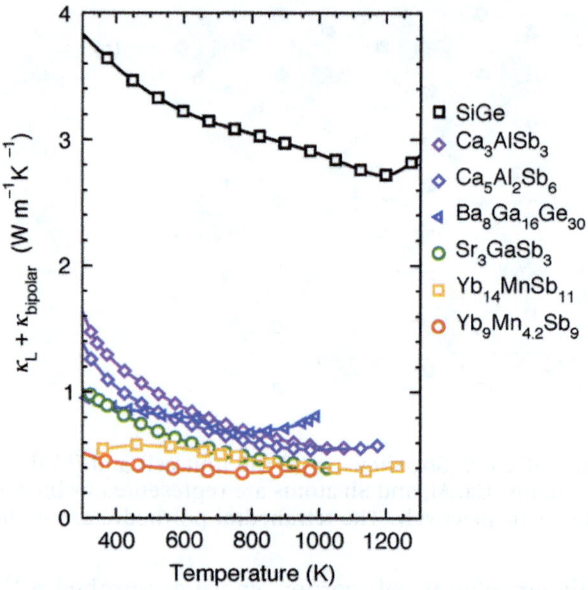

Figure 1.5 The experimental κ_L for many Zintl compounds is much lower than that observed in the heritage SiGe material used in radioisotope thermoelectric generators.

Figure 1.5 shows that the lattice thermal conductivity, κ_L, for many Zintl compounds is significantly below that of heritage $Si_{0.8}Ge_{0.2}$.[117,126–131] Near room temperature, many of these materials show a decaying lattice thermal conductivity due to increased phonon–phonon scattering, while others exhibit thermal conductivity values with little temperature dependence. The latter behaviour will be explored below. Many of these materials display a slight upturn in thermal conductivity near their maximum measurement temperature. At these high temperatures, minority carrier contributions to the conductivity become significant, leading to a bipolar contribution.[132] The standard Wiedemann–Franz approach to removing the electronic contribution does not capture this contribution.

1.2.1 Theory behind Low κ_L in Complex Materials

A simple kinetic theory approach to phonon transport and heat conduction reveals that the three primary descriptors of thermal conductivity are the volumetric heat capacity (C_v), phonon group velocity (v) and mean free path between collisions (l).[133,134] As Zintl compounds are largely dense materials, there is relatively minor variation in the heat capacity. However, the group velocity and phonon relaxation time ($\tau_{ph} = l/v$) are critically dependent on material composition and crystal structure.

The phonon relaxation time is determined by several scattering sources, including phonon–phonon, defect, and boundary scattering. In most materials at high temperature, Umklapp (phonon–phonon) scattering is the dominant scattering mechanism. Shown in Figure 1.5, κ_L in $Ca_5Al_2Sb_6$ provides an excellent example of the $1/T$ temperature dependence expected when Umklapp scattering dominates.[128] Umklapp scattering is governed by the anharmonicity of the lattice, typically parameterized by the Grüneisen parameter.[135] The Grüneisen parameter has been characterized using thermal expansion measurements for few Zintl compounds to date, and moderate values between 1.5–2 have been found.[136] However, in 'rattling' compounds, the mode-specific Grüneisen parameter can be quite high near the rattling frequency and leads to increased scattering of phonons in that frequency regime. Such mode-specific values are obtained from temperature dependent inelastic scattering measurements.

In addition to Umklapp scattering, point defect scattering[137] in Zintl compounds has been harnessed to scatter high frequency phonons. One early example is $Yb_{1-x}Ca_xZn_2Sb_2$.[138] Here, the thermal conductivity responds as predicted from classic thermal scattering theory (Figure 1.6), with the most substantial reduction to κ_L occurring at 300 K near $x = 0.5$. Likewise, $Yb_9Mn_{4.2}Sb_9$ has Mn interstitials that act as scattering sources and also reduce κ_L at room temperature.[131] To scatter low frequency phonons, boundary scattering from grain boundaries, nanostructures, for example, can be used.[139] Despite the success of nanostructures in reducing κ_L in other materials, little work has been devoted to nano-structuring Zintl

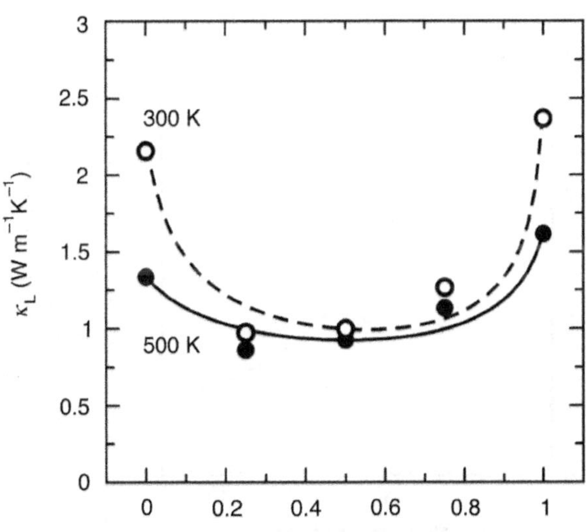

Figure 1.6 Alloying has a profound effect on the room temperature thermal conductivity of $Yb_{1-x}Ca_xZn_2Sb_2$; the effect is less pronounced at high temperature due to increased Umklapp scattering.
Reprinted with permission from E. S. Toberer, A. F. May and G. J. Snyder, *Chem. Mater.*, 2010, **22**, 624–634.[122] Copyright (2010) American Chemical Society.

compounds. Extended defects (stacking faults) have been observed in the layered Zintl compound $SrZnSb_2$, and have been shown to significantly reduce κ_L near room temperature.[140]

Phonon group velocity, ν_g, has been found to play a critical role in explaining the low thermal conductivity in Zintl compounds.[136] Like many thermoelectric materials, the low frequency limit of ν_g, (speed of sound) in many Zintl compounds is quite low, arising from a combination of heavy atoms and soft bonding between covalent moieties. Furthermore, Zintl compounds often possess large mass contrast, which increases the gap between phonon modes at the Brillouin zone edges and at the zone centre.[141] Such large gaps flatten the phonon modes and lead to reduced group velocity. When this phenomenon is combined with extremely complex structures (here, complexity refers to the number of atoms within the primitive cell), the group velocity throughout much of the dispersion can be significantly suppressed. As the size of the real-space cell grows, the Brillouin zone shrinks, and the fraction of optical phonon modes with low velocity increases rapidly. As the number of atoms in the primitive cell (N) increases, the acoustic contribution to κ_L shrinks, scaling roughly as $1/N^{1/3}$. Figure 1.7 shows the experimental κ_L in selected Zintl phases as a function of N, illustrating the benefit of a large unit cell and correspondingly small contribution from acoustic phonons. Further reduction in the acoustic group velocity can be achieved through the inclusion of 'rattling' modes that flatten the phonon dispersion near the resonant frequency of the rattling species.[142–144]

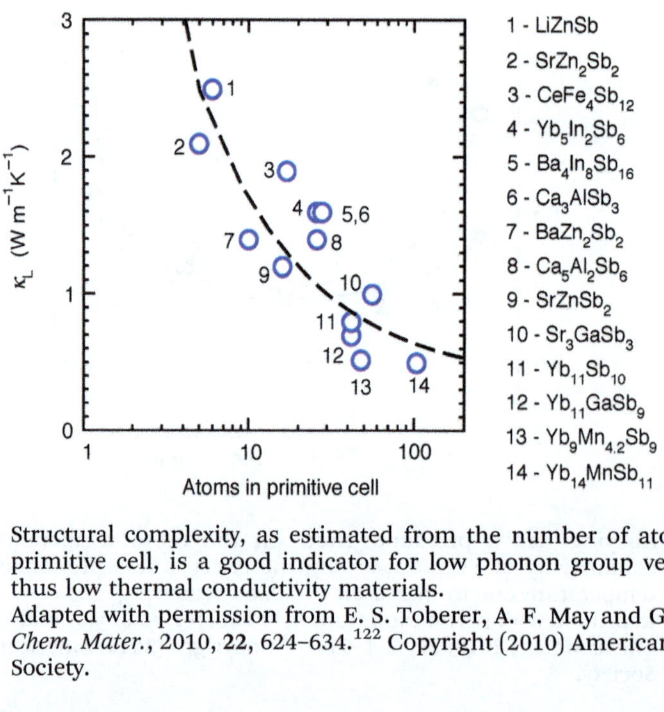

Figure 1.7 Structural complexity, as estimated from the number of atoms in the primitive cell, is a good indicator for low phonon group velocity, and thus low thermal conductivity materials.
Adapted with permission from E. S. Toberer, A. F. May and G. J. Snyder, *Chem. Mater.*, 2010, **22**, 624–634.[122] Copyright (2010) American Chemical Society.

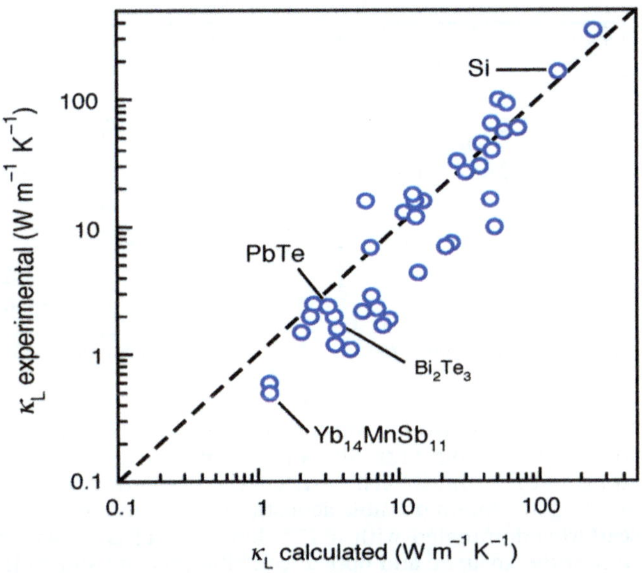

Figure 1.8 The experimental κ_L can be readily predicted (within a factor of 2) from knowledge of speed of sound, Grüneisen parameter, atomic density, cell volume and average atomic mass. The current approach thus successfully captures the underlying relevant physics.

At high temperature, the optical branches in complex materials can be approximated *via* Cahill's minimum thermal conductivity description (κ_{min}),[145] and the remaining phonons in the acoustic branches can be approximated *via* a Debye–Callaway model.[134] We have extensively considered the accuracy of this 'Debye + κ_{min}' approach in ref. 136. Experimental measurements of the thermal conductivity were compared with κ_L predicted using experimental measurements of the speed of sound, Grüneisen parameter, atomic density, cell volume and average atomic mass. The results of this approach are shown in Figure 1.8 for all materials where a complete property characterization was available from the literature. This approach was found to be predictive to a factor of two on average for the compounds in this study.

1.2.2 Case Studies

1.2.2.1 Complex Antimonides – $SrZn_2Sb_2$ vs. $Yb_{14}MnSb_{11}$

The role of complexity can be observed nicely by considering a pair of Zintl compounds: $SrZn_2Sb_2$ and $Yb_{14}MnSb_{11}$.[146,147] In Figure 1.9, we show the experimental lattice thermal conductivity for these two compounds, as well theoretical curves for the acoustic, optical, and total contributions to κ_L. From the Cahill theory of minimum thermal conductivity, we can estimate a lower bound on the optical phonon contribution to the lattice thermal

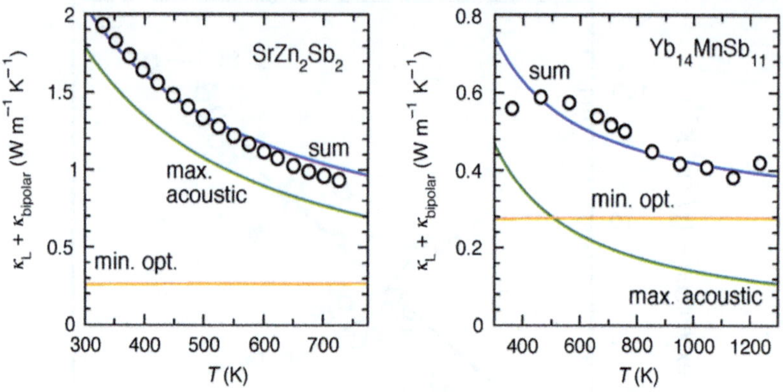

Figure 1.9 The minimum optical mode contribution to the thermal conductivity of structurally simple $SrZn_2Sb_2$ and complex $Yb_{14}MnSb_{11}$ (five and 104 atoms in the primitive cell, respectively) is shown in orange. The corresponding maximum possible acoustic contribution (modeled as Umklapp scattering-dominated with a T^{-1} dependence) is shown in green. The sum of the acoustic and optical contributions are shown in blue. While acoustic modes are typically considered the primary vehicle for thermal conduction, the optical modes represent the dominant source of thermal conduction at high temperature in structurally complex materials.

conductivity. In the case of $SrZn_2Sb_2$, which has only five atoms in the primitive cell, the optical contribution is fairly minor compared to the experimental κ_L. In contrast, $Yb_{14}MnSb_{11}$ has 104 atoms in the primitive cell and a corresponding large relative optical contribution. At high temperatures, the optical contribution accounts for more than half of the total thermal conductivity. We emphasize that this model predicts the minimum contribution from the optical modes and is likely an underestimate. From this understanding, the maximum acoustic contribution can be estimated, assuming Umklapp scattering dominates the phonon transport. As shown in Figure 1.9, κL in the relatively simple $SrZn_2Sb_2$ is comprised primarily of the acoustic contribution, while in $Yb_{14}MnSb_{11}$ the acoustic contribution becomes almost insignificant, particularly at very high temperatures. Thus, there is much greater reduction potential available for scattering the acoustic phonons of $SrZn_2Sb_2$ than in more structurally complex solids like $Yb_{14}MnSb_{11}$.

1.2.2.2 $SrZnSb_2$ vs. *$SrZn_2Sb_2$ – Stacking Faults*

The simplified consideration of acoustic and optical phonons presented above is a helpful way to discriminate between low- and high-frequency phonons. However, classical scattering theory provides a more nuanced view of phonon behaviour. In particular, it can be helpful to consider the frequency dependence of phonon scattering when more than one scattering source is present. The two closely related Zintl phases, $SrZnSb_2$ and

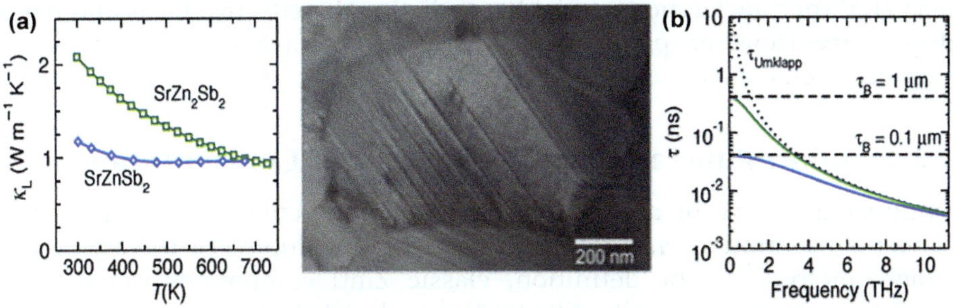

Figure 1.10 (a) The lattice thermal conductivity of $SrZnSb_2$ is significantly lower than $SrZn_2Sb_2$ (left). Bright-field TEM indicates planar defects in $SrZnSb_2$ (right) but not in $SrZn_2Sb_2$ (not shown). (b) This difference in κ_L can be successfully modeled using a Debye–Callaway approach that includes both Umklapp and boundary scattering of the acoustic branches. Here, the boundary scattering is limited by grain boundaries ($\sim$1 μm) for $SrZn_2Sb_2$ and stacking faults ($\sim$100 nm) for $SrZnSb_2$.

$SrZn_2Sb_2$, provide an excellent example. Despite extremely similar chemical structures, these two compounds exhibit a factor of two difference in thermal conductivity (Figure 1.10) at room temperature; by 600 K this difference disappears.[146,148] Here, $SrZnSb_2$ is found by transmission electron microscopy (TEM) to possess a high density of stacking faults.[140] Qualitatively, the expectation is that these stacking faults significantly decrease thermal conductivity at room temperature, but the effect is washed out by Umklapp scattering at high temperature.

To develop a more quantitative assessment of these stacking faults, we turn to classic phonon transport theory and the Debye–Callaway model. To begin, the Umklapp scattering is treated as an identical contribution in these two chemically similar materials. Based on the difference in grain structure, the boundary scattering mean free path is approximated as 0.1 and 1 μm for $SrZnSb_2$ and $SrZn_2Sb_2$, respectively. These calculations reveal that the decreased relaxation time due to boundary scattering reduces thermal conductivity by approximately 30% at room temperature. This impact can be seen in Figure 1.10(b), with the significantly lower relaxation time due to enhanced boundary scattering.

1.3 Electronic Transport

Electronic transport in Zintl compounds varies widely, stemming from large differences in the electronic band structures, carrier relaxation times and carrier concentration (n). While the relationship between the thermal properties of Zintl compounds and readily available crystallographic information is relatively well understood, the electronic properties are often more difficult to predict from the structure alone. Fortunately, calculated electronic structures are becoming increasingly available due to improvements in computational capabilities and can be used to gauge the potential of a new Zintl thermoelectric material by predicting the effective mass and band degeneracy. However, prediction of the electron relaxation time remains a significant challenge.

1.3.1 Controlling and Optimizing Carrier Concentration

Optimizing the zT of a thermoelectric material requires controlling the carrier concentration, n, through either electronic doping or by varying the stoichiometry.[122,123] By definition, classic Zintl compounds are valence precise, suggesting a fully filled valence band and empty conduction band.[149,150] Provided the bandgap is greater than zero, Zintl compounds generally behave as intrinsic semiconductors with very low n (less than 10^{19} carriers per cm^3), low electronic conductivity (σ) and high Seebeck coefficients (α). Doping with aliovalent elements leads to a valence imbalance and an increased concentration of free carriers, allowing for a transition from semiconducting to metallic behavior. By rationally optimizing the electronic properties, large improvements in zT relative to undoped Zintl phase material can be achieved.[117,127,128,147]

The $Ca_5M_2Sb_6$ system (M = Al, Ga, or In) provides a straightforward example of controlling and optimizing the carrier concentration *via* doping. $Ca_5M_2Sb_6$ compounds are valence precise when M is a trivalent element, and, consistent with the Zintl formalism, samples with M = Al, Ga or In have similar, low carrier concentrations ($\sim 10^{18}$ h^+ cm^{-3}).[151] As in most Zintl phases,[122] the $Ca_5M_2Sb_6$ system is naturally slightly p-type and apparently can only be doped p-type. Substituting divalent elements such as Mn or Zn on the trivalent M site leads to an electron deficiency and the resulting carrier concentration can be predicted as a function of dopant content (x) by assuming that each substitution yields one free hole (see dashed line in Figure 1.11(a)). Experimentally, we find that Zn-doping in the $Ca_5In_{2-x}Zn_xSb_6$ system increases the Hall carrier concentration as expected at low concentrations of Zn, while the doping effectiveness is reduced above $x = 0.05$ (shown as symbols in Figure 1.11(a)),[152] Similar trends were observed in Zn-doped $Ca_5Al_2Sb_6$ and $Ca_5Ga_2Sb_6$[153,154] and in the substitution of a monovalent element (Na^+) on the divalent cation site in $Ca_5Al_2Sb_6$.[127]

The transition from non-degenerate to degenerate semiconducting behaviour is illustrated for $Ca_5In_{2-x}Zn_xSb_6$ in Figure 1.11(b) by the decreasing α and increasing σ with increasing Zn content (x). As the Zn concentration increases, the peak zT increases from 0.1 to 0.7 when $x = 0.1$ (Figure 1.11(c)).

In several doped Zintl phases, a single parabolic band model (SPB model) has been used to approximate the dependence of α and σ on carrier concentration (see dashed curves in Figure 1.11(b)) by assuming that the band mass (m^*) is invariant with doping (rigid band approximation), and that acoustic phonons are the predominant scattering mechanism. The figure of merit can then be modeled as a function of carrier concentration if κ_L is assumed to be independent of doping, reducing the need to synthesize a large number of samples. In the $Ca_5In_{2-x}Zn_xSb_6$ system, a SPB model

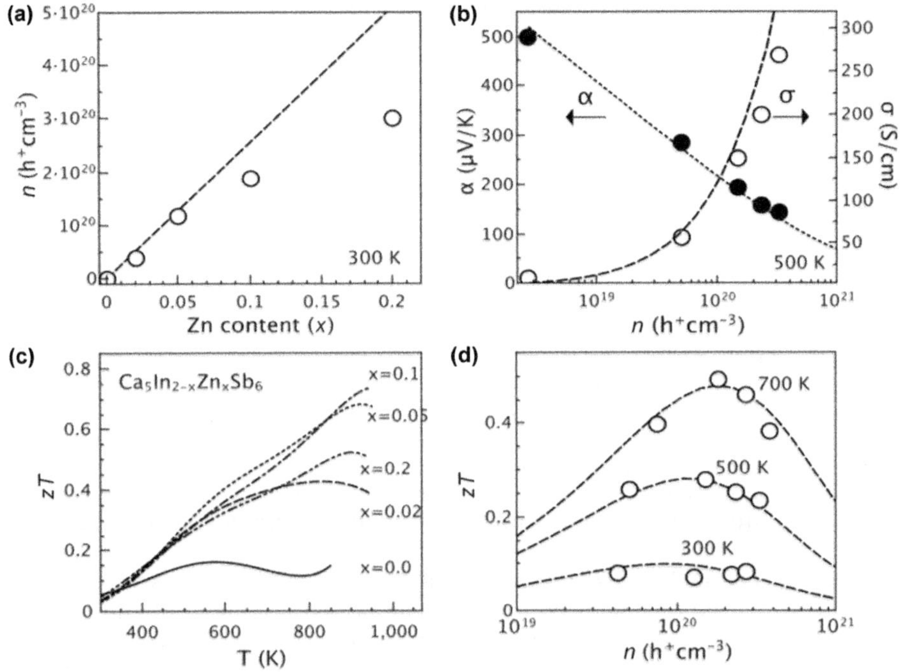

Figure 1.11 (a) Doping $Ca_5In_2Sb_6$ with Zn^{2+} on the In^{3+} site leads to an increase in p-type carrier concentration and (b) a transition from non-degenerate to degenerate semiconducting behaviour. Dashed lines were generated using an SPB model with $m^* = 2m_e$ and $m_o = 6.9$ cm^2 V^{-1}. (c) This leads to an increase from a peak zT of 0.2 to $zT = 0.7$ in extrinsically doped samples. (d) The optimum carrier concentration predicted using a SPB model at 700 K corresponds to $x = 0.05$.

accurately describes the relationship between zT and n, as shown in Figure 1.11(d) for several temperatures. A more detailed description of the SPB model and its limitations can be found in ref. 129.

1.3.2 Limits to Controlling Carrier Concentration

In traditional Zintl phases, the maximum carrier concentration that can be achieved is often limited by the solubility of the dopant. However, the chemistry and electronic structure of the Zintl phase also plays a role. The charge imbalance in a heavily doped material impacts the stability of the structure, potentially leading to the spontaneous formation of charge-compensating defects.[155]

Some Zintl phases are able to tolerate very large dopant concentrations and show highly degenerate behaviour, while in others, doping is very limited (Figure 1.12). For example, the $Yb_{14}Al_{1-x}Mn_xSb_{11}$ system, the most successful high temperature p-type thermoelectric material to date,[156] tolerates $n_H > 10^{21}$ h^+ cm^{-3}. $A_{14}MSb_{11}$ compounds are valence precise

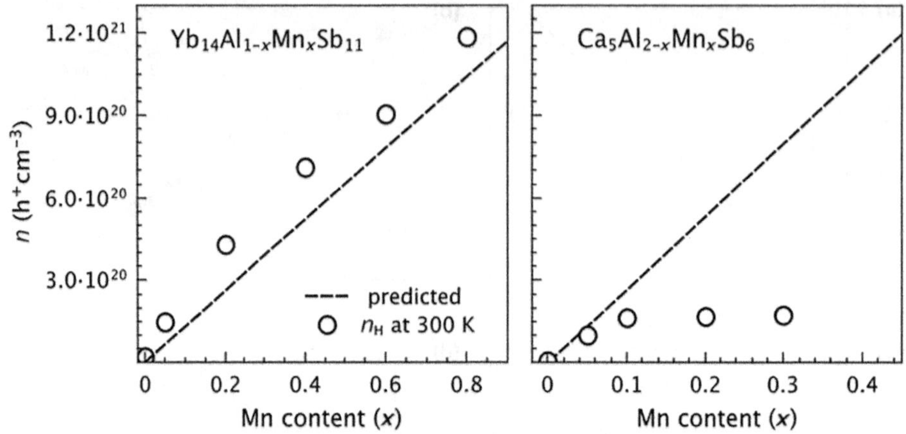

Figure 1.12 Left panel: The carrier concentration in $Yb_{14}Al_{1-x}Mn_xSb_{11}$ increases linearly with increasing x. Right panel: In contrast, the substitution of Mn on the Al site in $Ca_5Al_2Sb_6$ is limited to very low concentrations.

when the A ion is 2+ and the M ion is 3+. $Yb_{14}MnSb_{11}$ (in which Mn is divalent) can thus be regarded as a heavily doped analogue of the valence precise Zintl phase, $Yb_{14}AlSb_{11}$.[147] The carrier concentration in the Mn-analogue ($n_H = 1.2 \times 10^{21}$ h^+ cm^{-3}) is consistent with simple charge counting predictions, assuming that each Mn substitution leads to one free hole. This high doping tolerance in $A_{14}MSb_{11}$compounds may be partly explained by the presence of several heavy bands (and thus a large density of states) slightly below the valence band edge.[157] In contrast, the $Ca_5Al_2Sb_6$ system tolerates only 5 atomic % Mn on the Al site, leading to a maximum carrier concentration of 2×10^{20} h^+ cm^{-3}.[158]

Stretching the limit of the Zintl formalism are compounds that cannot be synthesized with a valence precise composition, such as $Yb_9Mn_{4+x}Sb_9$. The $Yb_9Mn_{4+x}Sb_9$ system is valence precise when $x = 0.5$. However, when x is greater than zero, the additional Mn atoms must occupy interstitial lattice sites.[159,160] The competing energetic requirements of charge balance and structural stability lead to an electron deficient composition of approximately $Yb_9Mn_{4.2}Sb_9$. The resulting high p-type carrier concentration (2×10^{20} h^+ cm^{-3}) is very close to the optimal n predicted by an SPB model, leading to excellent thermoelectric performance without the need for doping.[131]

1.3.3 Band Structure Requirements

When selecting and optimizing Zintl phases for thermoelectric applications, several key characteristics of the electronic band structure should be considered. The most critical requirement is the presence of a band gap ($E_g > 0$), which allows for a single carrier type to dominate transport. The magnitude of E_g determines the temperature at which minority carrier effects begin to negatively impact thermoelectric performance (*e.g.*, bipolar thermal conductivity and degradation of the Seebeck coefficient). Generally the band gap in Zintl phases arises due to the energy difference between the bonding and anti-bonding states of the covalently bonded anions.[161] Many Zintl compounds are semiconductors with small band gaps ($E_g < 1$ eV), although semi-metallic Zintl phases have also been reported ($E_g < 0$ eV).[162] Good thermoelectric performance is most often found in phases with band gaps in the range of 0.2 and 0.6 eV.

While a band gap is the only strictly necessary band structure criterion, the band mass and the number of bands involved in transport (band degeneracy, N_v) also play an important role in determining thermoelectric performance. In the optimization of thermoelectric materials, the effective mass presents a fundamental conflict: while a light inertial effective mass (m_I^*) along the direction of transport is desired for high mobility, a heavy 'density of states effective mass' (m_{DOS}^*) is necessary for a high Seebeck coefficient. The density of states effective mass is related to the band degeneracy and the inertial effective mass by $m_{DOS}^* = m_i^* N_v^{2/3}$. Thus, when N_v is large and m_i^* is small, a high Seebeck coefficient and high mobility can be obtained simultaneously.[163] When electrons are scattered primarily by acoustic phonons, as is the case in most known thermoelectric materials, the improvement in μ conferred by a light band mass outweighs the detrimental effect on α.

The presence of N_v and m_i^* in the thermoelectric quality factor highlights the value of calculated electronic structures. The electronic densities of states of many Zintl compounds have been reported, providing estimates of the magnitude of E_g. However, the electronic band structures from which both N_v and m_i^* can be obtained are often omitted in published computational studies. To date, band structures have been published for only a handful of Zintl compounds, including $Ca_{14}MnBi_{11}$, $Ca_5M_2Sb_6$ (M = Al, Ga, In), Sr_3GaSb_3, Ca_3AlSb_3, and several AZn_2Sb_2 compounds (A = Sr, Ca, Yb, Eu). Among these reported band structures, the number of bands involved in p-type transport is typically between one and three. This is far less than that of PbTe and its alloys, which have up to $N_v = 16$.[163] This disparity is partly explained by the higher symmetry of the latter compounds, which leads to multiple symmetrically equivalent carrier pockets.[164] However, there is no theoretical limitation on N_v in Zintl compounds, suggesting that a computational survey may reveal Zintl phases with high N_v.

Because electronic mobility is higher in covalent materials than in ionic materials, it is often assumed that charge transport in Zintl compounds occurs preferentially through the anionic framework. This is supported by the influence that the dimensionality and orientation of the covalently bonded substructure appears to have on the anisotropy of the electronic band structure in some Zintl phases. For example, in $Ca_5Al_2Sb_6$, transport parallel to the chains of corner-linked $AlSb_4$ tetrahedra exhibits the lightest band mass.[117] In AZn_2Sb_2, the lightest bands correspond to the directions within the plane of the covalent Zn_2Sb_2 slabs.[146] In contrast, $Ca_{14}MnBi_{11}$, which is characterized by isolated $MnBi_4$ tetrahedra, has a relatively isotropic band dispersion with no clearly preferred direction for charge transport.[165] It is not yet clear, however, whether transport along the anionic framework is controlled by the covalent bonds themselves, or by the 'non-bonding' states that comprise the valence band edge. Anisotropic transport has not been confirmed experimentally due to the difficulty of growing sufficiently large single crystal or oriented polycrystalline samples.

1.3.4 Carrier Relaxation Time

In cases when the band mass cannot account for the variations in mobility observed experimentally, the carrier relaxation time (τ) must be considered. τ a critical but often-overlooked transport parameter, and the relationship between τ and structure and chemistry of Zintl phases is not yet well understood.

The AZn_2Sb_2 series, where A = Sr, Ca, Eu, or Yb, provides a good example of the influence of the carrier relaxation time. The electronic mobility in AZn_2Sb_2 compounds is very high relative to most Zintl phases, in part due to relatively light m_i^*. Transport data suggests that m_{DOS}^* in AZn_2Sb_2 remains constant regardless of the choice of cation, as illustrated by the Pisarenko curve (Figure 1.13) generated using an SPB model and an effective mass of $m^* = 0.6\ m_e$.[146] Despite the similar m_{DOS}^* across the AZn_2Sb_2 series, the carrier mobility varies drastically. The mobility is found to be significantly higher in the rare earth containing analogues than in the alkaline earth analogues (Figure 1.13).[146] A recent investigation of AZn_2Sb_2 single crystals (included in Figure 1.13), by May *et al.* suggests that this trend is an intrinsic material property, rather than a result of processing.[166] May *et al.* observed that the Debye temperature is the only material parameter that correlates

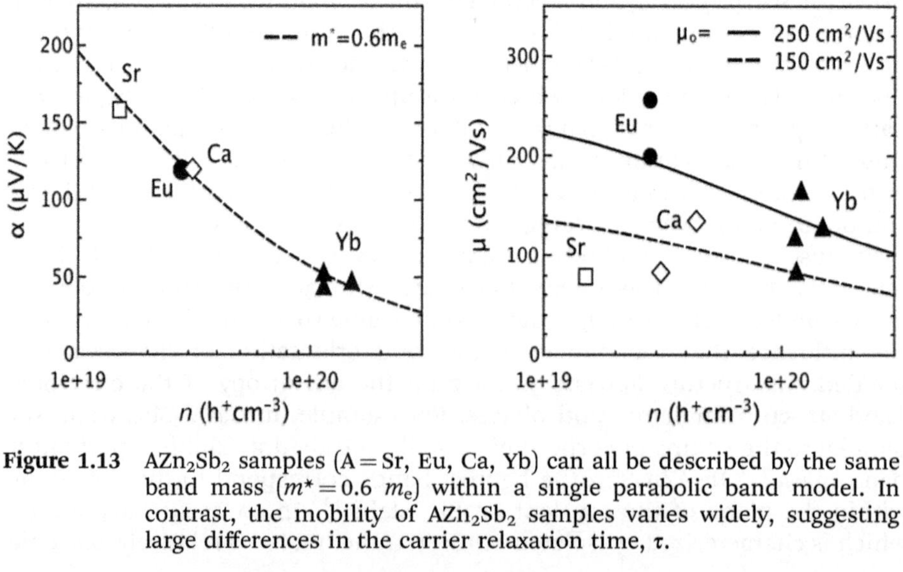

Figure 1.13 AZn_2Sb_2 samples (A = Sr, Eu, Ca, Yb) can all be described by the same band mass ($m^* = 0.6\ m_e$) within a single parabolic band model. In contrast, the mobility of AZn_2Sb_2 samples varies widely, suggesting large differences in the carrier relaxation time, τ.

with mobility, suggesting a potential route to understanding and perhaps controlling τ.

1.4 Future Opportunities for Zintl Thermoelectric Materials

Zintl phases are successful thermoelectric materials due to their complex structures and bonding, which leads to highly tunable properties, and at times exceptionally low, glass-like lattice thermal conductivity. To date, several Zintl compounds have been reported with zT values in excess of unity, including $Yb_{14}MnSb_{11}$, YbM_2Sb_2 (M = Zn, Cd), and many others have promising figures of merit. Among compounds that are known to have good thermoelectric properties, strategies such as band structure engineering through chemical substitutions should be pursued to further improve zT.

One of the most exciting aspects of the field is the vast potential for future discovery of new thermoelectric materials, not only for high temperature but also for moderate temperatures. The Zintl compounds optimized for thermoelectric applications described herein are only the tip of the iceberg. Many Zintl phases have promising electronic properties, but have not yet been optimized as thermoelectric materials. For example, $BaGa_2Sb_2$, $Ba_4In_8Sb_{16}$, $Eu_{11}Cd_6Sb_{12}$, $Eu_7Ga_6Sb_8$, $Eu_{10}Mn_6Sb_{13}$, $Ba_2Sn_3Sb_6$, $Ba_3Ga_4Sb_5$, and $Ba_2In_5As_5$ exhibit unique, complex structures and appear to have energy gaps according to their electronic structures.[167–172] Additionally, there is no reason to restrict these materials to high temperature applications; it is possible that high zT will be discovered at lower temperatures providing opportunities for refrigeration. Furthermore, the vast majority of known Zintl compounds have not been studied beyond the initial crystallographic characterization, providing a wealth of additional thermoelectric material candidates.[149]

Acknowledgements

The authors thank NSF DMR-1405973 (SMK) and NASA Science Missions Directorate's Radioisotope Power Systems Technology Advancement Program through NASA/JPL (AW, ET, and GJS).

References

1. R. Kniep, in *Chemistry, Structure, and Bonding of Zintl Phases and Ions*, ed. S. M. Kauzarich, VCH Publishers, Inc, New York, 1996, pp. xvii–xxx.
2. *Chemistry, Structure, and Bonding of Zintl Phases and Ions*, ed. S. M. Kauzlarich, VCH Publishers, Inc, New York, 1996.
3. H. Schäfer, B. Eisenmann and W. Müller, *Angew. Chem., Int. Ed. Engl.*, 1973, **12**, 694–712.
4. B. Eisenmann and G. Cordier, *Chemistry, Structure, and Bonding of Zintl Phases and Ions*, ed. S. M. Kauzlarich, VCH Publishers, New York, 1996, pp. 61–137.

5. H. Schäfer and B. Eisenmann, *Rev. Inorg. Chem.*, 1981, **3**, 29.
6. H. Schäfer, *Annu. Rev. Mater. Sci.*, 1985, **15**, 1–42.
7. R. Nesper, *Prog. Sol. State Chem.*, 1990, **20**, 1–45.
8. G. J. Miller, in *Chemistry, Structure, and Bonding of Zintl Phases and Ions*, ed. S. M. Kauzlarich, VCH, New York, 1996, pp. 1–59.
9. S. M. Kauzlarich, *Comments Inorg. Chem.*, 1990, **10**, 75–88.
10. W. B. Pearson, *The Crystal Chemistry and Physics of Metals and Alloys*, Wiley-Interscience, New York, 1972.
11. R. Nesper, *Angew. Chem., Int. Ed. Engl.*, 1991, **30**, 789.
12. E. Mooser and W. B. Pearson, in *Progress in Semiconductors*, ed. Gibson, Kröger and Burgess, Wiley & Sons Inc., New York, 1960, vol. 5, p. 103.
13. L. H. Yang, C. D. Consorte, C. Y. Fong, J. E. Pask, E. Nabighian, S. M. Kauzlarich and J. S. Nelson, *Chem. Mater.*, 1998, **10**, 4025–4029.
14. R. F. Gallup, C. Y. Fong and S. M. Kauzlarich, *Inorg. Chem.*, 1992, **31**, 115–118.
15. P. Alemany, S. Alvarez and R. Hoffmann, *Inorg. Chem.*, 1990, **29**, 3070–3073.
16. P. C. Schmidt, D. Stahl, B. Eisenmann, R. Kniep, V. Eyert and J. Kubler, *J. Solid State Chem.*, 1992, **97**, 93–104.
17. G. H. Grosch and K. J. Range, *J. Alloys .Compd.*, 1996, **235**, 250–255.
18. G. J. Miller, M. W. Schmidt, F. Wang and T.-S. You, *Struct. Bonding*, 2011, **139**, 1–55.
19. R. Hoffmann and C. Zheng, *J. Phys. Chem.*, 1985, **89**, 4175–4181.
20. S. M. Kauzlarich, in *Chemistry, Structure, and Bonding of Zintl Phases and Ions*, ed. S. M. Kauzlarich, VCH Publishers, Inc., New York, 1996, pp. 245–274.
21. B. R. Taylor, G. A. Fox, L. J. Hope-Weeks, R. S. Maxwell, S. M. Kauzlarich and H. W. H. Lee, *Mater. Sci. Eng.*, 2002, **B96**, 90–93.
22. D. Sánchez-Portal, R. M. Martin, S. M. Kauzlarich and W. E. Pickett, *Phys. Rev. B: Condens. Matter Mater. Phys.*, 2002, **65**, 144414.
23. S.-Q. Xia and S. Bobev, *J. Comput. Chem.*, 2008, **29**, 2125–2133.
24. S. M. Kauzlarich, T. Y. Kuromoto and M. M. Olmstead, *J. Am. Chem. Soc.*, 1989, **111**, 8041–8042.
25. S. L. Brock, J. E. Greedan and S. M. Kauzlarich, *J. Solid State Chem.*, 1994, **109**, 416–418.
26. Z. N. Chong and R. Hoffmann, *J. Solid State Chem.*, 1988, **72**, 58–71.
27. S. Jia and R. J. Cava, *Phys. Rev. B: Condens. Matter Mater. Phys.*, 2010, **82**, 180410.
28. S. Jia, P. Jiramongkolchai, M. R. Suchomel, B. H. Toby, J. G. Checkelsky, N. P. Ong and R. J. Cava, *Nat. Phys.*, 2011, **7**, 207–210.
29. A. P. Holm, M. M. Olmstead and S. M. Kauzlarich, *Inorg. Chem.*, 2003, **42**, 1973–1981.
30. F. Gascoin and S. C. Sevov, *Inorg. Chem.*, 2002, **41**, 5920–5924.
31. F. Gascoin and S. C. Sevov, *Inorg. Chem.*, 2003, **42**, 85567–88571.

32. F. Gascoin and S. C. Sevov, Ternary and Quaternary Niobium Arsenide Zintl Phases, in *Inorganic Chemistry in Focus: III*, ed. G. Meyer, D. Naumann and L. Wesemann, Wiley-VCH, 2006, pp. 195–206.
33. A. Rehr and S. M. Kauzlarich, *J. Alloys Compd.*, 1994, **207**, 424–426.
34. J. Y. Chan, S. M. Kauzlarich, P. Klavins, R. N. Shelton and D. J. Webb, *Chem. Mater.*, 1997, **9**, 3132–3135.
35. J. Y. Chan, M. E. Wang, A. Rehr, S. M. Kauzlarich and D. J. Webb, *Chem. Mater.*, 1997, **9**, 2131–2138.
36. J. Y. Chan, M. M. Olmstead, S. M. Kauzlarich and D. J. Webb, *Chem. Mater.*, 1998, **10**, 3583–3588.
37. A. P. Holm, S. M. Park, C. L. Condron, M. M. Olmstead, H. Kim, P. Klavins, F. Grandjean, R. P. Hermann, G. J. Long, M. G. Kanatzidis, S. M. Kauzlarich and S. J. Kim, *Inorg. Chem.*, 2003, **42**, 4660–4667.
38. S. R. Brown, S. M. Kauzlarich, F. Gascoin and G. J. Snyder, *Chem. Mater.*, 2006, **18**, 1873–1877.
39. S. M. Kauzlarich, S. R. Brown and G. J. Snyder, *Dalton Trans.*, 2007, 2099–2107.
40. E. S. Toberer, A. F. May and G. J. Snyder, *Chem. Mater.*, 2010, **22**, 624–634.
41. G. Cordier, H. Schäfer and M. Stelter, *Z. Anorg. Allg. Chem.*, 1984, **519**, 183–188.
42. T. Y. Kuromoto, S. M. Kauzlarich and D. J. Webb, *Mol. Cryst. Liq. Cryst.*, 1989, **181**, 349–357.
43. S. M. Kauzlarich and T. Y. Kuromoto, *Croat. Chem. Acta*, 1991, **64**, 343–352.
44. S. M. Kauzlarich, M. M. Thomas, D. A. Odink and M. M. Olmstead, *J. Am. Chem. Soc.*, 1991, **113**, 7205–7208.
45. D. J. Webb, T. Y. Kuromoto and S. M. Kauzlarich, *J. Magn. Magn. Mater.*, 1991, **98**, 71–75.
46. D. J. Webb, T. Y. Kuromoto and S. M. Kauzlarich, *J. Appl. Phys.*, 1991, **69**, 4825.
47. T. Y. Kuromoto, S. M. Kauzlarich and D. J. Webb, *Chem. Mater.*, 1992, **4**, 435–440.
48. D. P. Siemens, J. Del Castillo, W. Potter, D. J. Webb, T. Y. Kuromoto and S. M. Kauzlarich, *Solid State Commun.*, 1992, **84**, 1029–1031.
49. S. L. Brock, L. J. Weston, M. M. Olmstead and S. M. Kauzlarich, *J. Solid State Chem.*, 1993, **107**, 513–523.
50. J. Del Castillo, D. J. Webb, S. M. Kauzlarich and T. Y. Kuromoto, *Phys. Rev. B: Condens. Matter Mater. Phys.*, 1993, **47**, 4849–4852.
51. A. Rehr, T. Y. Kuromoto, S. M. Kauzlarich, J. Del Castillo and D. J. Webb, *Chem. Mater.*, 1994, **6**, 93–99.
52. D. M. Young, C. C. Torardi, M. M. Olmstead and S. M. Kauzlarich, *Chem. Mater.*, 1995, **7**, 93–101.
53. K. Vidyasagar, W. Honle and H. G. von Schnering, *Z. Anorg. Allg. Chem.*, 1996, **622**, 518–524.
54. J. T. Vaughey and J. D. Corbett, *Chem. Mater.*, 1996, **8**, 671–675.

55. H. Kim, Q. Huang, J. W. Lynn and S. M. Kauzlarich, *J. Solid State Chem.*, 2002, **168**, 162–168.
56. H. Kim and S. M. Kauzlarich, *J. Solid State Chem.*, 2005, **178**, 1935–1939.
57. J. Y. Chan, S. M. Kauzlarich, P. Klavins, R. N. Shelton and D. J. Webb, *Phys. Rev. B: Condens. Matter Mater. Phys.*, 1998, **57**, 8103–8106.
58. I. R. Fisher, T. A. Wiener, S. L. Bud'ko, P. C. Canfield, J. Y. Chan and S. M. Kauzlarich, *Phys. Rev. B: Condens. Matter Mater. Phys.*, 1999, **59**, 13829–13834.
59. A. P. Holm, J. Tobin and S. M. Kauzlarich, *Spin Polarized Photemission Study of Rare Earth Transition Metal Zintl Compounds, $Eu_{14}MnSb_{11}$, $Eu_{14}MnBi_{11}$, $Yb_{14}MnSb_{11}$, $Yb_{14}MnBi_{11}$*, Report ALS-00160, LBNL, 1999.
60. I. R. Fisher, S. L. Bud'ko, C. Song, P. C. Canfield, T. C. Ozawa and S. M. Kauzlarich, *Phys. Rev. Lett.*, 2000, **85**, 1120–1123.
61. A. C. Payne, M. M. Olmstead, S. M. Kauzlarich and D. J. Webb, *Chem. Mater.*, 2001, **13**, 1398–1406.
62. A. P. Holm, S. M. Kauzlarich, S. A. Morton, G. D. Waddill, W. E. Pickett and J. G. Tobin, *J. Am. Chem. Soc.*, 2002, **124**, 9894–9898.
63. H. Kim, J. Y. Chan, M. M. Olmstead, P. Klavins, D. J. Webb and S. M. Kauzlarich, *Chem. Mater.*, 2002, **14**, 206–216.
64. H. Kim, P. Klavins and S. M. Kauzlarich, *Chem. Mater.*, 2002, **14**, 2308–2316.
65. H. Kim, M. M. Olmstead, P. Klavins, D. J. Webb and S. M. Kauzlarich, *Chem. Mater.*, 2002, **14**, 3382–3390.
66. E. Ratai, P. Bruins, C. J. Hernandez, S. M. Kauzlarich and M. P. Augustine, *Chem. Mater.*, 2002, **14**, 2467–2475.
67. R. P. Hermann, F. Grandjean, S. M. Kauzlarich, J. Jiang, S. Brown and G. J. Long, *Inorg. Chem.*, 2004, **43**, 7005–7013.
68. A. P. Holm, T. C. Ozawa, S. M. Kauzlarich, S. A. Morton, G. D. Waddill and J. G. Tobin, *J. Solid State Chem.*, 2005, **178**, 262–269.
69. R. P. Hermann, F. Grandjean, D. Kafle, D. E. Brown, C. E. Johnson, S. M. Kauzlarich and G. J. Long, *Inorg. Chem.*, 2007, **46**, 10736–10740.
70. S. R. Brown, E. S. Toberer, T. Ikeda, C. A. Cox, F. Gascoin, S. M. Kauzlarich and G. J. Snyder, *Chem. Mater.*, 2008, **20**, 3412–3419.
71. E. S. Toberer, C. A. Cox, S. R. Brown, T. Ikeda, A. F. May, S. M. Kauzlarich and G. J. Snyder, *Adv. Funct. Mater.*, 2008, **18**, 2795–2800.
72. C. A. Cox, E. S. Toberer, A. A. Levchenko, S. R. Brown, G. J. Snyder, A. Navrotsky and S. M. Kauzlarich, *Chem. Mater.*, 2009, **21**, 1354–1360.
73. C. A. Cox, S. R. Brown, G. J. Snyder and S. M. Kauzlarich, *J. Electron. Mater.*, 2010, **39**, 1373–1375.
74. J. F. Rauscher, C. A. Cox, T. Yi, C. M. Beavers, P. Klavins, E. S. Toberer, G. J. Snyder and S. M. Kauzlarich, *Dalton Trans.*, 2010, **39**, 1055–1062.
75. S. Kastbjerg, C. A. Uvarov, S. M. Kauzlarich, E. Nishibori, M. A. Spackman and B. B. Iversen, *Chem. Mater.*, 2011, **23**, 3723–3730.

76. C. A. Uvarov, J. F. Rauscher and S. M. Kauzlarich, *Sci. Adv. Mater.*, 2011, **3**, 646–651.
77. C. A. Uvarov, F. Ortega-Alvarez and S. M. Kauzlarich, *Inorg. Chem.*, 2012, **51**, 7617–7624.
78. B. C. Sales, P. Khalifah, T. P. Enck, E. J. Nagler, R. E. Sykora, R. Jin and D. Mandrus, *Phys. Rev. B: Condens. Matter Mater. Phys.*, 2005, **72**, 205207.
79. E. S. Toberer, S. R. Brown, T. Ikeda, S. M. Kauzlarich and G. J. Snyder, *Appl. Phys. Lett.*, 2008, **93**, 062110.
80. T. Yi, M. N. Abdusalyamova, F. Makhmudov and S. M. Kauzlarich, *J. Mater. Chem.*, 2012, **22**, 14378–14384.
81. C. Yu, Y. Chen, H. Xie, G. J. Snyder, C. Fu, J. Xu, X. Zhao and T. Zhu, *Appl. Phys. Express*, 2012, **5**, 031801.
82. R. A. Ribeiro and M. A. Avila, *Philos. Mag.*, 2012, **92**, 2492–2507.
83. Z. Chong, R. Hoffmann, R. Nesper and H. G. Von Schnering, *J. Am. Chem. Soc.*, 1986, **108**, 1876–1884.
84. P. Klufers and A. Mewis, *Z. Naturforsch., B: J. Chem. Sci.*, 1977, **32**, 753–756.
85. A. Mewis, *Z. Naturforsch., B: J. Chem. Sci.*, 1978, **33**, 382–384.
86. A. Mewis, *Z. Naturforsch., B: J. Chem. Sci.*, 1978, **33**, 606–609.
87. P. Klufers, H. Neumann, A. Mewis and H. U. Schuster, *Z. Naturforsch., B: J. Chem. Sci.*, 1980, **35**, 1317–1318.
88. A. Mewis, *Z. Naturforsch., B: J. Chem. Sci.*, 1980, **35**, 939–941.
89. A. Mahan and A. Mewis, *Z. Naturforsch., B: J. Chem. Sci.*, 1983, **38**, 1041–1045.
90. P. Klufers and A. Mewis, *Z. Kristallogr.*, 1984, **169**, 135–147.
91. A. Artmann, A. Mewis, M. Roepke and G. Michels, *Z. Anorg. Allg. Chem.*, 1996, **622**, 679–682.
92. C. Kranenberg, D. Johrendt and A. Mewis, *Solid State Sci.*, 2002, **4**, 261–265.
93. F. Wartenberg, C. Kranenberg, R. Pocha, D. Johrendt, A. Mewis, R. D. Hoffmann, B. D. Mosel and R. Pottgen, *Z. Naturforsch., B: J. Chem. Sci.*, 2002, **57**, 1270–1276.
94. A. V. Morozkin, O. Isnard, P. Henry, S. Granovsky, R. Nirmala and P. Manfrinetti, *J. Alloys Compd.*, 2006, **420**, 34–36.
95. E. S. Toberer, A. F. May, B. C. Melot, E. Flage-Larsen and G. J. Snyder, *Dalton Trans.*, 2010, **39**, 1046–1054.
96. H. Zhang, M. Baitinger, M.-B. Tang, Z.-Y. Man, H.-H. Chen, X.-X. Yang, Y. Liu, L. Chen, Y. Grin and J.-T. Zhao, *Dalton Trans.*, 2010, **39**, 1101–1104.
97. P. E. R. Blanchard, S. S. Stoyko, R. G. Cavell and A. Mar, *J. Solid State Chem.*, 2011, **184**, 97–103.
98. A. F. May, M. A. McGuire, D. J. Singh, R. Custelcean and G. E. Jellison, Jr., *Inorg. Chem.*, 2011, **50**, 11127–11133.
99. A. T. Nientiedt, H. Lincke, U. C. Rodewald, R. Poettgen and W. Jeitschko, *Z. Naturforsch., B: J. Chem. Sci.*, 2011, **66**, 221–226.

100. A. F. May, M. A. McGuire, J. Ma, O. Delaire, A. Huq and R. Custelcean, *J. Appl. Phys.*, 2012, **111**, 033708.
101. A. F. May, M. A. McGuire, D. J. Singh, J. Ma, O. Delaire, A. Huq, W. Cai and H. Wang, *Phys. Rev. B: Condens. Matter Mater. Phys.*, 2012, **85**, 035202.
102. A. N. Nateprov, V. C. Kravtsov, V. Moshnyaga and S. Schorr, *Surf. Eng. Appl. Electrochem.*, 2012, **48**, 375–379.
103. V. Ponnambalam, S. Lindsey, W. Xie, D. Thompson, F. Drymiotis and T. M. Tritt, *J. Phys. D: Appl. Phys.*, 2011, **44**, 155406.
104. C. Zheng, R. Hoffmann, R. Nesper and H.-G. von Schnering, *J. Am. Chem. Soc.*, 1986, **108**, 1876–1884.
105. C. Zheng and R. Hoffmann, *J. Solid State Chem.*, 1988, **72**, 58–71.
106. F. Gascoin, S. Ottensmann, D. Stark, S. M. Haile and G. J. Snyder, *Adv. Funct. Mater.*, 2005, **15**, 1860–1864.
107. X.-J. Wang, M.-B. Tang, H.-H. Chen, X.-X. Yang, J.-T. Zhao, U. Burkhardt and Y. Grin, *Appl. Phys. Lett.*, 2009, **94**, 092106.
108. Q.-G. Cao, H. Zhang, M.-B. Tang, H.-H. Chen, X.-X. Yang, Y. Grin and J.-T. Zhao, *J. Appl. Phys.*, 2010, **107**, 053714.
109. H. Zhang, L. Fang, M. B. Tang, Z. Y. Man, H. H. Chen, X. X. Yang, M. Baitinger, Y. Grin and J. T. Zhao, *J. Chem. Phys.*, 2010, **133**, 194701.
110. K. Guo, Q.-G. Cao, X.-J. Feng, M.-B. Tang, H.-H. Chen, X. Guo, L. Chen, Y. Grin and J.-T. Zhao, *Eur. J. Inorg. Chem.*, 2011, 4043–4048.
111. G. Cordier, H. Schafer and M. Stelter, *Z. Naturforsch., B: J. Chem. Sci.*, 1987, **42**, 1268–1272.
112. G. Cordier, H. Schafer and M. Stelter, *Z. Naturforsch., B: J. Chem. Sci.*, 1985, **40**, 1100–1104.
113. G. Cordier, G. Savelsberg and H. Schäfer, *Z. Naturforsch.*, 1982, **37b**, 975–980.
114. A. Zevalkink, W. G. Zeier, G. Pomrehn, E. Schechtel, W. Tremel and G. J. Snyder, *Energy Environ. Sci.*, 2012, **5**, 9121–9128.
115. J. Jiang, A. C. Payne, M. M. Olmstead, H. O. Lee, P. Klavins, Z. Fisk, S. M. Kauzlarich, R. P. Hermann, F. Grandjean and G. J. Long, *Inorg. Chem.*, 2005, **44**, 2189–2197.
116. A. Zevalkink, E. S. Toberer, W. G. Zeier, E. Flage-Larsen and G. J. Snyder, *Energy Environ. Sci.*, 2011, **4**, 510–518.
117. A. Zevalkink, W. G. Zeier, G. Pomrehn, E. Schechtel, W. Tremel and G. J. Snyder, *Energy Environ. Sci.*, 2012, **5**, 9121–9128.
118. A. Zevalkink, G. S. Pomrehn, Y. Takagiwa, J. Swallow and G. Snyder, *ChemSusChem*, 2013, **6**, 2316–2321.
119. G. Cordier, E. Czech, M. Jakowski and H. Schäfer, *Rev. Chim. Miner.*, 1981, **18**, 9–18.
120. P. Verdier, P. L'Haridon, M. Mauaye and Y. Laurent, *Acta Crystallogr.*, 1976, **B32**, 726–728.
121. G. Cordier, H. Schafer and M. Stelter, *Z. Naturforsch., B: J. Chem. Sci.*, 1985, **40**, 5–8.

122. E. S. Toberer, A. F. May and G. J. Snyder, *Chem. Mater.*, 2010, **22**, 624–634.
123. G. J. Snyder and E. S. Toberer, *Nat. Mater.*, 2008, **7**, 105–114.
124. G. S. Nolas, J. L. Cohn, G. A. Slack and S. B. Schujman, *Appl. Phys. Lett.*, 1998, **73**, 178–180.
125. G. Nolas, G. A. Slack, D. Morelli, T. M. Tritt and A. C. Ehrlich, *J. Appl. Phys.*, 1996, **79**, 4002.
126. C. B. Vining, *J. Appl. Phys.*, 1991, **69**, 4333–4340.
127. A. Zevalkink, E. S. Toberer, W. G. Zeier, E. Flage-Larsen and G. J. Snyder, *Energy Environ. Sci.*, 2011, **4**, 510–518.
128. E. S. Toberer, A. Zevalkink, N. Crisosto and G. J. Snyder, *Adv. Funct. Mater.*, 2010, **20**, 4375–4380.
129. A. F. May, E. S. Toberer, A. Saramat and G. J. Snyder, *Phys. Rev. B: Condens. Matter Mater. Phys.*, 2009, **80**, 125205.
130. C. A. Cox, E. S. Toberer, A. A. Levchenko, S. R. Brown, G. J. Snyder, A. Navrotsky and S. M. Kauzlarich, *Chem. Mater.*, 2009, **21**, 1354–1360.
131. S. K. Bux, A. Zevalkink, O. Janka, D. Uhl, S. Kauzlarich, J. G. Snyder and J.-P. Fleurial, *J. Mater. Chem. A*, 2014, **2**, 215–220.
132. H. J. Goldsmid, *Applications of Thermoelectricity*, J. Wiley and Sons Inc., New York, 1960.
133. T. M. Tritt, *Thermal Conductivity Theory, Properties, and Application*, Plenum Publishers, New York, NY, 2004.
134. J. Callaway, *Phys. Rev.*, 1959, **113**, 1044–1051.
135. M. A. Black, *Am. J. Phys.*, 1972, **41**, 691–696.
136. E. S. Toberer, A. Zevalkink and G. J. Snyder, *J. Mater. Chem.*, 2011, **21**, 15843.
137. P. G. Klemens, *Proc. Phys. Soc., London, Sect. A*, 1955, **68**, 1113–1127.
138. F. Gascoin, S. Ottensmann, D. Stark, M. S. Haile and G. Snyder, *Adv. Funct. Mater.*, 2005, **15**, 1860–1864.
139. S. K. Bux, J.-P. Fleurial and R. B. Kaner, *Chem. Commun.*, 2010, **46**, 8311–8324.
140. O. Prytz, E. Flage-Larsen, E. S. Toberer, G. J. Snyder and J. Taftø, *J. Appl. Phys.*, 2011, **109**, 043509.
141. G. A. Slack, *Solid State Physics*, Academic Press, New York, 1979.
142. J. Cohn, G. Nolas, V. Fessatidis, T. H. Metcalf and G. A. Slack, *Phys. Rev. Lett.*, 1999, **82**, 779–782.
143. M. Zebarjadi, K. Esfarjani, J. Yang, Z. F. Ren and G. Chen, *Phys. Rev. B: Condens. Matter Mater. Phys.*, 2010, **83**, 195207.
144. M. Christensen, A. B. Abrahamsen, N. B. Christensen, F. Juranyi, N. H. Andersen, K. Lefmann, J. Andreasson, C. R. Bahl and B. B. Iversen, *Nat. Mater.*, 2008, **7**, 811–815.
145. D. Cahill, S. Watson and R. Pohl, *Phys. Rev. B: Condens. Matter Mater. Phys.*, 1992, **46**, 6131–6140.
146. E. S. Toberer, A. F. May, B. C. Melot, E. Flage-Larsen and G. J. Snyder, *Dalton Trans.*, 2010, **39**, 1046–1054.
147. E. S. Toberer, C. A. Cox, S. R. Brown, T. Ikeda, A. F. May, S. M. Kauzlarich and G. J. Snyder, *Adv. Funct. Mater.*, 2008, **18**, 2795–2800.

148. A. F. May, E. S. Toberer and G. J. Snyder, *J. Appl. Phys.*, 2009, **106**, 013706.
149. H. Schaefer, B. Eisenmann and W. Mueller, *Angew. Chem., Int. Ed. Engl.*, 1973, **12**, 694–712.
150. H. Schaefer, *Annu. Rev. Mater. Sci.*, 1985, **15**, 1–41.
151. A. Zevalkink, G. S. Pomrehn, S. Johnson, J. Swallow, Z. M. Gibbs and G. J. Snyder, *Chem. Mater.*, 2012, **24**, 2091–2098.
152. A. Zevalkink, J. Swallow and G. J. Snyder, *Dalton Trans.*, 2013, 9713–9179.
153. A. Zevalkink, E. S. Toberer, T. Bleith, E. Flage-Larsen and G. J. Snyder, *J. Appl. Phys.*, 2011, **110**, 013721.
154. S. I. Johnson, A. Zevalkink and G. J. Snyder, *J. Mater. Chem. A*, 2013, **1**, 4244–4249.
155. A. Zunger, *Appl. Phys. Lett.*, 2003, **83**, 57–59.
156. S. Brown, S. Kauzlarich, F. Gascoin and G. J. Snyder, *Chem. Mater.*, 2006, **18**, 1873–1877.
157. D. Sánchez-Portal, R. Martin, S. Kauzlarich and W. Pickett, *Phys. Rev. B: Condens. Matter Mater. Phys.*, 2002, **65**, 144414.
158. A. Zevalkink, J. Swallow and G. J. Snyder, *J. Electron. Mater.*, 2012, **41**, 813–818.
159. S.-Q. Xia and S. Bobev, *Chem. Mater.*, 2009, **22**, 840–850.
160. S. Bobev, J. D. Thompson, J. L. Sarrao, M. M. Olmstead, H. K. Hope and S. M. Kauzlarich, *Inorg. Chem.*, 2004, **43**, 5044–5052.
161. S. M. Kauzlarich, in *Chemistry, Structure, and Bonding of Zintl Phases and Ions*, VCH, New York, 1996, pp. 245–274.
162. I. Todorov, D. Y. Chung, L. Ye, A. J. Freeman and M. G. Kanatzidis, *Inorg. Chem.*, 2009, **48**, 4768–4776.
163. Y. Pei, X. Shi, A. LaLonde, H. Wang, L. Chen and G. J. Snyder, *Nature*, 2011, **473**, 66–69.
164. F. J. DiSalvo, *Science*, 1999, **285**, 703–706.
165. D. Sánchez-Portal, R. M. Martin, S. M. Kauzlarich and W. E. Pickett, *Phys. Rev. B: Condens. Matter Mater. Phys.*, 2002, **65**, 144414.
166. A. F. May, M. A. McGuire, J. Ma, O. Delaire, A. Huq and R. Custelcean, *J. Appl. Phys.*, 2012, **111**, 033708.
167. S.-M. Park, S.-J. Kim and M. G. Kanatzidis, *J. Solid State Chem.*, 2003, **175**, 310–315.
168. J. Mathieu, R. Achey, J.-H. Park, K. M. Purcell, S. W. Tozer and S. E. Latturner, *Chem. Mater.*, 2008, **20**, 5675–5681.
169. R. Lam and A. Mar, *Inorg. Chem.*, 1996, **35**, 6959–6963.
170. S.-J. Kim, S. Hu, C. Uher and M. G. Kanatzidis, *Chem. Mater.*, 1999, **11**, 3154–3159.
171. B. Saparov, S. Bobev, A. Ozbay and E. R. Nowak, *J. Solid State Chem.*, 2008, **181**, 2690–2696.
172. A. P. Holm, P. Seon-Mi, C. L. Condron, M. M. Olmstead, K. Hyungrak, P. Klavins, F. Grandjean, R. P. Hermann, G.-J. Long, M. G. Kanatzidis, S. M. Kauzlarich and K. S. Jin, *Inorg. Chem.*, 2003, **42**, 4660–4667.

CHAPTER 2

Chalcogenide Thermoelectric Materials

ANTHONY V. POWELL* AND PAZ VAQUEIRO

Department of Chemistry, University of Reading, Whiteknights, Reading RG6 6AD, UK
*Email: a.v.powell@reading.ac.uk

2.1 Introduction

There is growing recognition of the gains that may be made through the more efficient use of existing energy resources. In particular, the combustion of fossil fuels generates considerable quantities of waste heat, conversion of which into useful electrical energy may be achieved using a thermoelectric generator. There has been significant recent progress in the provision of thermoelectric materials with improved performance at elevated temperatures, appropriate to waste heat recovery in high-temperature (600–900 K) systems such as vehicle exhaust streams. However, a significant fraction of the waste heat generated in industrial processes is released at temperatures between 373 and 535 K, suggesting that there is also a role for thermoelectric energy recovery at low to intermediate temperatures. In general, thermoelectric materials exhibit optimum performance over a relatively narrow temperature range. Therefore, it is unlikely that a single material would be suitable for applications in the low, intermediate and high temperature regimes. This has motivated a worldwide search for new materials,[1] particularly those comprised of earth-abundant elements.

The efficiency of a thermoelectric device is directly related to the performance of the n- and p-type materials from which it is composed.

RSC Energy and Environment Series No. 17
Thermoelectric Materials and Devices
Edited by Iris Nandhakumar, Neil M. White and Stephen Beeby

Published by the Royal Society of Chemistry, www.rsc.org

The materials performance is embodied in a dimensionless figure-of-merit, zT, eqn (2.1)

$$zT = \frac{S^2 \sigma T}{\kappa} \tag{2.1}$$

where S, σ and κ are the Seebeck coefficient, electrical conductivity and thermal conductivity, respectively. The latter involves the sum of the contributions from charge carriers (κ_e) and from lattice vibrations (κ_L). High performance requires a high Seebeck coefficient and low-thermal conductivity, features typically found in non-metallic systems, to be coupled with a high electrical conductivity, characteristic of metallic materials. Therein lies the problem in designing high-performance materials, since each of the properties S, σ and κ are dependent on charge-carrier density and the three quantities cannot be independently optimised. Moreover, S shows the opposite dependence on carrier density to that of σ and κ_e (Figure 2.1) and a degree of compromise is therefore necessary. Consequently, heavily doped semiconductors with carrier densities in the range 10^{19}–10^{21} cm^{-3} attract the greatest interest for the design of new thermoelectric materials.

Commercially-available thermoelectric modules are comprised of Bi_2Te_3, suitably doped for the required n- and p-type variants, for which zT approaches unity at temperatures close to ambient. However, Bi_2Te_3 devices suffer from a number of disadvantages, including relatively low efficiencies, performance that falls off markedly at higher temperatures, materials

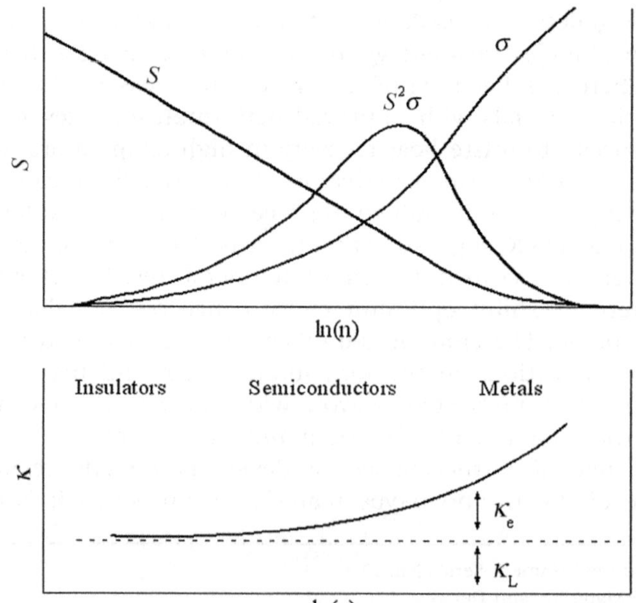

Figure 2.1 The dependence of the electrical conductivity (σ), Seebeck coefficient (S) and electronic (κ_e) and lattice contributions (κ_L) to the thermal conductivity on the charge-carrier concentration, n.

degradation above *ca.* 573 K and toxicity. In addition, tellurium is comparatively scarce, with a terrestrial abundance of only 1 ppb. It has been identified as one of the top 9 'at risk' elements.[2] Furthermore, it is obtained as a by-product of refining of copper ores and therefore its availability is also low. The shortcomings associated with Bi_2Te_3 have led to a resurgence of interest in the identification of new thermoelectric materials,[3,4] particularly those that are tellurium free. Whilst oxides are attractive candidates for high temperature applications,[5] their power factors are comparatively modest. However, power factors may be increased through replacement of oxygen by sulfur or selenium, where the more covalent metal–chalcogenide interaction leads to broader bands and hence increased charge-carrier mobility. This has led to an increasing focus on materials containing sulfur and selenium, including those in which these elements partially replace tellurium.

A number of design strategies for improved thermoelectric materials have emerged[6,7] that seek to achieve a degree of separation between the electrical ($S^2\sigma$) and thermal (κ) terms of zT. These include:

i. The phonon-glass electron-crystal (PGEC) approach, in which localised vibrational modes of a weakly bound atom serve to scatter phonons.[8] This achieves reductions in κ_L, whilst impacting little on the electrical transport properties.
ii. The exploitation of compositional inhomogeneities to create endotaxially embedded nanoinclusions within a semiconducting matrix.[9] The nanoinclusions scatter acoustic phonons with no significant impact on the charge carriers.
iii. Grain boundary engineering to increase phonon scattering.[10,11]
iv. Nanocompositing in which nanoparticles of a second (frequently unrelated) phase form a physical mixture with a thermoelectric material of proven performance, prior to consolidation.[12,13] Improvements in thermoelectric response are typically achieved through substantial reductions in thermal conductivity.
v. The use of appropriate impurity atoms to distort the density of states, (*DOS*), in a semiconductor through the creation of resonant states.[14]
vi. The phonon-liquid electron-crystal (PLEC) approach in which highly mobile ions assume a liquid-like state within an otherwise rigid crystalline matrix, thereby creating disorder which serves to reduce κ_L.[15]
vii. The utilization of reduced length scale, inherent in a structure of reduced dimensionality, to create a more highly structured *DOS*, in which there may be sharp discontinuities.[16] Significant enhancements in the Seebeck coefficient may be realised by tuning the Fermi level, E_F to such a discontinuity, as the Mott relationship,[17] eqn (2.2), reveals that the Seebeck coefficient is related to the derivative of the *DOS* at E_F.

$$S = \frac{\pi^2}{3}\frac{k^2 T}{e}\left\{\frac{\mathrm{d}[\ln \sigma(E)]}{\mathrm{d}E}\right\}_{E=E_F} \tag{2.2}$$

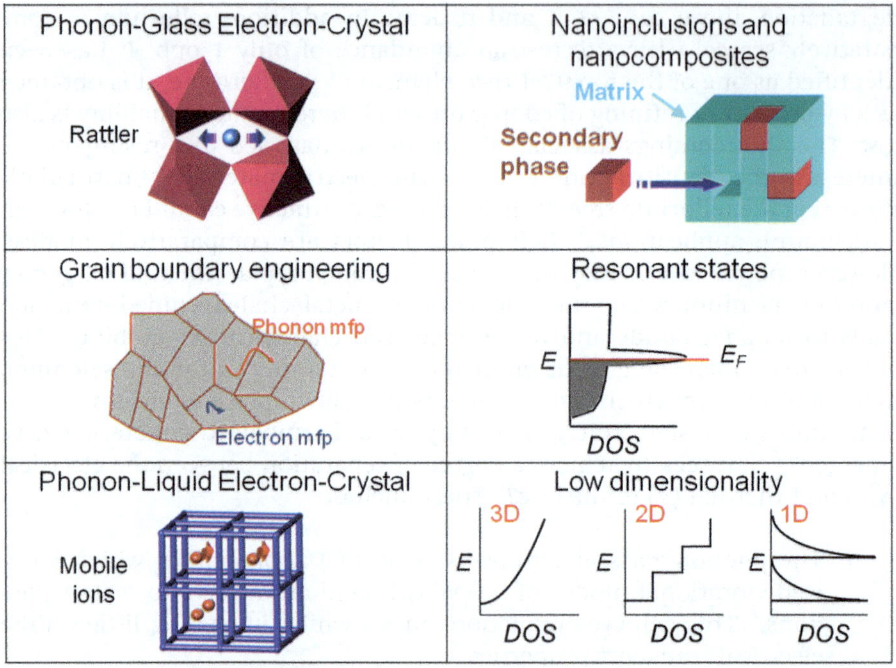

Figure 2.2 A schematic representation of the principal strategies that may be used to enhance the thermoelectric figure-of-merit, zT, through reductions in the thermal conductivity or increases in the power factor.

In this chapter, we describe how these strategies, which are summarised schematically in Figure 2.2, have been applied to a range of metal chalcogenides, particularly those involving earth abundant elements.

2.2 Synthesis

The synthesis of polycrystalline metal-chalcogenides is most commonly carried out by heating powdered mixtures of the elements and/or binary chalcogenides for prolonged periods at elevated temperatures. In many cases, one or more of the reactants may melt, making the process a solid–liquid reaction rather than the more conventional diffusion-limited solid–solid reaction encountered in oxide chemistry. Given the propensity for chalcogenides to convert to the corresponding oxides at high-temperatures in the presence of even trace amounts of oxygen, reactions are conducted either in an inert-gas stream or more generally in an evacuated sealed container. This is most conveniently made of fused silica. However, the more reactive metals may effect reduction of SiO_2 resulting in the formation of oxides. Attack of the container can be partially alleviated by carbon coating through pyrolysis of a hydrocarbon or by containing the reactants within a glassy-carbon crucible contained within the sealed silica ampoule. Fused silica imposes an upper limit of *ca.* 1300 K. on the reaction temperature. For higher temperatures, ampoules constructed from high-melting point metals such as tantalum may be used.

Sulfidation of an oxide precursor offers an alternative to high-temperature synthesis from the elements or binary sulfides. This may be effected by the passage of a gaseous sulfiding agent over the heated oxide. Whilst hydrogen sulfide has been commonly used, carbon disulfide provides a convenient alternative.[18] Passage of an inert carrier gas though liquid CS_2 prior to entry to the heated zone providing a sufficient concentration to effect reaction.

The introduction of nanocrystals of a second phase into a host matrix is dependent on the limited solubility of the former in the host matrix. Since the properties are heavily dependent on the resulting microstructure, both the rate of cooling from the reaction temperature and the consolidation method play a role in determining thermoelectric performance. The technique of Spark Plasma Sintering[19] (SPS) is generally preferred over hot-pressing for the consolidation of nanostructured materials as the short sintering times lead to preservation of the nanostructure. Matrix encapsulation through a form of arrested precipitation (Figure 2.3) has been used to introduce nanoscale inclusions into thermoelectric semiconductors.[20,21] This exploits the fact that a given material may be soluble in another in the liquid state but not in the solid state. In such a situation, rapid cooling of a multi-component liquid leads to precipitation of the minority (solute) phase in the form of nanoparticles embedded in the majority (solvent) phase.

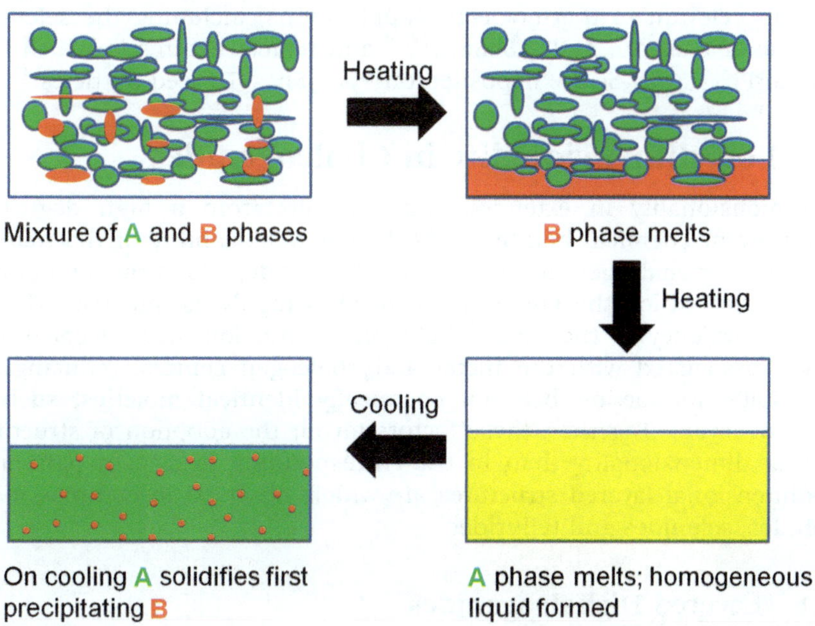

Figure 2.3 The process of arrested precipitation that may be used to introduce nanoprecipitates into a bulk phase.

Increasingly, the use of mechanical alloying[22] is being explored as an alternative to high-temperature synthesis for the production of thermoelectric materials. The method involves milling mixtures of reactant powders, usually the elements, contained in a jar with a number of grinding balls. These serve to create high-energy impacts between particles, leading typically to the formation of poorly crystalline products. Hardened steel, tungsten carbide or zirconia are often used for the construction of the jar and grinding balls. The method has been applied with success to complex tellurides[23] and sulfides.[24,25] It offers advantages in terms of scale-up of materials production and also leads to the formation of small particle size products, which may confer advantages in terms of increased interface scattering of phonons and concomitant reductions in thermal conductivity.

The application of alternative low-temperature processes for the production of chalcogenide thermoelectric materials has been limited to date. However, solution- and solvothermally-based methods have been used to produce chalcogenide thermoelectric materials, particularly in nanocrystalline form. Hot injection, whereby a solution of a chalcogen-containing precursor is introduced into a high-boiling point solvent containing metal salts, leads to precipitation of nanocrystals on cooling.[26,27] Such reactions can lead to the stabilization of metastable phases, such as the preferential formation of cubic $Pb_{2-x}Sn_xS_2$. The low synthesis temperature may also avoid the phase separation common at elevated temperatures, as observed in the nanocrystalline phases $Pb_mSb_{2n}Se_{m+3n}$ ($n=1$; $m=2$, 4, 6, 8). Hydrothermal synthesis in an autoclave has been used to prepare a range of materials relevant to thermoelectric applications, including the selenium analogue of LAST-10, $AgPb_{10}SbSe_{12}$,[28,29] and nanostructures in which thin layers of $Pb_{0.75}Sn_{0.25}Se$ are deposited on $Pb_{0.75}Sn_{0.25}Te$ seed particles.[30]

2.3 Low-dimensionality in Chalcogenides

Low-dimensionality in extended solids results from a high degree of anisotropy in chemical bonding. The higher polarizability of the dianion of the heavier chalcogenides, Q, compared to oxide, allows the adoption of structures in which the environment of Q is highly asymmetric. The increased covalency of the metal–chalcogen interaction also lowers formal charges associated with the metal and chalcogen centres, reducing the electrostatic interaction between chemically identical moieties, such as chains or layers. Together, these factors favour the adoption of structures of lower dimensionality than in the corresponding oxides. In particular, two-dimensional layered structures are widely observed in complex metal sulphides, selenides and tellurides.

2.3.1 Layered Dichalcogenides

Many transition-metal dichalcogenides adopt one of two structure types, in which transition-metal cations are located in sites of either octahedral or trigonal prismatic coordination, between pairs of close-packed chalcogenide layers. This gives rise to two-dimensional slabs of stoichiometry MQ_2, comprised of edge-sharing MQ_6 octahedra (Figure 2.4) or trigonal prisms. Stacking such slabs in a direction perpendicular to the close-packed layers produces structures in which adjacent MQ_2 blocks are separated by a van der Waals' gap. The 1T structure of TiS_2 (Figure 2.4(b)) containing octahedral blocks, has parallels with that of the thermoelectric material Bi_2Te_3, in which fusion of pairs of MQ_2 slabs through common octahedral faces

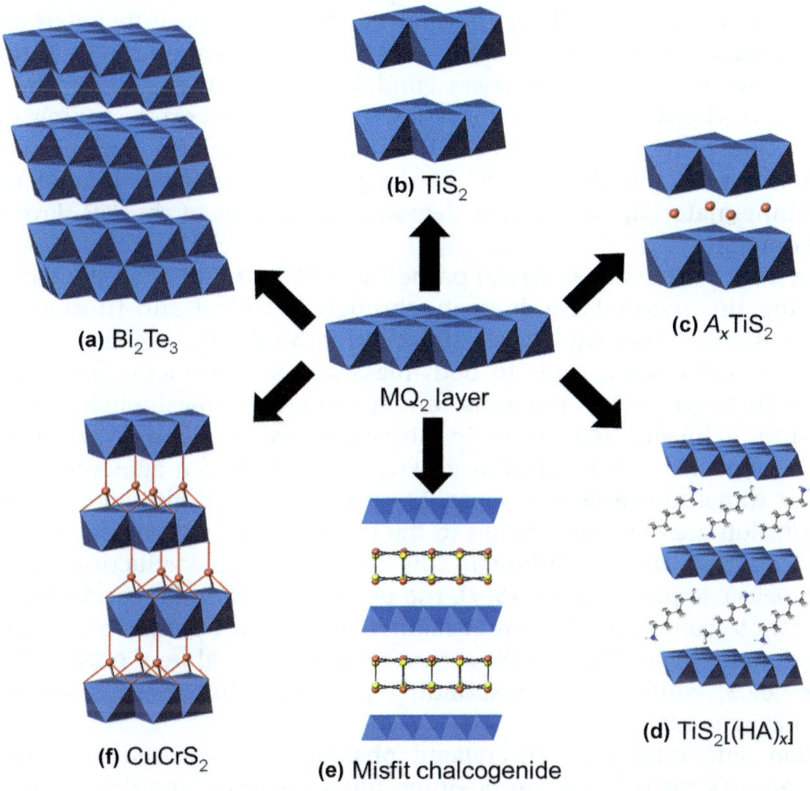

Figure 2.4 The octahedral metal dichalcogenide MQ_2 unit as a building block for the structures of low-dimensional materials. (a) The structure of Bi_2Te_3 derived from fusion of pairs of $BiTe_2$ units across a shared octahedral face. (b) The structure of TiS_2 in which neighbouring TiS_2 units are separated by a van der Waals' gap. (c) A stage 1 intercalate, A_xTiS_2 in which guest species, A, occupy octahedral sites in the van der Waals' gap of TiS_2. (d) Schematic representation of the structure of $TiS_2[(HA)_{0.08}(H_2O)_{0.22}(DMSO)_{0.03}]$ containing hexylammonium (HA) ions in the van der Waals' gap of TiS_2: H_2O and DMSO solvent molecules have been omitted for clarity. (e) Alternating MS_2 and rocksalt (MS) blocks in the misfit layered chalcogenides of idealised formula AMS_3. (f) Occupation by Cu^+ ions of 50% of the tetrahedral sites between pairs of CrS_2 blocks in the structure of $CuCrS_2$.

produces a double slab of M_2Q_3 stoichiometry (Figure 2.4(a)). The stacking of trigonal prismatic slabs may give rise to a variety of polytypes of different periodicity along the stacking direction, including those in which both octahedral and trigonal prismatic units are present.

Imai *et al.*[31] have shown that the electrical transport properties of single-crystals of TiS_2 are competitive with those of Bi_2Te_3. In particular, the in-plane thermoelectric power factor ($S^2\sigma$) is *ca.* 64% that of the telluride. The promising electrical behaviour of TiS_2 may be associated with the large *DOS* just above E_F and inter-valley scattering. The figure of merit, $zT = 0.16$, is however significantly reduced from that of Bi_2Te_3 owing to a relatively high in-plane thermal conductivity. As anticipated from the structure, the physical properties are highly anisotropic. The ratio of in-plane to out-of-plane electrical conductivity is 750 at 300 K. The anisotropy in thermal conductivity is less marked; the corresponding ratio being of the order of 1.6 at 300 K. This anisotropy in physical properties of a single crystal manifests itself in the properties of consolidated polycrystalline material,[18] where the degree of alignment of the TiS_2 layers is less complete.

Given the promising electrical properties of TiS_2, efforts have focused on reducing the thermal conductivity through chemical substitution. The introduction of impurity atoms, substituting randomly over the available crystallographic sites, leads to both mass and strain fluctuations, which scatter high frequency phonons, and provide a well-established mechanism for achieving reductions in lattice thermal conductivity, κ_L. Partial replacement of titanium in phases $Ti_{1-x}M_xS_2$, has been achieved for the heavier transition-series elements, M = Nb[32] and Ta[33] although levels of substitution are low. In addition to the impact on κ_L, the substituent acts as an electron donor. Although this leads to slight reductions in the thermoelectric power factor ($S^2\sigma$), the decrease in thermal conductivity is sufficient to compensate and modest increases in the figure of merit result. For example, in the $Ti_{1-x}Nb_xS_2$ series, zT reaches a value of 0.24 at 700 K for $x = 0.05$, whilst the corresponding maximum for Ta substitution is $zT = 0.28$ ($x = 0.05$).

Anion substitution in TiS_2-related phases has also been explored in $TiS_{2-x}Se_x$ ($0 \le x \le 2$).[34] The mixed-anion phases exhibit lattice thermal conductivities below that of either end-member phase, consistent with the introduction of mass fluctuation scattering. However, concomitant changes in the electrical properties, which have an impact on κ_e, lead to a more complex dependence of the total thermal conductivity on composition. Whilst $\sigma(T)$ and $S(T)$ are consistent with the retention of semi-metallic behaviour throughout the composition range $0 \le x \le 2$, with increasing selenium content, the absolute value of the Seebeck coefficient decreases and electrical conductivity increases. The increasing metallicity may be traced to the greater covalency of selenium compared to sulphur, which leads to broader bands as the selenium content increases. A maximum $zT = 0.41$ is observed for $TiS_{1.5}Se_{0.5}$ at 700 K.

2.3.2 Intercalated Phases

A further complication in the study of TiS_2 is the presence of non-stoichiometry. Excess titanium may occupy vacant octahedral sites between pairs of TiS_2 slabs, leading to $Ti_{1+\delta}S_2$. This has a marked effect on thermoelectric properties and is likely to be the origin of discrepancies between properties determined for TiS_2 prepared in different laboratories.[35] For example, for nominally stoichiometric TiS_2, values of the Seebeck coefficient ranging between $-40\ \mu V\ K^{-1}$ and $-275\ \mu V\ K^{-1}$ have been reported. Even when considering single-crystal samples only, variations can reach 70%. A recent investigation[36] of $Ti_{1+\delta}S_2$ in which the degree of non-stoichiometry was systematically varied over the range $0.0 \leq \delta \leq 0.05$ demonstrates that titanium self intercalation leads to a progressive increase in the carrier concentration. At low levels of non-stoichiometry, the absolute value of the negative Seebeck coefficient, is considerably higher than that reported by Imai *et al.*,[31] approaching $-300\ \mu V\ K^{-1}$ at 300 K. The thermoelectric power factor is therefore optimised at relatively low levels of non-stoichiometry, $\delta \leq 0.02$. However, titanium intercalation also has a pronounced effect on thermal conductivity, as disorder arising from the presence of titanium in inter-layer sites reduces κ_L by almost a factor of four across the range of composition $0 \leq \delta \leq 0.05$. Competing with this reduction in κ_L is the increase in the electronic contribution, κ_e, arising from the increased charge-carrier density. The lowest measured thermal conductivities are observed in the middle of the compositional range, $0.015 \leq \delta \leq 0.025$, and a maximum figure of merit of $zT = 0.48$ is attained for $\delta = 0.025$ at 700 K.

Intercalation of other elements (Figure 2.4(c)) has also been used to effect reductions in thermal conductivity of TiS_2 and related phases. Materials of the form A_xTiS_2 (A = Bi, Gd, Nd, Cu, Ag) are generally prepared through a two-step process, in which TiS_2 is prepared by high-temperature reaction and subsequently reacted with the guest species, A. The transfer of charge from guest to host that accompanies the intercalation of A, increases the electrical conductivity. However, this raises the charge-carrier contribution to the thermal conductivity, counteracting the decrease in lattice contribution arising from the effects of disorder, due to the introduction of the guest into the inter-layer space. The optimum thermoelectric properties are therefore generally found at relatively low levels of guest incorporation, where the increase in charge carrier density is insufficient to impact on the overall thermal conductivity.

Intercalates of Bi,[37] Gd[38] and Nd[39] exhibit large increases in the electrical conductivity, due to the increased charge-carrier density arising from electron transfer to the TiS_2 host. Although enhancements in the figure of merit are observed on intercalation, those of the pristine phase are significantly lower than reported elsewhere. This may be associated with densities of hot-pressed materials that are markedly lower than for materials consolidated by SPS. The beneficial effects of a high degree of densification produced by SPS are illustrated by Cu_xTiS_2.[40] Materials with compositions in the range $0.02 \leq x \leq 0.1$, exhibit increases in electrical conductivity of almost an order of magnitude over hot-pressed materials, whilst retaining similarly high Seebeck coefficients. The intercalation of copper ions effects a reduction in κ_L by almost 50%, with the result that a large increase in electrical conductivity is accompanied by a slight decrease in the overall thermal conductivity. This leads to significant improvements in the figure of merit, with $zT = 0.46$ reported for $Cu_{0.1}TiS_2$. Whilst the impact on thermoelectric properties of silver intercalation is broadly similar, the structural chemistry of Ag_xTiS_2 is more complex.[41] In the compositional range $0 \leq x \leq 0.2$, with increasing levels of silver substitution, the structure evolves from a stage 1 intercalate, in which silver ions are randomly distributed over sites between each pair of TiS_2 slabs, to a stage 2 structure above $x = 1/6$, in which 1/3 of the sites between alternate pairs of TiS_2 layers are occupied in an ordered fashion. However, even at levels of silver incorporation below $x = 1/6$, there is evidence that stage 1 and stage 2 phases co-exist. Despite the presence of staging, the decrease in κ_L is similar to that achieved at a comparable level of copper intercalation, and the maximum $zT = 0.45$ at 700 K for $0.02 < x < 0.1$, is similar to that achieved for copper intercalation.

In an effort to combine the beneficial effects of mass fluctuation scattering through anion substitution with those of inter-layer disorder through intercalation, Nunna *et al.*,[42] prepared the series $Cu_xTiS_{1.5}Se_{0.5}$ $(0 \leq x \leq 0.1)$. Whilst the overall thermal conductivity increases slightly on copper intercalation, due to a large increase in the κ_e term, thermoelectric power factors for samples with low copper contents $(x \leq 0.05)$ exceed those of the pristine material above *ca.* 380 K. Above 450 K, at these low doping levels, the figure of merit surpasses that of $TiS_{1.5}Se_{0.5}$ reaching $zT = 0.54$ at 700 K for $Cu_{0.05}TiS_{1.5}Se_{0.5}$.

Investigations have recently been extended to organic guests (Figure 2.4(d)). Electrochemical intercalation of hexylammonium (HA) in the presence of the solvent dimethylsulfoxide (DMSO), produces $TiS_2[(HA)_{0.19}(DMSO)_{0.35}]$.[43] This contains a bilayer of HA in the inter-layer space, together with co-intercalated solvent molecules. The latter may be exchanged with water to yield $TiS_2[(HA)_{0.08}(H_2O)_{0.22}(DMSO)_{0.03}]$, which exhibits an increase in the in-plane electrical conductivity and a decrease in absolute value of the Seebeck coefficient. These observations are consistent with charge transfer from the organic guest and reduction of the TiS_2 host. The power factor is reduced by around a factor of two, whilst the reduction in the in-plane thermal conductivity is more marked, with $\kappa = 0.69\ \mathrm{W\,m^{-1}\,K^{-1}}$, as a result of an exceptionally low $\kappa_L = 0.12\ \mathrm{W\,m^{-1}\,K^{-1}}$. The in-plane figure of merit is tripled to $zT = 0.28$ at 373 K. Calculations indicate a decrease in the phonon relaxation time, suggesting an increase in phonon scattering in the hybrid material, due to coupling between organic and inorganic components. Remarkably, the hybrid material shows a high degree of mechanical flexibility; exhibiting a flexural modulus lower than that of a typical plastic. This feature may open up possibilities for flexible thermoelectric devices.

2.3.3 Structurally-related Phases

The octahedral MQ_2 slab can serve as a building block for a variety of complex sulfides. Higher levels of incorporation of cations in the inter-layer space than is typical for the intercalation compounds discussed above, leads to a range of two-dimensional superstructures for A_xMS_2, when $x<1.0$.[44] Electrical transport properties appear to be sensitively dependent on consolidation conditions.[45] For example, the electrical conductivity of n-type $NiCr_2S_4$ ($Ni_{0.5}CrS_2$), changes by more than an order of magnitude, as a result of changes in hot-pressing temperature. This may be due to a change in composition during consolidation as a result of sulfur volatilization. The higher electrical conductivities are also associated with reduced values of $|S|$. Therefore, although the thermal conductivity is relatively low at $\kappa = 1.4\ W\,m^{-1}\,K^{-1}$, $zT < 0.1$ for $NiCr_2S_4$ at 373 K. A similarly low $\kappa = 1.73\ W\,m^{-1}\,K^{-1}$, and $zT = 0.12$ at 700 K, is found for trigonal Cr_2S_3 ($Cr_{1/3}CrS_2$).[46]

Occupation in an ordered fashion of 50% of the tetrahedral sites between CrS_2 slabs leads to the trigonal structure adopted by $CuCrS_2$[47] (Figure 2.4(f)). An early report of $zT = 2$ at room temperature[48] generated considerable interest. However, this high figure of merit has not been reproduced by other workers[45,49] and appears to be due to an anomalously large electrical conductivity. It is possible this is associated with ex-solution of mobile copper ions as elemental copper. However, the isostructural $CuCrSe_2$ and $AgCrSe_2$ reach $zT = 1$ at 773 and 800 K, respectively.[50,51] These materials, which are superionic conductors, are good examples of the PLEC approach, with extremely low thermal conductivities of $\sim 0.5\ W\,m^{-1}\,K^{-1}$.

The misfit layered chalcogenides, in which MQ_2 slabs are interleaved with a rocksalt (MQ) building block (Figure 2.4(e)),[52] have attracted attention as potential PGEC materials. A structural mismatch between the two sub-units leads to an irrational ratio of one of the in-plane axial lengths and the structures are incommensurate. The potential to disrupt thermal transport due to the resulting loss of long-range periodicity in three dimensions, has motivated the investigation of the thermoelectric properties of such materials.[53] Detailed investigations of rare-earth phases $(RS)_{1+m}NbS_2$, $(LaS)_{1.2}CrS_2$, and the higher-order phases $[MS]_{1+m}[TiS_2]_2$ (M = Pb, Bi, Sn) reveal thermal conductivities are indeed reduced from that of the corresponding dichalcogenide, in line with expectations from the PGEC approach. The degree of anisotropy in the properties is greater in the niobium containing phases than in the chromium analogues, whilst evidence has been presented that the physical properties are sensitive to microstructure. Figures of merit at room temperature are generally <0.1, reaching 0.15 at elevated temperatures (≥ 900 K).

2.4 Shandite-related Phases

The beneficial effects of the structured *DOS* in a low-dimensional system may extend to three-dimensional structures in which low-dimensional structural motifs can be identified. One such example is that of materials of general formula $A_3M_2Q_2$ (A = Ni, Co, Rh, Pd; M = Pb, In, Sn, Tl; Q = S, Se),[54–56] related to the mineral shandite. These materials adopt a rhombohedral structure (Figure 2.5) in which kagome-like sheets of A atoms, containing additional M atoms located in six-coordinate sites, confer two-dimensional character. Electronegative atoms, Q, are located above and below triangular A_3 units, whilst the A_3M sheets are linked into a three-dimensional structure through additional M atoms located in trigonal antiprismatic inter-layer sites. The formal oxidation states of the constituent elements is a matter of some debate, with both zero,[57,58] and mixed[59] valence states being proposed for the transition-series element and an intermediate valence state for tin.[60] Recent DFT calculations coupled with comprehensive spectroscopic investigations[61] suggest $Co_3Sn_2S_2$ and its indium-substituted derivatives have characteristics of intermetallic phases with formal oxidation states that are either zero (Co) or closely approach this value (Sn, In S).

Much of the interest from a thermoelectric perspective has focused on the half-metallic ferromagnet, $Co_3Sn_2S_2$, and its derivatives. Despite its metallic nature, $Co_3Sn_2S_2$ possesses an unusually large Seebeck coefficient (*ca.* $-50\ \mu V\ K^{-1}$ at 300 K).[62] Band structure calculations[63] reveal states in the

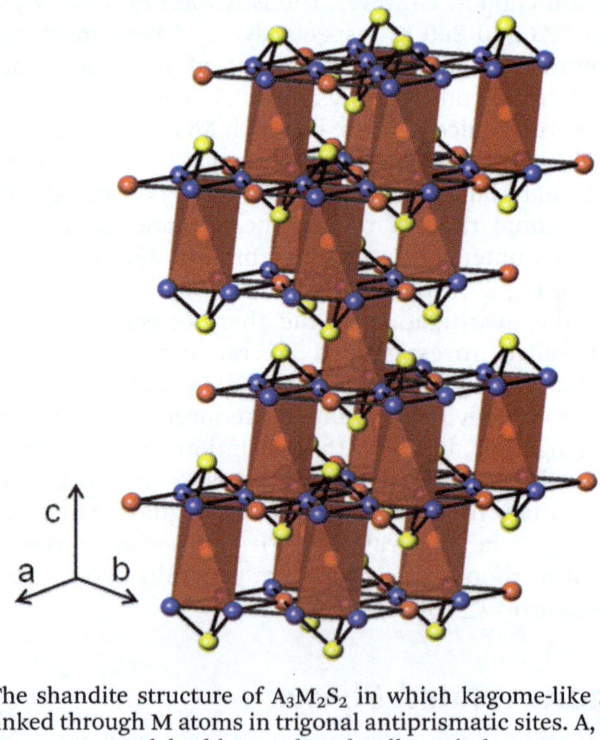

Figure 2.5 The shandite structure of $A_3M_2S_2$ in which kagome-like A_3M layers are linked through M atoms in trigonal antiprismatic sites. A, M and S atoms are represented by blue, red and yellow circles, respectively, and the trigonal antiprismatic sites by shaded polyhedra.

vicinity of E_F to be predominantly of Co d-character. This results in narrow bands in the *DOS* in the vicinity of E_F, offering opportunities to increase the Seebeck coefficient, by tuning E_F to one of the discontinuities in the *DOS*.

In the series $Co_3Sn_{2-x}In_xS_2$ ($0.0 \leq x \leq 2.0$), progressive replacement of tin by indium, provides a means of continuously tuning E_F (Figure 2.6).[61] DFT calculations suggest the preferential occupation of inter-layer sites by indium is associated with the different bonding preferences of the two main-group elements.[61,64] Both end member phases ($x = 0.0$, 2.0) exhibit metallic behaviour, which is consistent with the electron populations and calculated band structures. However, these considerations also predict a transition to semiconducting behaviour at $x = 1.0$. This is confirmed by experimental data which reveal an unusual double metal to semiconductor to metal transition as E_F is lowered through the introduction of indium.[65] As the semiconducting composition of $x = 1.0$ is approached, $|S|$ increases, leading to a substantial improvement in zT over that of the unsubstituted ($x = 0$) phase. The value $zT = 0.32$ for $Co_3Sn_{1.6}In_{0.4}S_2$ at 673 K is competitive with other sulphide materials at this temperature. At lower temperatures of 425 K, the figures of merit for $Co_3Sn_{2-x}In_xS_2$ ($0.8 \leq x \leq 0.85$) are amongst the highest reported for sulphides in this temperature region, whilst the power factors of up to 20 $\mu W\ cm^{-1}\ K^{-2}$ are of the same order as those of bismuth telluride.[66]

Substitution at the transition-metal site offers an alternative means of tuning the electron transport properties of $Co_3Sn_2S_2$ (Figure 2.6). Whilst much of the focus on materials, $Co_{3-x}A_xSn_2S_2$ (A = Ni, Fe),[67,68] has been on magnetic properties, the present authors have recently investigated the thermoelectric behaviour of such phases.[69] Electron doping, through nickel substitution, yields single-phase materials, $Co_{3-x}Ni_xSnS_2$, over the entire range of composition $0.0 \leq x \leq 3.0$. The metallic character increases with nickel content, with the result that the thermoelectric properties are progressively degraded. By contrast, hole doping through iron substitution in

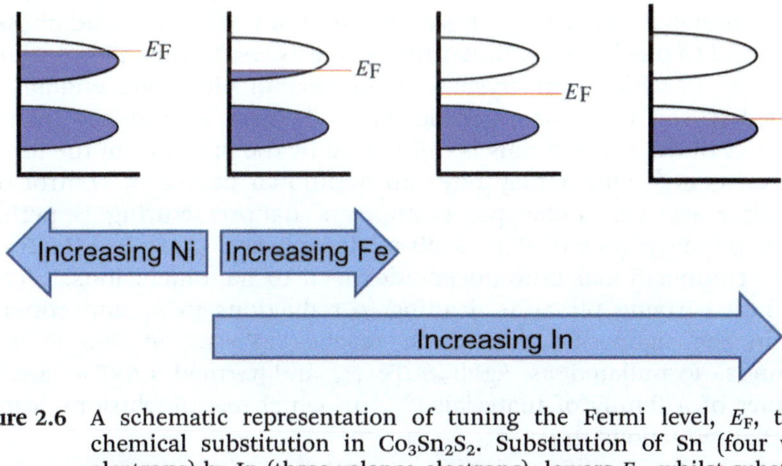

Figure 2.6 A schematic representation of tuning the Fermi level, E_F, through chemical substitution in $Co_3Sn_2S_2$. Substitution of Sn (four valence electrons) by In (three valence electrons), lowers E_F, whilst substitution at the Co site can be used to add (Ni) or remove (Fe) electrons.

the series $Co_{3-x}Fe_xSn_2S_2$, leads to a more limited solid solution range ($0 \leq x \leq 0.6$), which may be extended to $x \approx 1.0$[68] by high-pressure synthesis, albeit with an increase in the level of impurity phases. Iron substitution leads to a marked reduction in the electrical conductivity between $x = 0$ and $x = 0.6$. However, a concomitant increase in the magnitude of $|S|$ results in power factors that are comparable with or exceed that of $Co_3Sn_2S_2$. When coupled with the decrease in thermal conductivity, arising from reductions in the charge-carrier contribution, this produces enhancements in the thermoelectric figure-of-merit. The maximum $zT = 0.2$ at 523 K, for the composition $Co_{2.6}Fe_{0.4}Sn_2S_2$, is comparable with that for an equivalent level of hole doping through indium substitution at the main-group atom site ($ZT(Co_3Sn_{1.6}In_{0.4}S_2) = 0.26$ at 528 K).[61]

Anion substitution in the series $Co_3SnInS_{2-x}Se_x$[70] leads to a solid solution over the range $0 \leq x \leq 0.8$, with trace impurities present at $x = 1.0$. Semiconducting behaviour persists to $x = 0.4$ but a sharp drop in resistivity occurs at $x = 0.6$ and $d\rho/dT$ changes sign as x increases further. Calculations suggest the transition to metallic behaviour may be associated with a change in the distribution of indium over the two available M sites in the shandite structure. Remarkably, the transition to a metallic state at $x = 0.6$ is accompanied by an increase in $|S|$. Although thermal conductivity is increased on substitution, presumably due to the increased charge-carrier contribution, the figure of merit is also improved, zT reaching *ca.* 0.06 for $Co_3SnInS_{1.2}Se_{0.8}$ at 300 K. Since sample densities are only 75% of the crystallographic values, further improvements may be possible through better consolidation.

2.5 Rocksalt-derived Chalcogenides

Lead chalcogenides PbQ (Q = Te, Se, S) adopt the cubic rocksalt structure. In principle, a variety of solid solutions involving alloying with a second phase may be formulated. However, limits on the solubility of this second phase in PbQ can lead to precipitation under the synthesis conditions. This produces a composite in which nanocrystals of the second phase are endotaxially embedded in the lead chalcogenide matrix.[9] There is evidence that the morphology of the nanocrystals is influenced by the presence of the ternary dopant element,[71] which may offer an additional degree of control over the precipitates. The archetypal example of nanostructuring is perhaps the pseudo-binary system of rocksalt-related phases, $(AgSbTe_2)_x(PbTe)_{1-x}$, in which compositional inhomogeneities lead to nanoinclusions.[72] These scatter heat carrying phonons, leading to reductions in κ_L and enhancements in the figure-of-merit, which reaches $zT = 1.7$ at 700 K. Such compounds, formulated as $AgPb_mSbTe_{2+m}$ and termed LAST-m are the forerunner of a family of materials,[73–75] in which nanoinclusions lead to substantial reductions in κ.

Nanostructuring results in significant scattering of phonons whose mean free path (mfp) is below *ca.* 100 nm: phonons with long mfp remaining largely unaffected by the nanostructure. However, it has been demonstrated[76] that control of the microstructure over longer length scales enables scattering of phonons over a wider range of the phonon spectrum, through inter-granular scattering. Achieving the highest thermoelectric performance therefore requires consideration of phonon scattering over a wide range of lengths scales in a hierarchical fashion (Figure 2.7),[77,78] spanning the atomic scale (defects and disorder), through endotaxial precipitates (nanoinclusions) to the mesoscale (grain boundaries). The success of a hierarchical approach is exemplified by p-type PbTe alloyed with SrTe, which exhibits the exceptional $zT = 2.2$ at 915 K.[76]

The growing appreciation of the importance of microstructure in thermoelectric materials has stimulated efforts to manipulate microstructure to effect control over interface scattering over a range of length scales.[79] This work has demonstrated that the formation of nanoinclusions or nanoprecipitates is ubiquitous in chalcogenides with the rocksalt structure. However, the atomic disorder that reduces the thermal conductivity through increased phonon scattering, may also increase charge-carrier scattering, thereby reducing the carrier mobility. The deleterious impact on carrier transport is reduced when the energy difference between the valence band of the parent matrix and that of the nanocrystalline phase is minimised.[80] Recent work[81] reveals that alloying is most effective at low temperatures and

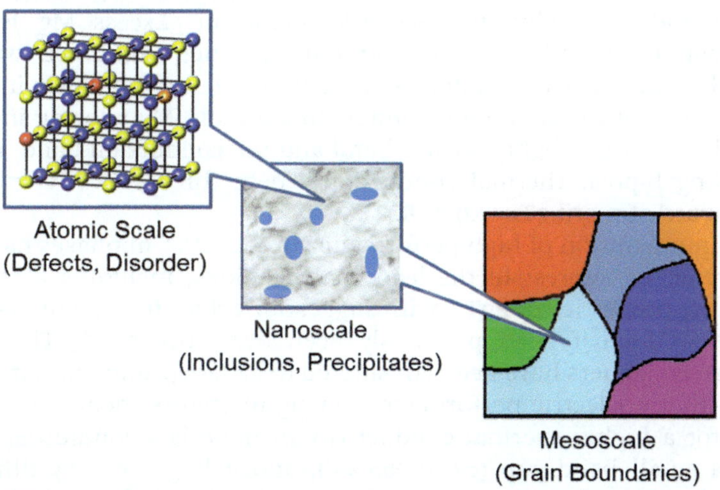

Figure 2.7 The hierarchical approach to the manipulation of phonon scattering in thermoelectric materials. At the atomic length scale, defects corresponding to impurity atoms in the rocksalt structure result in alloy or mass-fluctuation scattering of phonons with short ($\leq$4 nm) mean-free paths (mfp), through the effects of disorder (blue circles, cation; yellow circles, anion; red/orange circles, impurity atoms at the cation/anion site). At longer length scales, nanoinclusions (shown in blue) scatter phonons with longer mfp ($\leq$100 nm), whilst grain boundaries scatter phonons with mfp up to 1 μm.
Adapted from Ref. 76.

for materials for which the unalloyed forms have a relatively high thermal conductivity.

Improvements in the thermoelectric power factor $S^2\sigma$, have also been targeted through band engineering.[82,83] The electronic structure of PbTe[84,85] consists of a 'light' upper valence band (L-band) at the L points of the Brillouin zone, with degeneracy, $N_v = 4$, and a second 'heavy' lower valence band (Σ-band) along the Σ line, *ca.* 0.2 eV below the L-band with $N_v = 12$. A large valley degeneracy increases the density of states effective mass, m^* ($\propto N_v^{2/3}$), resulting in a higher mobility and enhanced thermoelectric properties. The degeneracy is increased by high symmetry, which results in isolated pockets in the Fermi surface, or by a small energy separation between bands. In the case of the lead chalcogenides, N_v may be increased significantly if convergence between the L and Σ valence bands can be achieved, through chemical substitution.[83]

Increasingly, a combination of approaches is adopted that enables electronic and thermal contributions to the figure of merit to be targeted simultaneously.[86] This has led to spectacular increases in zT at high temperatures and values in the range 1 to 1.5 are now commonplace amongst rocksalt phases. An excellent example of the benefits of the complementary effects on electronic and thermal properties is provided by the incorporation of MgTe into p-type PbTe.[87] Within the Mg solubility limit, the gap in energy between the L and Σ valence bands is reduced, promoting charge-carrier injection and enhancing the Seebeck coefficient. Excess Mg leads to the formation of endotaxial nanostructures, which combined with the mesoscale grain structure, reduces the lattice thermal conductivity to *ca.* 0.5 W m^{-1} K^{-1} at elevated temperatures. In addition, Mg alloying increases the gap between the (light) valence band and the conduction band, thereby suppressing bipolar thermal conduction. These three effects combine to yield a figure of merit $zT = 2.0$ at 823 K.

The demonstration of high performance in the LAST-m phases has led to a resurgence of interest in the lead chalcogenides, including a reassessment of the properties of PbTe: with suggestions that the high temperature thermal conductivity had previously been overestimated.[88] The lighter chalcogen congeners have been considered to be less promising candidates for high thermoelectric performance, owing to factors including the propensity for a higher thermal conductivity in materials containing lighter atoms, a small band gap (*ca.* 0.065 eV), and a larger energy difference (0.3 eV) between the L and Σ valence bands ($\Delta E_{L\Sigma}$) than in PbTe.[89,90] However, recent calculations[91] suggest that heavily hole doped PbSe could exhibit $zT \approx 2$ at elevated temperatures as a result of flattening of a heavy, high *DOS* band *ca.* 0.4 eV below the band gap. Contrary to expectations, the measured thermal conductivity of PbSe is comparable with that of PbTe.[92] This may be traced to the smaller cubic lattice parameter and higher degree of anharmonicity in the lattice vibrations. In addition, the band gap of PbSe is strongly temperature dependent, exceeding that of PbTe above 500 K,[90] whilst band engineering allows the magnitude of $\Delta E_{L\Sigma}$ to be reduced.[93] These results have led to a re-examination of materials derived from PbSe and PbS.

A technologically-important feature of the lead chalcogenides, PbQ (Q = S, Se, Te), is the capacity to create n- and p-type derivatives through suitable doping, generally with halogens and electropositive species respectively, and to control the charge-carrier density. PbTe doped with 2% Na, exhibits a sharp increase in *DOS* in the region of E_F, leading to holes with heavy mass that induce an increase in the positive Seebeck coefficient, raising zT to 1.4 at 750 K.[94] When combined with nanostructuring, through the inclusion of Ag_2Te precipitates,[95] a maximum $zT > 1.5$ is attained above 650 K. Significantly from a device efficiency perspective, a high figure of merit is maintained over a wide temperature range, resulting in a 30% higher efficiency than for sodium-doped PbTe. Silver doping of PbSe is qualitatively different,[96] owing to the limited solubility. The variation in carrier concentration is non-monotonic and beyond the solubility limit of 1.0%, excess Ag acts as a donor. Coupled with a band gap that increases with temperature, thereby avoiding the saturation and/or local maximum of the Seebeck coefficient at higher temperatures that occurs in undoped and lightly doped PbSe, this leads to enhanced high temperature power factors and zT, which reaches 1.0 at 770 K for 1.5% Ag doping.

Although n-type PbQ phases have a lower valley degeneracy ($N_v = 4$) in the conduction band than p-type analogues ($N_v \geq 12$), weaker electron–phonon interactions in the conduction band lead to higher electron mobility.[97] Consequently, zT can achieve values comparable with that of p-type variants; that for PbSe doped with Br reaching $zT = 1.1$ at 850 K for example. Interestingly when the dopant is Cl, optical phonon scattering of charge carriers reduces the electron mobility at high temperatures, whilst also providing an additional heat carrying path. Together these limit zT to 0.8 at 700 K.[98] The existence of n- and p-type variants with comparable performance, suggest that PbSe is a viable, more abundant alternative to PbTe for energy harvesting in the mid- to high-temperature range.

Introduction of post transition-series elements such as Tl, Al, In, into PbQ has a more marked effect on the *DOS*, through the introduction of resonant states, which can produce substantial improvements in the figure-of-merit.[99] For example, doping of n-type PbSe with Al, leads to $zT = 1.3$ at 850 K.[100] Jaworski *et al.*[101] have sought to combine the effects of resonant levels with the reduction in the lattice component of thermal conductivity that arises from alloy scattering of phonons, through the incorporation of Tl in p-type $PbTe_{1-x}Q_x$ (Q = S, Se). In the case of Q = S, the high *DOS*, arising from a resonant state, is retained to $x = 0.08$, and charge carrier mobility is increased, leading to a maximum $zT = 1.6$ in $PbTe_{0.92}S_{0.08}$. By contrast the selenium analogues show a reduction in the *DOS*. Calculations suggest the loss of resonant behaviour is due to a change in overlap between Tl(6s) and Te(5p) levels, associated with a decrease in lattice constant.

Amongst the pseudo binary series $PbQ_{1-x}Q'_x$ (where Q and Q′ are different chalcogens), p-type $PbTe_{1-x}S_x$, exhibits both alloy scattering, arising from a mass fluctuation effect and nanostructuring. At $x \approx 0.25$ spinodal decomposition[102] leads to a nanocomposite[103] of a PbTe-rich phase and a partially decomposed phase of $x \approx 0.5$.[104] *In situ* growth of nanocrystals of PbS within a PbTe matrix reveals that nanostructuring reduces the thermal conductivity by over a factor of two from that of a disordered solid solution phase of the same composition.[105] Mass fluctuation scattering due to S/Te disorder combines with nanostructuring to reduce κ_L significantly, leading to an exceptionally high figure-of-merit, $zT = 2.3$ at 923 K.[106] Although the performance of n-type PbS containing nanoprecipitates of PbTe is less striking[107] ($zT \approx 0.8$ at 910 K), this provides an example of a material that is largely composed of earth-abundant elements. In the corresponding p-type telluride-selenide phases, the increase in zT (1.8 at 850 K for $PbTe_{0.85}Se_{0.15}$)[83] appears to be primarily due to a large enhancement in the power factor. This has been attributed to band tuning that leads to convergence of valence and conduction bands and increased valley degeneracy. High zT is also achieved in n-type $PbSe_{1-x}S_x$ phases,[108,109] for which it has been suggested that the performance deviates little from the expected average for a mixture of the two binary phases.[108] Remarkably, despite the apparent absence of nanostructuring in the pseudo ternary p-type system $(PbTe_{0.86})(PbS_{0.07})(PbSe_{0.07})$, multiple types of point defect scattering produce significant reductions in κ_L and a particularly high figure of merit ($zT = 2.0$ at 823 K).[110]

Alloying of lead telluride with other binary tellurides, including those of the alkaline earth metals,[111–113] Ag[114] and Cd,[115] has been investigated and the now ubiquitous nanoinclusions identified. Alloying of Ag_2Te with PbTe leads to relatively large precipitates of up to 200 nm,[114] and perhaps surprisingly their impact on thermal conductivity appears to be greater than in the case of the more commonly observed nanoparticles. The thermal conductivity approaches the theoretical minimum and zT exceeds 1.5 at 775 K. Large inclusions also play a role in scattering heat-carrying phonons in the pseudo-binary PbTe–$PbSnS_2$ system.[116,117] The layered $PbSnS_2$, segregates into lamellar structures of 50–100 nm thickness and up to 100 μm in length. Despite these large dimensions, κ_L is reduced to <0.65 W m^{-1} K^{-1}, leading to a maximum figure of merit, $zT = 1.1$ in n-type $(PbTe)_{0.94}(PbSnS_2)_{0.06}$ at the comparatively low temperature of 625 K.

Whilst nanostructuring is perhaps less explored in lead sulphide and selenide, a variety of binary sulphides (Bi_2S_3, Sb_2S_3, SrS and CaS) has been shown to induce a high degree of phonon scattering in an n-type PbS matrix, leading to $zT \approx 1.1$ at 923 K.[118] Similarly, in p-type $(PbSe)_{0.98}(SrSe)_{0.02}$, scattering of phonons due to features at both the nano- and mesoscale, increases ZT to 1.3 at 923 K. This behaviour contrasts with that of p-type $(PbSe)_{0.97}(CdS)_{0.03}$, where the increase in zT to 1.6 at 923 K is primarily due to an enhancement in the Seebeck coefficient.[86] This has been attributed to tailoring of the electronic structure, leading to a decrease in $\Delta E_{L\Sigma}$.

The limited work on lead-free rocksalt phases has focused on tin analogues. Preliminary results suggest that many of the phenomena identified in the lead phases occur in the tin congeners. For example, SnTe alloyed with low levels of CdTe and nanostructured with CdS results in p-type phases with $zT = 1.3$ at 873 K.[119] Similarly p-type $(SnTe)_{0.97}(InTe)_{0.015}(CdTe)_{0.015}$ shows evidence for the operation of multiple effects.[120] Resonant states due to In incorporation enhance $|S|$ at lower temperatures, whilst Cd promotes band convergence, enhancing $|S|$ at higher temperatures, both of which increase the power factor. In addition, nanostructuring with CdS leads to a reduction in κ_L, resulting in an increase in the figure of merit to $zT = 1.4$ at 923 K.

2.6 Tin Selenide and Related Materials

SnSe adopts an orthorhombic layered structure (Figure 2.8) derived from that of rocksalt through a three-dimensional distortion. The structure consists of strongly-bonded corrugated SnSe slabs, two atoms thick, aligned parallel to the crystallographic *ab* plane. Adjacent slabs are linked along the crystallographic *c* direction through weaker Sn–Se bonds. Physical property measurements on a single crystal of SnSe[121] reveal an effectively isotropic Seebeck coefficient but an electrical conductivity that

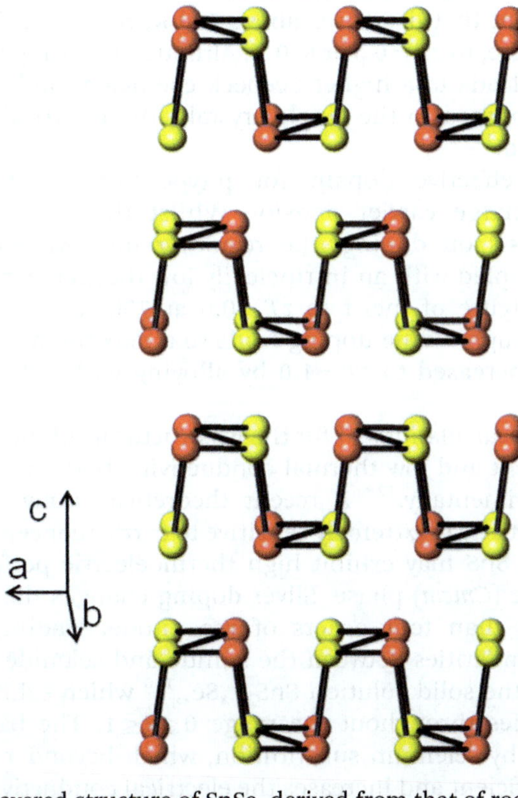

Figure 2.8 The layered structure of SnSe, derived from that of rocksalt, showing the distorted $SnSe_7$ coordination polyhedra in which there are four long and three short Sn–Se distances.

exhibits a high degree of anisotropy. Along each of the unit cell directions, conduction is metallic to 525 K, where it changes to activated semiconduction. This persists to 800 K, where a displacive phase transition (*Pnma* to *Cmcm*)[122] occurs and the conductivity becomes almost temperature independent. The highest power factor (10.1 μW cm^{-1} K^{-2}) is along *b*, which is *ca.* 35% higher than along *c*. Thermal conductivity, which is dominated by the lattice contribution, is remarkably low for a crystalline solid, lying in the range 0.23–0.34 W m^{-1} K^{-1} at 973 K. This may be associated with the pronounced anharmonicity of the bonding. The combination of an exceptionally low thermal conductivity and a moderately high power factor, lead to an unparalleled $zT = 2.6$ at 923 K along the *b* direction of single crystals of SnSe.

Given the strongly anisotropic physical properties, it is perhaps unsurprising that the figure of merit of polycrystalline SnSe does not reach the exceptional values of the single crystals.[123] Differences between electrical measurements made parallel and perpendicular to the pressing direction suggest a degree of anisotropy is carried through into the consolidated powder, consistent with the preferred orientation evident in powder X-ray diffraction data. A significant reduction in electrical conductivity and a Seebeck coefficient that decreases above 500 K, reduce *zT* from that of the single crystal phase, to $zT = 0.5$ at 820 K. Although texturing in an arc-melted sample of SnSe leads to a higher Seebeck coefficient and a lower thermal conductivity compared to the single crystals, the electrical conductivity is markedly reduced.[124]

Silver is an effective dopant for p-type SnSe[125] and significantly increases the charge carrier density. Whilst the anisotropic electrical behaviour persists on doping, the resulting improvement in electrical conductivity, coupled with an intrinsically low thermal conductivity raises the maximum figure of merit to $zT = 0.6$ at 750 K. Extension to n-type materials[126] through iodine doping leads to a maximum $zT = 0.8$ at 773 K, which may be increased to $zT = 1.0$ by alloying with 10% SnS, to reduce κ further.

Band structure calculations[127] for the isostructural sulphide predict a high Seebeck coefficient and low thermal conductivity, both of which have been confirmed experimentally.[128] A recent theoretical study[129] indicates the electrical conductivity is extremely sensitive to carrier concentration and that optimally doped SnS may exhibit high thermoelectric performance in the high-temperature (*Cmcm*) phase. Silver doping changes the carrier concentration by more than four orders of magnitude, leading to $zT = 0.6$ at 923 K.[128] The similarities between the sulfide and selenide have prompted investigation of the solid solution $SnS_{1-x}Se_x$,[130] which exhibits anisotropic physical properties throughout the range $0 \leq x \leq 1$. The band gap is continuously tuned by selenium substitution, which beyond $x = 0.2$ decreases the Seebeck coefficient and increases the electrical conductivity. A maximum of $zT = 0.82$ for $SnS_{0.2}Se_{0.8}$ at 823 K, represents a four-fold increase over that of the binary sulfide.

2.7 Oxychalcogenides

Limited attention has been given to mixed-anion compounds as potential thermoelectric materials, and in particular to those in which ionic oxide anions and more covalent chalcogenide anions coexist. The preference of 'hard' non-polarisable cations to coordinate to smaller oxide anions and 'soft' more polarisable cations to coordinate to larger chalcogenide anions, leads to the adoption by quaternary mixed-anion materials of structures in which oxide and chalcogenide anions are segregated, each coordinating preferentially to one type of cation. The low-dimensional character of the resulting structures, in which ionic and covalent bonding co-exist, leads to highly anisotropic electronic band structures, and potentially, a high *DOS* at the Fermi level. In addition, covalent character promotes high-mobility semiconduction, whereas low thermal conductivity is favoured by interactions that are more ionic in character. With a judicious choice of cations to ensure markedly different coordination preferences, a building block approach for the construction of layered oxychalcogenides can be implemented. Certain commonly encountered inorganic units, such as perovskite, fluorite, or rocksalt blocks, can be considered as two-dimensional building blocks for layered structures, in which two or more types of such unit, with contrasting electronic properties, are stacked along a given direction.

Layered oxychalcogenides often have unusually low thermal conductivities, in the range normally exhibited by amorphous materials. To date, the most promising thermoelectric performances have been reported for the family of [MO][CuQ] oxychalcogenides (M = rare earth element, Bi; Q = S, Se, Te), which are composed of alternating fluorite $[M_2O_2]^{2+}$ and antifluorite $[Cu_2Q_2]^{2-}$ layers stacked along the *c*-axis (Figure 2.9(a)). The first report of the thermoelectric performance of these materials focused on $[La_{1-x}Sr_xO][CuSe]$,[131] but interest in these materials grew considerably after the report of $zT = 0.76$ at 873 K for the bismuth analogue, $[Bi_{1-x}Sr_xO][CuSe]$.[132] As there is a relationship between the magnitude of the band gap and the maximum thermoelectric figure of merit,[133] higher zT values are found for the oxychalcogenides with smaller band gaps. Bismuth-containing phases have significantly lower band gaps than those containing rare earths, due to the contribution of Bi 6p states to the bottom of the conduction band.[134] Although the lowest band gap is found for the oxytelluride, there is a very limited number of doping studies,[135,136] and most of the effort in optimising the thermoelectric performance has centred on the bismuth oxyselenide. Given that the thermal conductivity of [BiO][CuSe] is remarkably low ($\kappa = 0.99$ W m^{-1} K^{-1}), research has focused on improving the power factor ($S^2\sigma$) through doping. Although charge carrier mobilities in these materials are rather small (~20 cm^2 V^{-1} s^{-1} for [BiO][CuSe]),[137] it has been shown that aliovalent substitution at the Bi^{3+} site in [BiO][CuSe] can lead to p-type thermoelectric materials with excellent performance. For instance, high values of the thermoelectric figure of merit, zT, have been obtained by doping with divalent (Sr^{2+}, Ca^{2+}, Pb^{2+}, Mg^{2+})[132,138–140] or monovalent

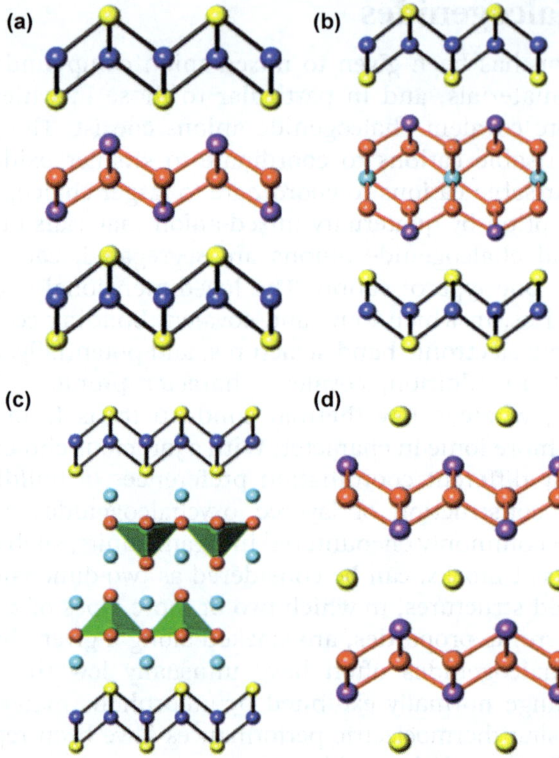

Figure 2.9 Crystal structure of the oxychalcogenides viewed along [010]. (a) [BiO][CuSe] consisting of fluorite and antifluorite blocks; (b) [Bi_2YO_4][Cu_2Se_2], which contains thicker fluorite-type oxide blocks; (c) [Sr_2FeO_3][CuSe] containing perovskite and antifluorite blocks and (d) [Bi_2O_2][Se] in which fluorite blocks alternate with square nets of chalcogen. Key: copper = blue circles; selenium = yellow circles; oxygen = red circles; bismuth = purple circles; strontium/yttrium = cyan circles. Iron-centered square pyramids in (c) are shown in green.

(Na^+, K^+) cations.[141,142] Alternatively, p-type doping can be achieved by introducing vacancies at the copper site, which leads to $zT = 0.81$ at 923 K for [BiO][$Cu_{0.975}$Se].[143] Substitution of Bi^{3+} ions with Ba^{2+}, combined with a reduction in grain size to 200–400 nm, which decreases the thermal conductivity by *ca.* 40%, results in even higher zT values of 1.1 at 923 K.[144] Similar reductions in thermal conductivity have been found for ball milled [BiO][CuSe],[145] suggesting that nanostructuring may be an effective approach to enhance the thermoelectric response of these materials. Texturing of $Bi_{0.875}Ba_{0.125}OCuSe$ by hot forging also increases zT to 1.4 at 923 K.[146] More recently, dual vacancies at the bismuth and the copper site have been exploited to reduce the thermal conductivity and control the charge carrier concentration, leading to a zT value of 0.84 at only 750 K.[147] Although these materials achieve high values of zT typically at temperatures above 800 K, thermogravimetric studies under an O_2 atmosphere have shown that oxidation occurs above 653 and 543 K for [BiO][CuSe] and [BiO][CuTe], respectively.[135,139]

Attempts to obtain the n-type doped analogues by doping at the copper site have been unsuccessful.[148] This may be a consequence of charge carrier compensation due to the ease of formation of Cu^+ vacancies in this family of oxychalcogenides. Isoelectronic substitution at the chalcogen site, to form $[BiO][CuSe_{1-x}Te_x]$ leads to improved electrical transport properties due to band-gap tuning,[149] whilst the analogous $[BiO][CuS_{1-x}Se_x]$ series exhibits a marked increase in electrical resistivity with rising sulfur content, which degrades the thermoelectric performance.[150] We have recently observed that the electrical transport properties of [BiO][CuS] are strongly dependent on the consolidation conditions, due to the volatility of sulfur at elevated temperatures.

The thermoelectric performance of the structurally-related $[Bi_2YO_4]$-$[Cu_2Se_2]$ (Figure 2.9(b)), which contains thicker oxide layers, has also been explored.[151] In this material, the nominal oxidation state of copper is increased to +1.5 and band structure calculations are consistent with the observed metallic behaviour. The large charge-carrier density associated with metallic conduction leads to a significantly reduced Seebeck coefficient (~25 μV K^{-1} at room temperature), and a value of zT of only 0.03 at 673 K.[152] Materials consisting of antifluorite $[Cu_2Se_2]^-$ layers alternating with perovskite-type oxide layers (Figure 2.9(c)) have been considered as potential thermoelectric materials, although their properties have not been investigated in any detail.[153] The electron-transport properties of $[Bi_2YO_4][Cu_2Se_2]$ and [BiO][CuSe] are dominated by the antifluorite $[Cu_2Se_2]$ layers, as the top of the valence band is formed by Cu 3d and Se 4p states.[154] It is therefore likely that the charge carrier mobility of perovskite-type oxychalcogenides such as $[Sr_2FeO_3][CuSe]$,[153] is also low, and this will result in similarly low values of the power factor to those found for [BiO][CuSe].

There are reports of thermoelectric oxychalcogenides which do not contain copper chalcogenide layers. For example, fluorite-type $[Bi_2O_2]^{2+}$ layers can be combined with Q^{2-} square-nets (Q = Se, Te), to form phases $[Bi_2O_2][Q]$, which crystallise in the anti-$ThCr_2Si_2$ structure type (Figure 2.9(d)). The n-type semiconductor $[Bi_2O_2][Se]$ exhibits $zT = 0.007$ at 300 K,[155] whilst the analogous oxy-telluride, which is a narrow band-gap semiconductor with $E_g \approx 0.23$ eV, attains $zT = 0.13$ at 573 K.[156] Rocksalt $[M_4S_6]$ layers (M = Bi, Pb) and fluorite-type $[La_2O_2]$ layers combine in $[LaO][BiPbS_3]$, which exhibits a promising $zT = 0.23$ at room temperature.[157]

2.8 Copper-containing Chalcogenides with Low Thermal Conductivities

The observation[15] of a remarkable $zT = 1.5$ (at 100 K) in the superionic conductor $Cu_{2-x}Se$, has led to the investigation of a wide range of copper-containing chalcogenides in which an ultralow thermal conductivity may be linked to a liquid-like state of highly mobile copper ions. Examples of this category of materials, termed phonon-liquid electron-crystals (PLECs), include $Cu_{2-x}S$,[158] Cu_3SbSe_3,[159] and argyrodite-type Cu_7PSe_6.[160] Whilst highly mobile cations lead to thermal conductivities below the glass limit, ionic conductivity has a detrimental effect on device performance.[161] There are however examples of copper-containing sulfo-minerals in which copper mobility is substantially reduced, and low thermal conductivity is associated with low-energy vibrations of the copper atoms.[162] In particular, p-type materials $A_{10}B_2C_4Q_{13}$ (A = Cu, Ag; B = Mn, Fe, Co, Ni, Zn, Cu, Hg, Cd; C = As, Sb, Bi; Q = S, Se).[163] with the tetrahedrite structure (Figure 2.10(a)) have attracted considerable attention.[164–170] The tetrahedrite structure is often described as a complex defect derivative of sphalerite, although it can also be considered as a collapsed sodalite-type tetrahedral framework.[171] Remarkably, the naturally-occurring mineral tetrahedrite can be used as a direct source of thermoelectric material.[170] Synthetic efforts have focused on the optimisation of the power factor in metallic $Cu_{12}Sb_4S_{13}$,[164] through partial substitution of copper by a divalent transition metal in materials $Cu_{12-x}TM_xSb_4S_{13}$ (TM = Mn, Fe, Co, Ni, and Zn, $0 \leq x \leq 2$).[164–167] $Cu_{10.4}Ni_{1.6}Sb_4S_{13}$ exhibits $zT = 0.7$ at 660 K,[165] whilst the doubly-substituted phase, $Cu_{10.5}NiZn_{0.5}Sb_4S_{13}$, attains $zT = 1.03$ at 723 K.[168] Partial substitution at the pnictogen site with tellurium results in a figure of merit, $zT = 0.92$ at 723 K for $Cu_{12}Sb_3TeS_{13}$.[172]

Other sulfur-containing minerals recently identified as potential thermoelectric materials include colusites $Cu_{26}V_2M_6S_{32}$ (M = Ge, Sn),[173,174] and bornite Cu_5FeS_4,[25] both of which have thermal conductivities below 1 W m^{-1} K^{-1} at room temperature. $zT = 0.73$ has been reported for $Cu_{26}V_2Ge_6S_{32}$ at 663 K,[173] while stoichiometric bornite reaches $zT = 0.55$ at 543 K.[25] Bornite adopts a structure related to antifluorite (Figure 2.10(b)) that

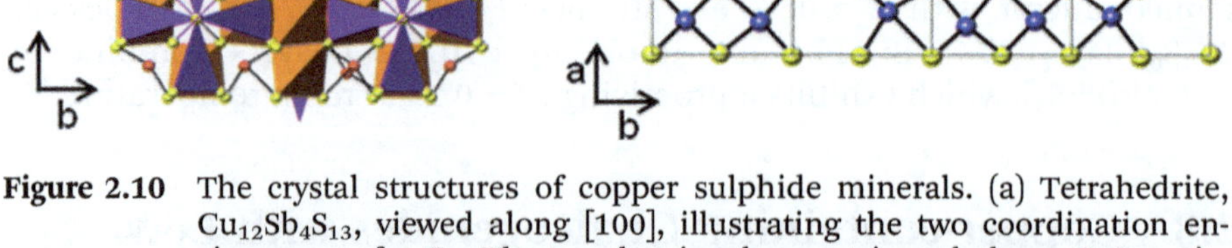

Figure 2.10 The crystal structures of copper sulphide minerals. (a) Tetrahedrite, $Cu_{12}Sb_4S_{13}$, viewed along [100], illustrating the two coordination environments for Cu. The purple triangles are trigonal-planar CuS_3 units and orange tetrahedra, CuS_4. Antimony and sulphur atoms are represented by red and yellow circles, respectively. (b) The low-temperature phase of bornite, Cu_5FeS_4 showing the vacancy ordering and alternating sphalerite and antifluorite subcells. Key: Cu/Fe = blue circles; S = yellow circles.

undergoes complex cation order–disorder transitions on heating.[175–177] At elevated temperatures a solid solution, formulated as $m\mathrm{Cu_8S_4}\cdot(1-m)$-$\mathrm{Cu_5FeS_4}$ is formed with antifluorite-type Cu_2S and gives rise to a figure of merit, $zT = 1.2$ at 900 K for $m = 0.8$.[178]

2.9 Concluding Remarks

Considerable recent effort has gone into the development of new high-performance thermoelectric materials for energy harvesting across a range of temperature scales. The hitherto perceived 'limit' of $zT = 1.0$ is now routinely exceeded in metal chalcogenides at elevated temperatures, principally through nano-structuring and manipulation of the band structure. Judicious application of a combination of approaches that target both the lattice component of the thermal conductivity and the thermoelectric power factor components of the figure-of-merit have led to materials with unprecedented performance, with $zT > 2$ having been achieved in more than one material. However, even with the tremendous advances in understanding of thermal and electron-transport properties that have been achieved, chalcogenides continue to surprise. The current record of $zT = 2.6$ (at 923 K) is found in the apparently simple binary phase, SnSe. This remarkable behaviour is observed in a particular direction in a single crystal of the material and therefore the challenge is to translate this level of performance into polycrystalline materials that can be produced in large scale.

The majority of chalcogenides appropriate to energy recovery at high temperature are derived from PbTe, which given the low terrestrial abundance of tellurium, presents challenges for large-scale implementation. However, high-performance is increasingly being realised in materials in which tellurium is wholly or partially replaced by its more earth-abundant congeners, sulfur and selenium. Sulfides and selenides are also beginning to show promise for applications in lower temperature regimes, appropriate to energy recovery from low-grade waste heat. Whilst in general, for reasons outlined above, high-performance in rocksalt-type phases is more frequently encountered in p-type than n-type materials, the advances in understanding of the relationship between the detailed electronic structure and thermoelectric response are enabling even this disparity to be addressed. The concerted efforts to enhance the performance of chalcogenide thermoelectrics have led to an evolution from what has been termed a first generation ($zT \approx 1.0$) of material through a second generation ($zT \approx 1.5$) that exploits nanostructuring to the current third generation ($zT \approx 2.0$), in which multiple approaches to enhance zT are simultaneously applied.[77] Third generation materials potentially enable energy conversion efficiencies of up to 20% to be realised and offer prospects for the creation of wholly chalcogenide-based devices. However, the fabrication of such devices will require a new set of problems to be addressed, including identification of suitable barrier layer and contact materials. Given the nanostructured nature of many of the most promising systems, the assessment of long-term stability under high-temperature operating conditions should also be a high priority. The susceptibility of chalcogenides to aerial oxidation may also prove to be a factor on prolonged operation at elevated temperatures. This is likely to necessitate the development of suitable packaging or protective coatings for the fabricated thermoelements. Providing technical issues of this nature can be successfully addressed, chalcogenide-based thermoelectrics offer genuine opportunities to make a significant contribution to energy sustainability.

References

1. G. J. Snyder and E. S. Toberer, *Nat. Mater.*, 2008, **7**, 105.
2. E. Davies, *Chem. World*, 2011, 50.
3. J. R. Sootsman, D. Y. Chung and M. G. Kanatzidis, *Angew. Chem.*, 2009, **48**, 8616.
4. A. P. Gonçalves and C. Godart, *Eur. Phys. J. B*, 2014, **87**, 42.
5. J. W. Fergus, *J. Eur. Ceram. Soc.*, 2012, **32**, 525.
6. P. Vaqueiro and A. V. Powell, *J. Mater. Chem.*, 2010, **20**, 9577.
7. A. J. Minnich, M. S. Dresselhaus, Z. F. Ren and G. Chen, *Energy Environ. Sci.*, 2009, **2**, 466.
8. G. A. Slack, in *CRC Handbook of Thermoelectrics*, ed. D. M. Rowe, CRC Press, 1995, ch. 34.
9. M. G. Kanatzidis, *Chem. Mater.*, 2010, **22**, 648.
10. J. D. Sugar and D. L. Medlin, *J. Alloys Compd.*, 2009, **478**, 75.
11. S. Kim II, K. H. Lee, H. A. Mun, H. S. Kim, S. W. Hwang, J. W. Roh, D. J. Yang, W. H. Shin, X. S. Li, Y. H. Lee, G. J. Snyder and S. W. Kim, *Science*, 2015, **348**, 109.
12. L. D. Zhao, B. P. Zhang, J. F. Li, M. Zhou, W. S. Liu and J. Liu, *J. Alloys Compd.*, 2008, **455**, 259.
13. J. L. Mi, X. B. Zhao, T. J. Zhu and J. P. Tu, *Appl. Phys. Lett.*, 2007, **91**, 172116.
14. J. P. Heremans, V. Jovovic, E. S. Toberer, A. Saramat, K. Kurosaki, A. Charoenphakdee, S. Yamanaka and G. J. Snyder, *Science*, 2008, **321**, 554.
15. H. Liu, X. Shi, F. Xu, L. Zhang, W. Zhang, L. Chen, Q. Li, C. Uher, T. Day and G. J. Snyder, *Nat. Mater.*, 2012, **11**, 422.
16. L. D. Hicks and M. S. Dresselhaus, *Phys. Rev. B*, 1993, **47**, 16631.
17. N. F. Mott and E. A. Davis, *Electronic Processes in Non-Crystalline Materials*, Clarendon Press, Oxford, 1979.
18. M. Ohta, S. Satoh, T. Kuzuya, S. Hirai, M. Kunii and A. Yamamoto, *Acta. Mater.*, 2012, **60**, 7232.
19. Z. A. Munir, U. Anselmi-Tamburini and M. Ohyanagi, *J. Mater. Sci.*, 2006, **41**, 763.
20. J. R. Sootsman, H. Kong, C. Uher, J. J. D'Angelo, C. I. Wu, T. P. Hogan, T. Caillat and M. G. Kanatzidis, *Angew. Chem., Int. Ed.*, 2008, **47**, 8618.

21. J. R. Sootsman, R. J. Pcionek, H. J. Kong, C. Uher and M. G. Kanatzidis, *Chem. Mater.*, 2006, **18**, 4993.
22. P. Baláž, M. Achimovičova, M. Baláž, P. Billik, Z. Cherkezova-Zheleva, J. M. Criado, F. Delogu, E. Dutková, E. Gaffet, F. J. Gotor, R. Kumar, I. Mitov, T. Rojac, M. Senna, A. Streleskii and K. Wieczorek-Ciurowa, *Chem. Soc. Rev.*, 2013, **42**, 7571.
23. H. Wang, J. F. Li, C. W. Nan, M. Zhou, W. S. Liu, B. P. Zhang and T. Kita, *Appl. Phys. Lett.*, 2006, **88**, 092104.
24. T. Barbier, S. Rollin-Martinet, P. Lemoine, F. Gascoin, A. Kaltzoglou, P. Vaqueiro, A. V. Powell and E. Guilmeau, *J. Am. Ceram. Soc.*, 2016, **99**, 51.
25. G. Guélou, A. V. Powell and P. Vaqueiro, *J. Mater. Chem. C.*, 2015, **3**, 10624.
26. R. B. Soriano, C. D. Malliakas, J. Wu and M. G. Kanatzidis, *J. Am. Chem. Soc.*, 2012, **134**, 3228.
27. R. B. Soriano, J. Wu and M. G. Kanatzidis, *J. Am. Chem. Soc.*, 2015, **137**, 9937.
28. K. F. Cai and X. R. He, *Mater. Lett.*, 2006, **60**, 2461.
29. K. F. Cai, X. R. He, M. Avdeev, D. H. Yu, J. L. Cui and H. Li, *J. Solid State Chem.*, 2008, **181**, 1434.
30. B. Zhang, J. He, X. Ji, T. M. Tritt and A. Kumbhar, *Appl. Phys. Lett.*, 2006, **89**, 163114.
31. H. Imai, Y. Shimakawa and Y. Kubo, *Phys. Rev. B.*, 2001, **64**, 241104.
32. M. Beaumale, T. Barbier, Y. Bréard, B. Raveau, Y. Kinemuchi, R. Funahashi and E. Guilmeau, *J. Electron. Mater.*, 2014, **43**, 1590.
33. M. Beaumale, T. Barbier, Y. Bréard, S. Hebert, Y. Kinemuchi and E. Guilmeau, *J. Appl. Phys.*, 2014, **115**, 043704.
34. F. Gascoin, R. Nunna, E. Guilmeau and Y. Bréard, *J. Alloys. Compd.*, 2012, **521**, 121.
35. E. Guilmeau, A. Maignan, C. Wan and K. Koumoto, *Phys. Chem. Chem. Phys.*, 2015, **17**, 24541.
36. M. Beaumale, T. Barbier, Y. Bréard, G. Guélou, A. V. Powell, P. Vaqueiro and E. Guilmeau, *Acta Mater.*, 2014, **78**, 86.
37. D. Li, X. Y. Qin, J. Zhang, L. Wang and H. J. Li, *Solid State Commun.*, 2005, **135**, 237.
38. D. Li, X. Y. Qin and J. Zhang, *J. Mater. Res.*, 2006, **21**, 480.
39. D. Li, X. Y. Qin, J. Zhang and H. J. Li, *Phys. Lett. A*, 2006, **348**, 379.
40. E. Guilmeau, Y. Bréard and A. Maignan, *Appl. Phys. Lett.*, 2011, **99**, 052107.
41. T. Barbier, O. I. Lebedev, V. Roddatis, Y. Bréard, A. Maignan and E. Guilmeau, *Dalton Trans.*, 2015, **44**, 7887.
42. R. Nunna, F. Gascoin and E. Guilmeau, *J. Alloys Compd.*, 2015, **634**, 32.
43. C. Wan, X. Gu, F. Dang, T. Itoh, Y. Wang, H. Sasaki, M. Kondo, K. Koga, K. Yabuki, G. J. Snyder, R. Yang and K. Koumoto, *Nat. Mater.*, 2015, **14**, 622.
44. P. Vaqueiro and A. V. Powell, *Chem Mater*, 2000, **12**, 2705.

45. A. Kaltzoglou, P. Vaqueiro, T. Barbier, E. Guilmeau and A. V. Powell, *J. Eletron. Mater.*, 2014, **43**, 2029.
46. A. Maignan, Y. Bréard, E. Guilmeau and F. Gascoin, *J. Appl. Phys.*, 2012, **112**, 013716.
47. G. C. Tewari, T. S. Tripathi and A. K. Rastogi, *Z. Krist.*, 2010, **225**, 471.
48. G. C. Tewari, T. S. Tripathi and A. K. Rastogi, *J. Electron. Mater.*, 2010, **39**, 1133.
49. Y. X. Chen, B. P. Zhang, Z. H. Ge and P. P. Shang, *J. Solid State Chem.*, 2012, **186**, 109.
50. S. Bhattacharya, R. Basu, R. Bhatt, S. Pitale, A. Singh, D. K. Aswal, S. K. Gupta, M. Navaneethan and Y. Hayakawa, *J. Mater. Chem. A*, 2013, **1**, 11289.
51. F. Gascoin and A. Maignan, *Chem. Mater.*, 2011, **23**, 2510.
52. G. A. Wiegers, *Prog. Solid State Chem.*, 1996, **24**, 1.
53. P. Jood and M. Ohta, *Materials*, 2015, **8**, 1124.
54. K. J. Range, F. Rau, M. Zabel and H. Paulus, *Z. Krist.*, 1997, **212**, 50.
55. W. S. Brower, H. S. Parker and R. S. Roth, *Am. Mineral.*, 1974, **59**, 296.
56. M. A. Peacock and J. McAndrew, *Am. Mineral.*, 1950, **35**, 425.
57. P. Gütlich, K. J. Range, C. Felser, C. Schultz-Münzenberg, W. Tremel, D. Walcher and M. Waldeck, *Angew. Chem., Int. Ed.*, 1999, **38**, 2381.
58. W. M. Skinner, G. Qian and A. N. Buckley, *J. Solid State Chem.*, 2013, **206**, 32.
59. A. Umetani, E. Nagoshi, T. Kubodera and M. Matoba, *Phys. B*, 2008, **403**, 1356.
60. J. Rothballer, F. Bachhuber, F. Pielnhofer, F. M. Schappacher, R. Pöttgen and R. Weihrich, *Eur. J. Inorg. Chem.*, 2013, 248.
61. J. Corps, P. Vaqueiro, A. Aziz, R. Grau-Crespo, W. Kockelmann, J.-C. Jumas and A. V. Powell, *Chem. Mater.*, 2015, **27**, 3946.
62. P. Vaqueiro and G. G. Sobany, *Solid State Sci.*, 2009, **11**, 513.
63. R. Weihrich and I. Anusca, *Z. Anorg. Allg. Chem.*, 2006, **632**, 1531.
64. J. Rothballer, F. Bachhuber, S. M. Rommel, T. Söhnel and R. Weihrich, *RSC Adv.*, 2014, **4**, 42183.
65. J. Corps, P. Vaqueiro and A. V. Powell, *J. Mater. Chem. A*, 2013, **1**, 6553.
66. M. Saleemi, M. S. Toprak, S. Li, M. Johnsson and M. Muhammed, *J. Mater. Chem.*, 2012, **22**, 725.
67. T. Kubodera, H. Okabe, Y. Kamihara and M. Matoba, *Phys. B*, 2006, **378–380**, 1142.
68. Y. Sakai, R. Tanakadate, M. Matoba, I. Yamada, N. Nishiyama, T. Irifune, K. Funakoshi, T. Kunimoto, Y. Higo and Y. Kamihara, *J. Phys. Soc. Jpn.*, 2015, **84**, 044705.
69. P. Mangelis, A. V. Powell and P. Vaqueiro, unpublished results.
70. M. Fujioka, T. Shibuya, J. Nakai, K. Yoshiyasu, Y. Sakai, Y. Takano, Y. Kamihara and M. Matoba, *Solid State Commun.*, 2014, **199**, 56.
71. J. He, I. D. Blum, H. Q. Wang, S. N. Girard, J. Doak, L. D. Zhao, J. C. Zheng, G. Casillas, C. Wolverton, M. Jose-Yacaman, D. N. Seidman, M. G. Kanatzidis and V. P. Dravid, *Nano Lett.*, 2012, **12**, 5979.

72. K. F. Hsu, S. Loo, F. Guo, W. Chen, J. S. Dyck, C. Uher, T. Hogan, E. K. Polychroniadis and M. G. Kanatzidis, *Science*, 2004, **303**, 818.
73. J. Androulakis, K. F. Hsu, R. Pcionek, H. Kong, C. Uher, J. J. D'Angelo, A. Downey, T. Hogan and M. G. Kanatzidis, *Adv. Mater.*, 2006, **18**, 1170.
74. P. F. R. Poudeu, J. D'Angelo, A. D. Downey, J. L. Short, T. P. Hogan and M. G. Kanatzidis, *Angew. Chem., Int. Ed.*, 2006, **45**, 3835.
75. A. V. Powell, F. Guinet, P. Vaqueiro, I. M. Wilcock and R. L. Jones, in *Thermoelectric Power Generation*, ed. T. P. Hogan, J. Yang, R. Funahashi and T. Tritt, *Mater. Res. Soc. Symp. Proc.*, Warrendale, PA, 2008, vol. 1044, p. 343.
76. K. Biswas, J. He, I. D. Blum, C. I. Wu, T. P. Hogan, D. N. Seidman, V. P. Dravid and M. G. Kanatzidis, *Nature*, 2012, **489**, 414.
77. L. D. Zhao, V. P. Dravid and M. G. Kanatzidis, *Energy Environ. Sci.*, 2014, **7**, 251.
78. Y. Lee, S. H. Lo, J. Androulakis, C. I. Wu, L. D. Zhao, D. Y. Chung, T. P. Hogan, V. P. Dravid and M. G. Kanatzidis, *J. Am. Chem. Soc.*, 2013, **135**, 5152.
79. D. L. Medlin and G. J. Snyder, *Curr. Opin Colloid Interface Sci.*, 2009, **14**, 226.
80. L. D. Zhao, J. He, S. Hao, C. I. Wu, T. P. Hogan, C. Wolverton, V. P. Dravid and M. G. Kanatzidis, *J. Am. Chem. Soc.*, 2012, **134**, 16327.
81. H. Wang, A. D. LaLonde, Y. Pei and G. J. Snyder, *Adv. Funct. Mater.*, 2013, **23**, 1586.
82. Y. Pei, H. Wang and G. J. Snyder, *Adv. Mater.*, 2012, **24**, 6125.
83. Y. Pei, X. Shi, A. LaLonde, H. Wang, L. Chen and G. J. Snyder, *Nature*, 2011, **473**, 66.
84. R. S. Allgaier, *J. Appl. Phys.*, 1961, **32**, 2185.
85. D. J. Singh, *Phys. Rev. B*, 2010, **81**, 195217.
86. L. D. Zhao, S. Hao, S. H. Lo, C. I. Wu, X. Zhou, Y. Lee, H. Li, K. Biswas, T. P. Hogan, C. Uher, C. Wolverton, V. P. Dravid and M. G. Kanatzidis, *J. Am. Chem. Soc.*, 2013, **135**, 7364.
87. L. D. Zhao, H. J. Wu, S. Q. Hao, C. I. Wu, X. Y. Zhou, K. Biswas, J. Q. He, T. P. Hogan, C. Uher, C. Wolverton, V. P. Dravid and M. G. Kanatzidis, *Energy Environ. Sci.*, 2013, **6**, 3346.
88. A. D. LaLonde, Y. Pei and G. J. Snyder, *Energy Environ. Sci.*, 2011, **4**, 2090.
89. T. R. Globus and A. O. Olesk, *Sov Phys. Semicond.*, 1985, **19**, 385.
90. A. N. Veis, R. F. Kuteinkov, S. A. Kumzerov and Y. I. Ukhanov, *Sov. Phys. Semicond.*, 1976, **10**, 1320.
91. D. Parker and D. J. Singh, *Phys. Rev. B*, 2010, **82**, 035204.
92. H. Wang, Y. Pei, A. D. LaLonde and G. J. Snyder, *Adv. Mater.*, 2011, **23**, 1366.
93. H. Wang, Z. M. Gibbs, Y. Takagiwa and G. J. Snyder, *Energy Environ. Sci.*, 2014, **7**, 804.
94. Y. Pei, A. LaLonde, S. Iwanaga and G. J. Snyder, *Energy Environ. Sci.*, 2011, **4**, 2085.

95. Y. Pei, N. A. Heinz, A. LaLonde and G. J. Snyder, *Energy Environ. Sci.*, 2011, **4**, 3640.
96. S. Wang, G. Zheng, T. Luo, X. She, H. Li and X. Tang, *J. Phys. D: Appl. Phys.*, 2011, **44**, 475304.
97. H. Wang, Y. Pei, A. D. LaLonde and G. J. Snyder, *Proc. Natl. Acad. Sci. U. S. A.*, 2012, **109**, 9705.
98. J. Androulakis, D. Y. Chung, X. Su, L. Zhang, C. Uher, T. C. Hasapis, E. Hatzikraniotis, K. M. Paraskevopoulos and M. G. Kanatzidis, *Phys. Rev. B*, 2011, **84**, 155207.
99. J. Androulakis, Y. Lee, I. Todorov, D. Y. Chung and M. G. Kanatzidis, *Phys. Rev. B*, 2011, **83**, 195209.
100. Q. Zhang, H. Wang, W. Liu, H. Wang, B. Yu, Q. Zhang, Z. Tian, G. Ni, S. Lee, K. Esfarjani, G. Chen and Z. Ren, *Energy Environ. Sci.*, 2012, **5**, 5246.
101. C. M. Jaworski, B. Wiendlocha, V. Jovovic and J. P. Heremans, *Energy Environ. Sci.*, 2011, **4**, 4155.
102. J. He, S. N. Girard, M. G. Kanatzidis and V. P. Dravid, *Adv. Funct. Mater.*, 2010, **20**, 764.
103. S. N. Girard, K. Schmidt-Rohr, T. C. Chasapis, E. Hatzikraniotis, B. Njegic, E. M. Levin, A. Rawal, K. M. Paraskevopoulos and M. G. Kanatzidis, *Adv. Funct. Mater.*, 2013, **23**, 747.
104. H. Lin, E. S. Božin, S. J. L. Billinge, J. Androulakis, C. D. Malliakas, C. H. Lin and M. G. Kanatzidis, *Phys. Rev. B*, 2009, **80**, 045204.
105. S. N. Girard, J. He, C. Li, S. Moses, G. Wang, C. Uher, V. P. Dravid and M. G. Kanatzidis, *Nano Lett.*, 2010, **10**, 2825.
106. D. Wu, L. D. Zhao, X. Tong, W. Li, L. Wu, Q. Tan, Y. Pei, L. Huang, J. F. Li, Y. Zhu, M. G. Kanatzidis and J. He, *Energy Environ. Sci.*, 2015, **8**, 2056.
107. S. Johnsen, J. He, J. Androulakis, V. P. Dravid, I. Todorov, D. Y. Chung and M. G. Kanatzidis, *J. Am. Chem. Soc.*, 2011, **133**, 3460.
108. H. Wang, J. Wang, X. Cao and G. J. Snyder, *J. Mater. Chem. A*, 2014, **2**, 3169.
109. J. Androulakis, I. Todorov, J. He, D. Y. Chung, V. Dravid and M. Kanatzidis, *J. Am. Chem. Soc.*, 2011, **133**, 10920.
110. R. J. Korkosz, T. C. Chasapis, S. H. Lo, J. W. Doak, Y. J. Kim, C. I. Wu, E. Hatzikraniotis, T. P. Hogan, D. N. Seidman, C. Wolverton, V. P. Dravid and M. G. Kanatzidis, *J. Am. Chem. Soc.*, 2014, **136**, 3225.
111. K. Biswas, J. He, G. Wang, S. H. Lo, C. Uher, V. P. Dravid and M. G. Kanatzidis, *Energy Environ. Sci.*, 2011, **4**, 4675.
112. L. D. Zhao, J. He, C. I. Wu, T. P. Hogan, X. Zhou, C. Uher, V. P. Dravid and M. G. Kanatzidis, *J. Am. Chem. Soc.*, 2012, **134**, 7902.
113. K. Biswas, J. He, Q. Zhang, G. Wang, C. Uher, V. P. Dravid and M. G. Kanatzidis, *Nat. Chem.*, 2011, **3**, 160.
114. Y. Pei, J. Lensch-Falk, E. S. Toberer, D. L. Medlin and G. J. Snyder, *Adv. Funct. Mater.*, 2011, **21**, 241.

115. K. Ahn, M. K. Han, J. He, J. Androulakis, S. Ballikaya, C. Uher, V. P. Dravid and M. G. Kanatzidis, *J. Am. Chem. Soc.*, 2010, **132**, 5227.
116. S. N. Girard, T. C. Chasapis, J. He, X. Zhou, E. Hatzikraniotis, C. Uher, K. M. Paraskevopoulos, V. P. Dravid and M. G. Kanatzidis, *Energy Environ. Sci.*, 2012, **5**, 8716.
117. J. He, S. N. Girard, J. C. Zheng, L. Zhao, M. G. Kanatzidis and V. P. Dravid, *Adv. Mater.*, 2012, **24**, 4440.
118. L. D. Zhao, S. H. Lo, J. He, H. Li, K. Biswas, J. Androulakis, C. I. Wu, T. P. Hogan, D. Y. Chung, V. P. Dravid and M. G. Kanatzidis, *J. Am. Chem. Soc.*, 2011, **133**, 20476.
119. G. Tan, L. D. Zhao, F. Shi, J. W. Doak, S. H. Lo, H. Sun, C. Wolverton, V. P. Dravid, C. Uher and M. G. Kanatzidis, *J. Am. Chem. Soc.*, 2014, **136**, 7006.
120. G. Tan, F. Shi, S. Hao, H. Chi, L. D. Zhao, C. Uher, C. Wolverton, V. P. Dravid and M. G. Kanatzidis, *J. Am. Chem. Soc.*, 2015, **137**, 5100.
121. L. D. Zhao, S. H. Lo, Y. Zhang, H. Sun, G. Tan, C. Uher, C. Wolverton, V. P. Dravid and M. G. Kanatzidis, *Nature*, 2014, **508**, 373.
122. T. Chattopadhyay, J. Pannetier and H. G. von Schnering, *J. Phys. Chem Solids*, 1986, **47**, 879.
123. S. Sassi, C. Candolfi, J. B. Vaney, V. Ohorodniichuk, P. Masschelein, A. Dauscher and B. Lenoir, *Appl. Phys. Lett.*, 2014, **104**, 212105.
124. F. Serrano-Sánchez, M. Gharsallah, N. M. Nemes, F. J. Mompean, J. L. Martinez and J. A. Alonso, *Appl. Phys. Lett.*, 2015, **106**, 083902.
125. C. L. Chen, H. Wang, Y. Y. Chen, T. Day and G. J. Snyder, *J. Mater. Chem. A*, 2014, **2**, 11171.
126. Q. Zhang, E. K. Chere, J. Sun, F. Cao, K. Dahal, S. Chen, G. Chen and Z. Ren, *Adv. Energy Mater.*, 2015, **5**, 1500360.
127. D. Parker and D. J. Singh, *J. Appl. Phys.*, 2010, **108**, 083712.
128. Q. Tan, L. D. Zhao, J. F. Li, C. F. Wu, T. R. Wei, Z. B. Xing and M. G. Kanatzidis, *J. Mater. Chem. A*, 2014, **2**, 17302.
129. B. Z. Sun, Z. Ma, C. He and K. Wu, *RSC Adv.*, 2015, **5**, 56382.
130. Y. M. Han, J. Zhao, M. Zhou, X. X. Jiang, H. Q. Leng and L. F. Li, *J. Mater. Chem. A*, 2015, **3**, 4555.
131. M. Yasukawa, K. Ueda and H. Hosono, *J. Appl. Phys.*, 2004, **95**, 3594.
132. L. D. Zhao, D. Berardan, Y. L. Pei, C. Byl, L. Pinsard-Gaudart and N. Dragoe, *Appl. Phys. Lett.*, 2010, **97**, 092118.
133. C. Wood, *Rep. Prog. Phys.*, 1988, **51**, 459.
134. S. Sallis, L. F. J. Piper, J. Francis, J. Tate, H. Hiramatsu, T. Kamiya and H. Hosono, *Phys. Rev. B*, 2012, **85**, 085207.
135. P. Vaqueiro, G. Guélou, M. Stec, E. Guilmeau and A. V. Powell, *J. Mater. Chem. A*, 2013, **1**, 520.
136. T. H. An, Y. S. Lim, H. S. Choi, W. S. Seo, C. H. Park, G. R. Kim, C. Park, C. H. Lee and J. H. Shim, *J. Mater. Chem. A*, 2014, **2**, 19759.
137. L. D. Zhao, J. He, D. Berardan, Y. Lin, J. F. Li, C. W. Nan and N. Dragoe, *Energy Environ. Sci.*, 2014, **7**, 2900.

138. F. Li, T. R. Wei, F. Kang and J. F. Li, *J. Mater. Chem A*, 2013, **1**, 11942.
139. S. D. N. Luu and P. Vaqueiro, *J. Mater. Chem. A*, 2013, **1**, 12270.
140. J. L. Lan, B. Zhan, Y. C. Liu, B. Zheng, Y. Liu, Y. H. Lin and C. W. Nan, *Appl. Phys. Lett.*, 2013, **102**, 123905.
141. J. Li, J. Sui, Y. Pei, X. Meng, D. Berardan, N. Dragoe, W. Cai and L. D. Zhao, *J. Mater. Chem. A*, 2014, **2**, 4903.
142. D. S. Lee, T. H. An, M. Jeong, H. S. Choi, Y. S. Lim, W. S. Seo, C. H. Park, C. Park and H. H. Park, *Appl. Phys. Lett.*, 2013, **103**, 232110.
143. Y. Liu, L. D. Zhao, Y. Liu, J. Lan, W. Xu, F. Li, B. P. Zhang, D. Berardan, N. Dragoe, Y. H. Lin, C. W. Nan, J. F. Li and H. Zhu, *J. Am. Chem. Soc.*, 2011, **133**, 20112.
144. J. Li, J. Sui, Y. Pei, C. Barreteau, D. Berardan, N. Dragoe, W. Cai, J. He and L. D. Zhao, *Energy Environ. Sci.*, 2012, **5**, 8543.
145. F. Li, J. F. Li, L. D. Zhao, K. Xiang, Y. Liu, B. P. Zhang, Y. H. Lin, C. W. Nan and H. M. Zhu, *Energy Environ. Sci.*, 2012, **5**, 7188.
146. J. Sui, J. Li, J. He, Y. L. Pei, D. Berardan, H. Wu, N. Dragoe, W. Cai and L. D. Zhao, *Energy Environ. Sci.*, 2013, **6**, 2916.
147. Z. Li, C. Xiao, S. Fan, Y. Deng, W. Zhang, B. Ye and Y. Xie, *J. Am. Chem. Soc.*, 2015, **137**, 6587.
148. S. D. N. Luu and P. Vaqueiro, *Semicond. Sci. Technol.*, 2014, **29**, 064002.
149. Y. Liu, J. Lan, W. Xu, Y. Liu, Y. L. Pei, B. Cheng, D. B. Liu, Y. H. Lin and L. D. Zhao, *Chem. Commun.*, 2013, **49**, 8075.
150. D. Berardan, J. Li, E. Amzallag, S. Mitra, J. Sui, W. Cai and N. Dragoe, *Materials*, 2015, **8**, 1043.
151. S. G. Tan, D. F. Shao, W. J. Lu, B. Yuan, Y. Liu, J. Yang, W. H. Song, H. Lei and Y. P. Sun, *Phys. Rev. B*, 2014, **90**, 085144.
152. S. D. N. Luu, PhD thesis, Heriot-Watt University, UK, 2015.
153. D. Berthebaud, O. I. Lebedev, D. Pelloquin and A. Maignan, *Solid State Sci.*, 2014, **36**, 94.
154. H. Hiramatsu, H. Yanagi, T. Kamiya, K. Ueda, M. Hirano and H. Hosono, *Chem. Mater.*, 2008, **20**, 326.
155. P. Ruleova, C. Drasar, P. Lostak, C. P. Li, S. Ballikaya and C. Uher, *Mater. Chem. Phys.*, 2010, **119**, 299.
156. S. D. N. Luu and P. Vaqueiro, *J. Solid State Chem.*, 2015, **226**, 219.
157. Y. L. Sun, A. Ablimit, H. F. Zhai, J. K. Bao, Z. T. Tang, X. B. Wang, N. L. Wang, C. M. Feng and G. H. Cao, *Inorg. Chem.*, 2014, **53**, 11125.
158. Y. He, T. Day, T. S. Zhang, H. Liu, X. Shi, L. Chen and G. J. Snyder, *Adv. Mater.*, 2014, **26**, 3974.
159. W. Qiu, L. Xi, P. Wei, X. Ke, J. Yang and W. Zhang, *Proc. Natl. Acad. Sci. U. S. A.*, 2014, **111**, 15031.
160. K. S. Weldert, W. G. Zeier, T. W. Day, M. Panthofer, G. J. Snyder and W. Tremel, *J. Am. Chem. Soc.*, 2014, **136**, 12035.
161. G. Dennler, R. Chmielowski, S. Jacob, F. Capet, P. Roussel, S. Zastrow, K. Nielsch, I. Opahle and G. K. H. Madsen, *Adv. Energy Mater.*, 2014, **4**, 1301581.

162. E. Lara-Curzio, A. F. May, O. Delaire, M. A. McGuire, X. Lu, C. Y. Liu, E. D. Case and D. T. Morelli, *J. Appl. Phys.*, 2014, **115**, 193515.
163. E. Makovicky, *Rev. Mineral. Geochem.*, 2006, **61**, 7.
164. X. Lu, D. T. Morelli, Y. Xia, F. Zhou, V. Ozolins, H. Chi, X. Zhou and C. Uher, *Adv. Energy Mater.*, 2013, **3**, 342.
165. T. Barbier, P. Lemoine, S. Gascoin, O. I. Lebedev, A. Kaltzoglou, P. Vaqueiro, A. V. Powell, R. I. Smith and E. Guilmeau, *J. Alloys Compd.*, 2015, **634**, 253.
166. K. Suekuni, Y. Tomizawa, T. Ozaki and M. Koyano, *J. Appl. Phys.*, 2014, **115**, 143702.
167. K. Suekuni, K. Tsuruta, T. Ariga and M. Koyano, *Appl. Phys. Express*, 2012, **5**, 051201.
168. X. Lu, D. T. Morelli, Y. Xia and V. Ozolins, *Chem. Mater.*, 2015, **27**, 408.
169. X. Fan, E. D. Case, X. Lu and D. T. Morelli, *J. Mater. Sci.*, 2013, **48**, 7540.
170. X. Lu and D. T. Morelli, *Phys. Chem. Chem. Phys.*, 2013, **15**, 5762.
171. N. E. Johnson, J. R. Craig and J. D. Rimstidt, *Am. Mineral.*, 1988, **73**, 389.
172. X. Lu and D. Morelli, *J. Electron. Mater.*, 2014, **43**, 1983.
173. K. Suekuni, F. S. Kim, H. Nishiate, M. Ohta, H. I. Tanaka and T. Takabatake, *Appl. Phys. Lett.*, 2014, **105**, 132107.
174. K. Suekuni, F. S. Kim and T. Takabatake, *J. Appl. Phys.*, 2014, **116**, 063706.
175. A. J. Frueh, *Am. Mineral.*, 1950, **35**, 185.
176. N. Morimoto and G. Kullerud, *Am. Mineral.*, 1961, **46**, 1270.
177. B. A. Grguric, A. Putnis and R. J. Harrison, *Am. Mineral.*, 1998, **83**, 1231.
178. P. Qiu, T. Zhang, Y. Qiu, X. Shi and L. Chen, *Energy Environ. Sci.*, 2014, **7**, 4000.

CHAPTER 3

Thermoelectric Oxides

COLIN NORMAN, FERIDOON AZOUGH AND ROBERT FREER*

School of Materials, University of Manchester, Oxford Road, Manchester M13 9PL, UK
*Email: robert.freer@manchester.ac.uk

3.1 Introduction

Electrical energy is the most versatile form of energy as it used in many ways for driving machines, communications, lighting, sensors and much more. The generation of electricity from other forms of energy are generally inefficient with the most efficient converting 35% (fuel cells are potentially around 60%) of the input into electricity with the majority of the rest of the input energy lost as heat. If this heat is used then further efficiency gains can be made. The utilisation of this heat is one of the aims of developments in the field of thermoelectricity. The electricity can be used for both general power generation[1–4] as in satellites and vehicles but can also be for sensors[5,6] that are separated from a power supply. The use of thermoelectric devices for cooling is also a widespread application for this technology.[7,8]

The description and definitions of thermoelectric power conversion based on the Seebeck effect are usually based on the dimensionless number, the figure-of-merit (zT) and the conversion efficiency (η), which are given by the following equations.[9]

$$zT = \frac{S^2\sigma}{\kappa}T \tag{3.1}$$

RSC Energy and Environment Series No. 17
Thermoelectric Materials and Devices
Edited by Iris Nandhakumar, Neil M. White and Stephen Beeby

Published by the Royal Society of Chemistry, www.rsc.org

Where S is the Seebeck coefficient, σ is the electrical conductivity and κ is the thermal conductivity

$$\eta = \frac{T_h - T_c}{T_h} \cdot \frac{M - 1}{M + (T_c/T_h)} \tag{3.2}$$

where $M = \sqrt{1 + Z(T_h + T_c)/2}$, T_h and T_c are the hot and cold temperatures, respectively.

In order to improve the thermoelectric properties of a material it is generally considered that it is necessary to increase zT by increasing the Seebeck coefficient and the electrical conductivity whilst at the same time decreasing the thermal conductivity. However, the thermal conductivity has two components; lattice (phonon) (κ_l) and electron (carrier) thermal conductivity (κ_e).

$$\kappa = \kappa_l + \kappa_e \tag{3.3}$$

The ratio of the electronic-thermal and electrical conductivities is proportional to temperature as given by the Wiedemann Franz law

$$\kappa_e = L\sigma T \tag{3.4}$$

where L is the Lorenz number (2.4×10^{-8} V^2 K^{-2} for free electrons).

It can thus be seen that increasing electrical conductivity will also increase the thermal conductivity of the material. Therefore, in order to minimise the thermal conductivity, the emphasis must be on the lattice thermal conductivity as will be shown later.

An increase in the numerator of the expression for zT known as the power factor (PF, $S\sigma^2$) would also be useful for improved performance. Classical physics shows that the electrical conductivity is proportional to the carrier concentration and that the Seebeck coefficient decreases with increased carrier concentration. This leads to the result that there is a maximum value for the power factor and consequently z, which corresponds to a carrier concentration of approximately 10^{19} cm^{-3} as is shown in Figure 3.1.[10,11] As well as carrier concentration it is also important for the carrier mobility to be high and also to reduce any effect of minority carriers acting in the reverse direction. The close interrelationships between the material properties mean that compromises need to be made in order to produce good thermoelectric materials.

Improvement, that is reduction, in the lattice thermal conductivity as stated above is required. Heat transfer occurs through the passage of phonons through the ceramic. These phonons are the result of lattice vibrations. Modification of the lattice will produce variations in the vibration frequencies and thus affect the passage of phonons through the system. This can be achieved by the presence of vacancies, dopants or the presence of ions which 'rattle' within a cage in the crystal structure.[12] In addition to changes in the atomic structure, the introduction of interfaces in the form of grain boundaries and layered composites also can induce phonon

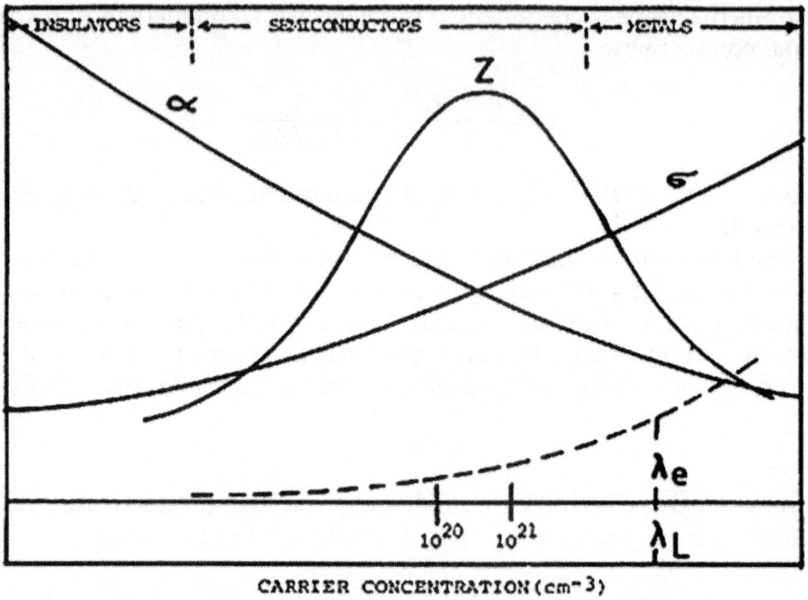

Figure 3.1 Schematic diagram showing the dependence of Seebeck coefficient α, electrical conductivity σ and thermal conductivity (electronic and lattice) λ_e λ_L and Z on carrier concentration.
Reprinted from Z. H. Dughaish, Lead telluride as a thermoelectric material for thermoelectric power generation, *Phys. B*, **322**, 205–223, Copyright (2002) with permission from Elsevier.[10]

scattering.[13,14] The thermal conductivity of a material is given by the following expression[9]

$$\kappa = \frac{1}{3} C v l \tag{3.5}$$

where C is the heat capacity, v the velocity of sound and l the mean free path of the phonons. As the grain size or layer thickness approaches l then the amount of scattering increases and conductivity falls. This effect can also be achieved through the presence of second phases. The difference between the phonon and charge carrier mean free paths allows the variation of lattice thermal conductivity independently to the electrical conductivity.

Thermoelectric material development tended to ignore the possibilities that oxides would be useful as they have low conductivity, and their ionic nature would lead to localized electrons that would be relatively immobile. However, the temperature ranges over which the existing materials could be used together with their stability, cost and toxicity led to the investigation of oxide materials. High temperature stability of oxides in air allows use at temperatures found in combustion engine exhaust systems and power stations. The availability, relatively low cost and reduced toxicology concerns of oxides have been fundamental drivers for research in this large group of materials. Many different classes of oxide materials have been investigated and include $SrTiO_3$, $CaMnO_3$, $Ca_3Co_4O_9$, $Na_xCo_2O_3$, $Bi_2Sr_2Co_2O_x$, ZnO, In_2O_3 and NiO. These systems have been well reviewed in recent years.[14–22]

The approaches used to develop oxide based thermoelectric materials have been similar over a wide range of different systems. These approaches include for a given composition *e.g.* $SrTiO_3$ to add dopants or to change the oxidation state of the cations by reduction to improve electrical conductivity, and to change the microstructure, add second phases or modify the crystal structure to reduce thermal conductivity.

The manufacture, composition and atomic structure, and microstructure of specific thermoelectric oxides will be discussed. This will be followed by a discussion of the production of thermoelectric modules containing thermoelectric oxides.

3.2 Manufacture

Oxide thermoelectric ceramics are made by conventional ceramic routes.[23] The typical routes employed for the preparation of the powders are:

1. Solid-state reaction in which the constituent oxides are mixed together and then calcined at temperatures >700 °C to form the desired phase. The starting powders can include oxides, hydroxides and carbonates.
2. Precipitation during which an intermediate is formed by the addition of a precipitant to a solution of usually mixed nitrates. The precipitants include hydroxides, carbonates and oxalates. The resulting powder is then calcined at lower temperatures than the solid-state reaction.
3. Sol–gel preparation in which organic reactants are added to the metal solution to form a homogeneous gel that is then dried and calcined.
4. Spray drying or roasting in which a solution of the metal components is sprayed into hot air to form, depending on the air temperature, either an intermediate powder or the desired oxide.

In all of these methods the aim is to produce a powder that has the correct chemical and phase composition and also suitable physical properties (crystallite size, particle size and surface area). These are required so that the sintering stage will produce a dense ceramic with the correct phase assemblage, grain size distribution, density and, ultimately, thermoelectric properties.

The powders are formed into ceramics using a variety of processing routes. There are two main groups of manufacture. The first in which a shape is formed and then sintered, and the second in which forming and sintering is one process. Alternatively, they can be classified into methods for bulk ceramics or thin films.

1. Dry pressing used to produce simple shapes
2. Slip casting used to make intricate shapes and thin films

3. Tape casting used for making thin films and laminated structures
4. Pressure-less sintering process can be adapted to take place in a variety of atmospheres and vacuum
5. Hot pressing, quicker than pressure-less sintering, giving a smaller final grain size
6. Hot isostatic pressing allows high density to be obtained
7. Spark plasma processing, very quick process allowing no reaction during sintering and little grain growth
8. Processes to produce thin films of single and multiple phases.[24]

The choice of all these methods is to ensure that the desired microstructures are formed.

In many studies of thermoelectric oxides, it has been shown that the final ceramic has contained minor phases with compositions close to that of the desired product. The reasons for this behaviour have, in general, been ascribed to inhomogeneity in the starting powder. The particle sizes of the constituents were too large to allow complete inter-diffusion of the components. The initial reaction of two of the components prior to a third can also contribute to the lack of homogeneity. One study[25] on $Ca_3Co_4O_9$ looked at four different synthesis routes. There were two types of milling differing in the energy supplied to the mixed powders and two precipitants, ammonium carbonate and oxalic acid. The grain sizes of the calcined powders were quite different varying from 100 nm to 500 nm agglomerates in the ammonium carbonate precipitated precursor to >5 μm for the lightly milled method. However, on sintering the ceramic grain sizes were similar apart from the lightly milled precursor. There were differences in the phase composition with the low energy milled route having a minor second phase after sintering as is shown in Figure 3.2. The power factors for the three most homogenous materials were higher than the other material. As $Ca_3Co_4O_9$ has a layered structure (see later) the texturing of this material through the production of platelets has also given rise to improved properties because of the anisotropy of the system.[26,27]

Sol–gel chemistry has been extensively studied as a way to form a homogeneous oxide powder prior to ceramic sintering. Superficially this looks to be the case as the initially formed gel is dried and calcined to form a single phase material. However detailed study of these systems shows that the homogeneous gel on drying frequently forms a number of different phases,[28] which subsequently react during calcination to form the single phase oxide. The formation of the pre-calcined mixture *via* a sol–gel route is on a much finer and intimate scale than the co-milling of the oxide precursors and so the problems of incomplete reaction are avoided. The one common feature of most sol–gel routes[29,30] is the formation of a carbon-rich powder, which is oxidised during drying and or calcination. This has been shown to be liable to auto-combustion,[31–33] spontaneous combustion[34] and ignition to flame.[35] As the system prior to drying contains organic material, ammonium and nitrate ions, there is the possibility of forming explosive

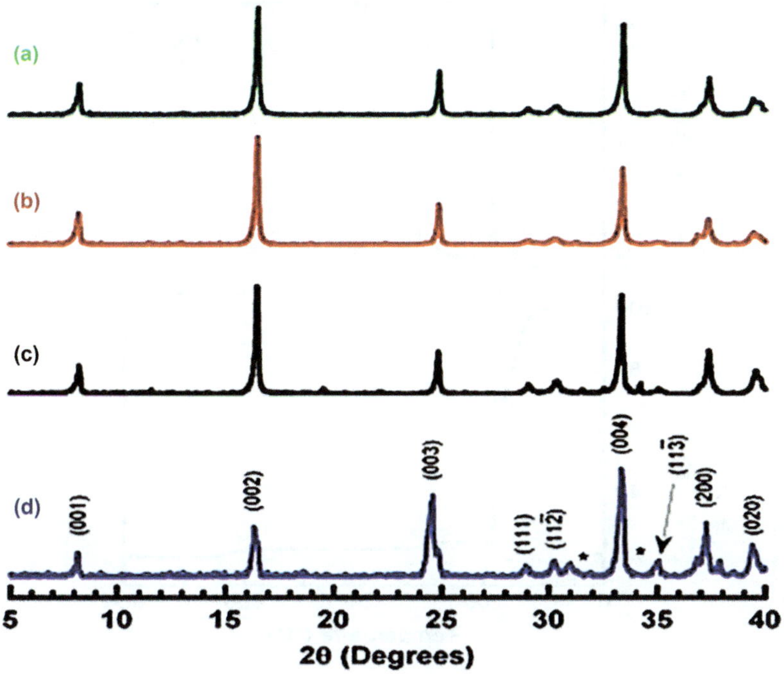

Figure 3.2 X-Ray diffraction spectra for the $Ca_3Co_4O_9$ sintered specimens prepared by four different synthesis methods (a) co-precipitation with ammonium carbonate, (b) co-precipitation with oxalic acid, (c) solid state with attrition milling, and (d) classical solid state. Crystallographic planes indicate the peaks for the $Ca_3Co_4O_9$ phase and the * shows those belonging to the $Ca_3Co_2O_6$ phase.
Reprinted from A. Sotelo *et al.*, Effect of synthesis methods on the $Ca_3Co_4O_9$ thermoelectric ceramic performances, *J. Solid State Chem.*, **221**, 247–254, Copyright (2015) with permission from Elsevier.[36]

mixtures, the nature of which is dependent on the materials used.[29,30] Though no explosions have been reported, the very rapid weight loss, sudden end to the weight loss and the exotherm reported[28,31] in the preparation of $Ca_3Co_4O_9$ suggests that some material has been ejected from the sample holder (Figure 3.3). In a variation on the sol–gel route a water soluble polymer, polyethylimine (PEI) was added to form on drying a paste was formed which on further heating underwent slow combustion with the evolution of brown fumes (nitrogen oxides).[36]

Dry pressing is the simplest of the forming methods but has the disadvantage of generally requiring the highest temperature to obtain high density which leads to larger grain size in the final ceramic. Isostatic pressing can help in densification and final grain size reduction as a higher green density is formed prior to sintering which also can remove the larger defects in the microstructure as a result of the higher pressures (200 MPa) generally used. Hot pressing and hot isostatic pressing involve the

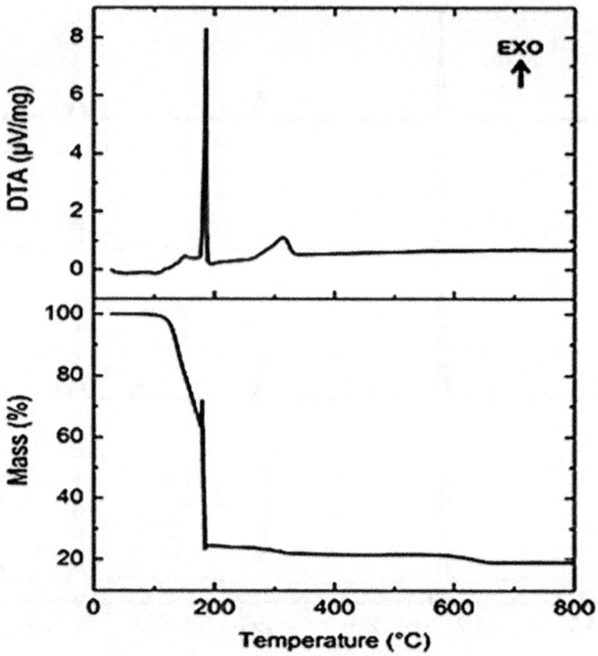

Figure 3.3 Thermogravimetric and differential thermal analysis curves of dried $Ca_3Co_4O_{9+\delta}$ citric-nitrate gel showing a rapid weight loss and associated exotherm.
Reprinted from N. Wu, T. C. Holgate, N. Van Nong, N. Pryds and S. Linderoth, High temperature thermoelectric properties of $Ca_3Co_4O_{9+\delta}$ by auto-combustion synthesis and spark plasma sintering, *J. Eur. Ceram. Soc.*, **34**, 925–931, Copyright (2014), with permission from Elsevier.[31]

application of pressure during the sintering process and give more rapid densification and because of the shorter time needed for densification lead to smaller final grain sizes. An extension of hot pressing, spark plasma sintering[37] in which pulses of direct current are passed through the sample during the pressing and sintering allows very quick sintering and little time for grain growth.

The use of slip, tape casting or spin coating is generally only used in thermoelectric oxides for the manufacture of materials of very specific shape often thin films. The production of stable suspensions of particles is critical for this process as agglomeration and settling can lead to the formation of defects within the ceramic, which are capable of degrading the properties. The dispersion of a ceramic into a polymer and the subsequent casting of a flexible tape that can be layered with similar tapes or ones containing different compositions allows the formation of layered structures that can reduce the thermal conductivity of the final ceramic. The choice of the appropriate polymer binder, lubricant dispersant combination for the ceramic powder is necessary to allow the formation of the tape, its handling, polymer removal and ceramic sintering.

Thin films have been shown to have good thermoelectric properties and have been produced by the many methods. These include physical and chemical vapour deposition, molecular beam epitaxy and pulsed laser deposition. The development of systems with an increasing number of components has led to difficulties in maintaining the desired composition. However the ability to control the position of formation of the film on a substrate has often made the fabrication of modules with p- and n-type legs close to each other easier. The production of superlattice structures in which there are alternating layers of two different phases is also possible.

3.3 Composition and Atomic Structure

The discovery of useful thermoelectric properties in semiconducting oxides has led to many investigations into them. The initial work on relatively simple oxides such as sodium cobaltite (Na_xCoO_2) and strontium titanate ($SrTiO_3$) developed in two parallel streams as the need for improved electrical conductivity and Seebeck coefficient looked at changing carrier properties through the use of dopants whilst the need for a reduction in the lattice thermal conductivity changed the local atomic structure to form phonon scattering centres. The oxide thermoelectric ceramics exist in both p-type and n-type which are necessary for optimum performance in a thermoelectric module. Common p-type materials include sodium cobaltite (Na_xCoO_2), calcium cobaltite ($Ca_3Co_4O_9$) and bismuth strontium cobaltite ($Bi_{2.1}Sr_2Co_2O_y$, BISCO). N-type materials include doped strontium titanate ($SrTiO_3$, STO), calcium manganite ($CaMnO_3$) and zinc oxide (ZnO).

3.3.1 Strontium Titanate Based Materials

Strontium titanate has the cubic perovskite structure (ABO_3) with octahedrally coordinated titanium ions and 12 coordinate strontium ions. This structure is very stable, provided that the tolerance factor (t) is kept close to 1, and can accommodate a large number of dopant ions on both cation sites, see eqn (3.6)

$$t = \frac{R_A + R_O}{\sqrt{2}(R_B + R_O)} \tag{3.6}$$

where R_x represents the atomic radii of the ions.

As the tolerance factor deviates more from the value of 1, other structures become more stable. The cubic structure becomes distorted into tetragonal and orthorhombic phases. Furthermore, changes in the ratio of filled A- and B-site occupancy can lead to the formation of Ruddlesden-Popper phases[38] (Figure 3.4) which comprise layers of perovskite structure separated by AO layers, they have a generalised formula of $AO(ABO_3)_n$.

Pure strontium titanate has reasonable power factors which coupled with its high thermal conductivity yield poor overall thermoelectric properties.

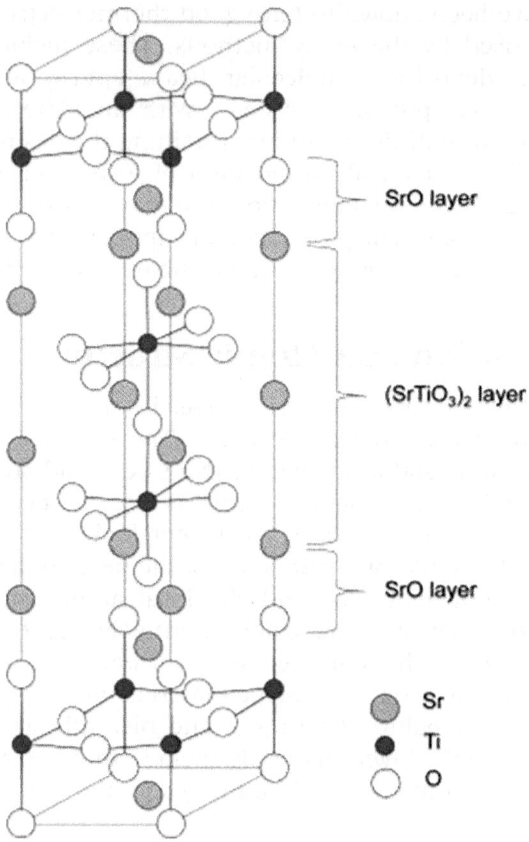

Figure 3.4 Schematic structure of Ruddlesden-Popper $SrO(SrTiO_2)_n$ where $n=2$. Reprinted from J. W. Fergus, Oxide materials for high temperature thermoelectric energy conversion, *J. Eur. Ceram. Soc.*, **32**, 525–540, Copyright (2012), with permission from Elsevier.[44]

The direction of research has been to improve the *zT via* a combination of increasing the electrical conductivity through doping or reduction of the material and lowering the thermal conductivity by introducing phonon scattering centres.

As strontium titanate is widely used for its dielectric and other ferroelectric properties, it has been doped with many different combinations of elements. For the development of improved thermoelectric properties lanthanides have mostly been used for A-site doping and niobium and tungsten on the B-site. Doping of strontium titanate has been carried out in two ways which are similar though distinct. The first is direct replacement of some of the strontium with another ion on a one for one basis. Thus the A-site to B-site cation ratio remains 1 : 1. If the dopant has the same charge as strontium (*e.g.* Ca or Ba) then simple substitution occurs. However, aliovalent substitution can be accommodated in a number of different possible ways. The addition of lanthanum, which is an electron donor, means that the extra charge has to be accommodated somewhere in the structure. This accommodation can be accomplished by either formation of oxygen interstitials, strontium vacancy formation or reduction of the titanium as shown in the following Kroger-Vink equations

$$La_2O_3 \rightarrow 2La_{Sr}^{\cdot} + O_i' \quad (3.7)$$

$$La_2O_3 \rightarrow 2La_{Sr}^{\cdot} + V_{Sr}'' \quad (3.8)$$

$$La_2O_3 \rightarrow 2La_{Sr}^{\cdot} + 2Ti_{Ti}' \quad (3.9)$$

The perovskite lattice is not an open structure so the formation of interstitial oxygen is unlikely but could at high dopant levels give rise to the formation of oxygen rich layers as found in Ruddlesden-Popper phases. The introduction of vacancies on the strontium sites can then give rise to order disorder changes in the materials based on both the two cations and the cations and vacancies. The reduction of titanium is widely considered to be present as has been observed in X-ray photoelectron studies.

The alternative is to partially replace the strontium and or the titanium whilst maintaining electrical neutrality. In this way three strontium ions are replaced by two lanthanum ions and a vacancy. Doping on both sites can be achieved by using a solid solution of strontium titanate and lanthanum niobate ($x\mathrm{SrTiO_3}\cdot(1-x)\mathrm{La_{1/3}NbO_3}$). This again has vacancies on the A-sites. Reduction of strontium titanate by sintering in argon/hydrogen atmospheres to introduce vacancies on the oxygen sites has also been widely used.

$$O_O \rightarrow V_O^{\cdot\cdot} + \frac{1}{2}O_O + 2e'' \quad (3.10)$$

As strontium titanate is widely used for its dielectric and other ferroelectric properties, it has been doped with many different combinations of elements. The lanthanides have mostly been used for A-site doping and niobium and tungsten on the B-site. The effect on the atomic structure when doping is carried out to generate A-site vacancies, depends on the doping level; at low levels there will be a random distribution of the dopant cations when $x < 0.2$; at higher levels when $x > 0.5$ the A-site cations are ordered[39,40] as shown in Figure 3.5, which also shows the presence of high concentrations of cation vacancies. In the intermediate region there is no long range order but some evidence for local ordering. This effect gives rise to a change in the properties so that there is an optimum concentration of the dopant and a maximum value of zT of 0.3 at 1000 K was achieved.[41] The thermal conductivity was reduced with increasing values of x with the appearance of a change from crystalline to glass-like behaviour. The presence of heavier dopants and vacancies produces phonon scattering and consequently reduces the thermal conductivity.

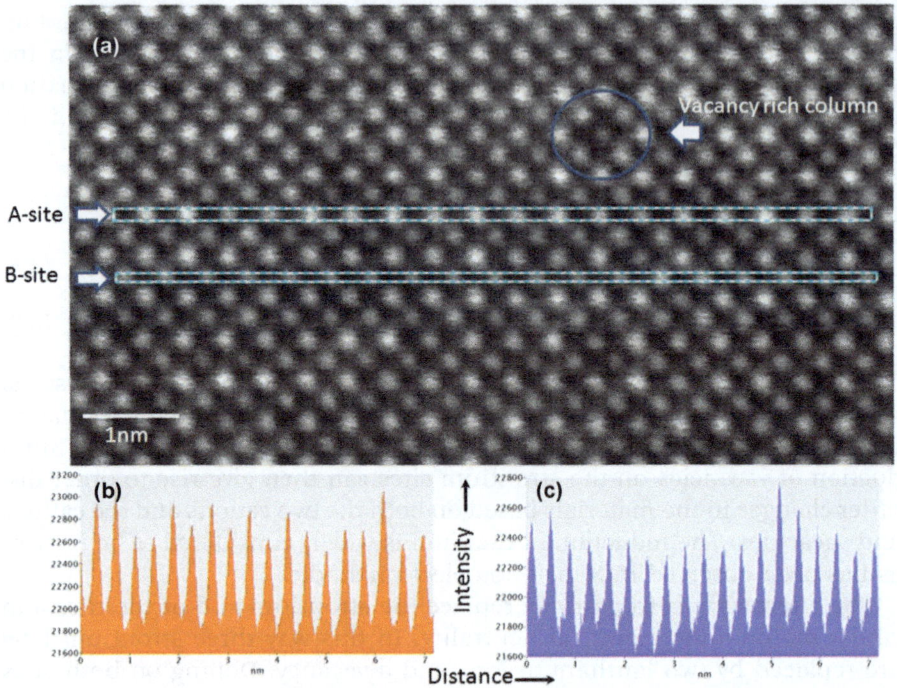

Figure 3.5 (a) HAADF (Zone axis: ⟨1 0 0⟩) image for $Sr_{0.7}La_{0.1}Ti_{0.7}Nb_{0.3}O_3$ showing both A-site and B-site atomic columns. The variation in the optical intensity along the rows indicates random distribution of Sr, La and vacancies in A-site and Ti and Nb in B-sites; (b) detailed intensity scan along A-sites; (c) detailed intensity scan along B-sites.[40]
Reprinted from D. Srivastava, F. Azough, M. Molinari, S. C. Parker and R. Freer, High-Temperature Thermoelectric Properties of $(1-x)$ $SrTiO_3-(x)$ $La_{1/3}NbO_3$ Ceramic Solid Solution, *J. Electr. Mater.*, 2014, **44**, 1803–1808 (© 2014 The Minerals, Metals & Materials Society) with permission of Springer.

The formation of a material with 65 nm grains in the final ceramic enabled a zT of 0.37 at 973 K to be obtained.[42] The material was prepared by a combination of a colloidal preparation of the oxide combined with spark plasma sintering at 1400 °C to give a 92% dense ceramic.

A study of doping with lanthanum and sintering in a reducing atmosphere showed that covering the sample with carbon 'felt' enhanced the electrical conductivity from room temperature to 900 K.[43] Further doping with niobium did not enhance the conductivity but did improve the overall zT with the best result of 0.22 at 473 K in $La_{0.05}Sr_{0.95}Nb_{0.05}Ti_{0.95}O_3$.

3.3.2 Cobaltites

These materials have a very different structure to the perovskite based oxides. The structures are based on two types of layers as is shown in Figure 3.6.[44]

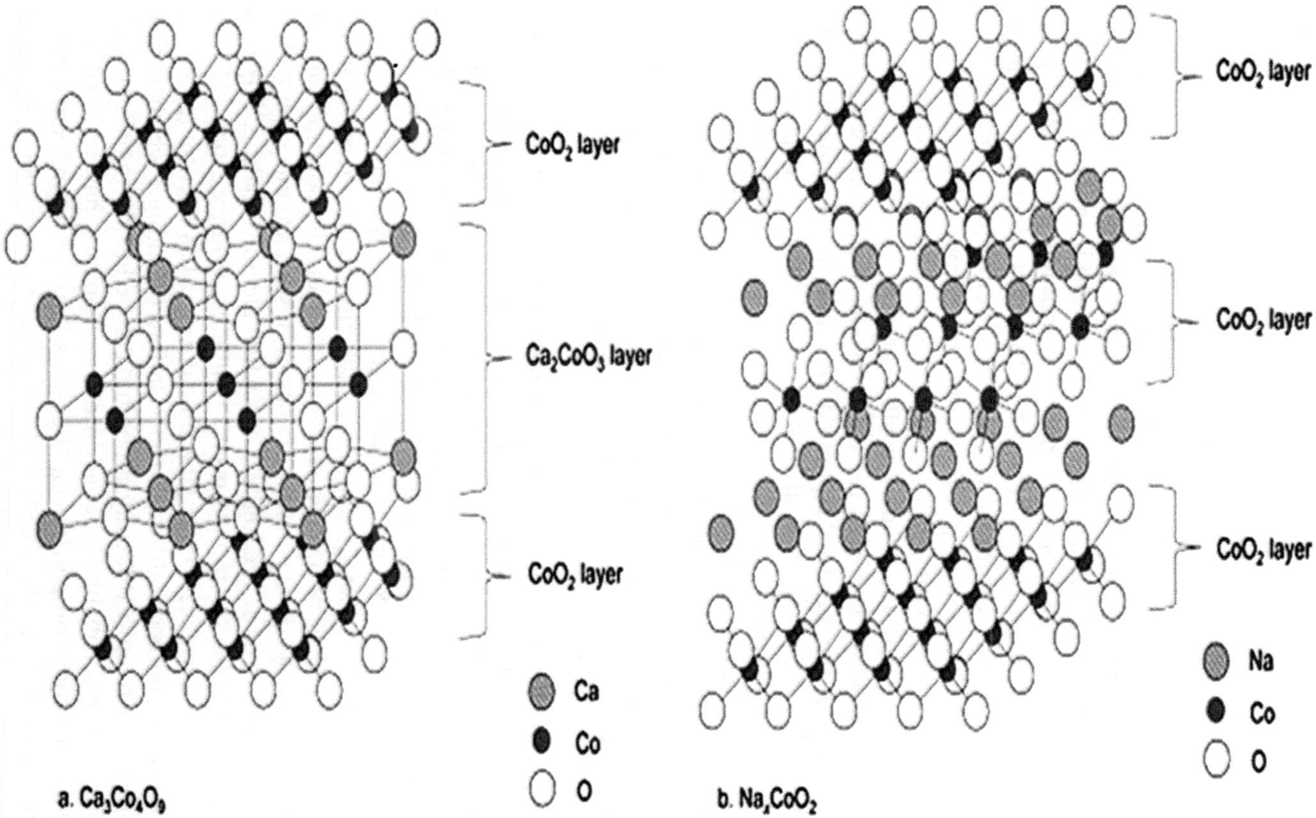

Figure 3.6 Schematic structures of (a) $Ca_3Co_4O_9$ and (b) Na_xCoO_2.
Reprinted from J. W. Fergus, Oxide materials for high temperature thermoelectric energy conversion, *J. Eur. Ceram. Soc.*, **32**, 525–540. Copyright (2012), with permission from Elsevier.[44]

Calcium cobaltite contains a CoO_2 layer and a Ca_2CoO_3 layer. Sodium cobaltite similarly has CoO_2 layers but this time they are separated by sodium layers. The two layers form misfit or incommensurate structures[45] in which they are not exactly aligned with each other or have a non-integral offset. Thus when doped with other cations there are number of extra distortions that can occur in the structure which may enhance the thermoelectric properties. One of the areas of uncertainty in cobaltite systems is the oxidation and spin states of the cobalt ions within the structure.[46,47] The low spin character of the cobalt ions has been proposed as a reason for the large Seebeck coefficients.

Calcium cobaltite, $Ca_3Co_4O_9$, can be doped with a wide range of metals. In general the use of higher atomic number dopants has given better properties. Bismuth[48–50] and silver[51,52] are two of the most widely used dopants and have been shown to improve the electrical conductivity and Seebeck coefficient. The improvement of carrier mobility through the addition of bismuth has given rise to the properties as increase in carrier concentration leads to a lower value of the Seebeck coefficient.

The addition of silver has been carried out so as to produce two distinct systems those with the silver incorporated into the calcium cobaltite structure and those with a metallic silver second phase. As a dopant, silver increases the conductivity but as it increases in amount the carrier concentration as well as its mobility has been shown to both increase and decrease the Seebeck coefficient.[51,52] Its presence as a heavy ion in the crystal structure helps to reduce the thermal conductivity. Together this has given an overall increase in the figure of merit. Silver as a metallic second phase gave an increase in electrical conductivity but a definite decrease in Seebeck coefficient.

The effect of oxygen concentration on sputtered layers showed that 20% oxygen was optimum for thermoelectric properties even though lower concentrations gave the desired phase.[53] The variation of grain size (18 nm to 2 μm) and associated increase in density did not alter the Seebeck coefficient but reduced both the electrical and thermal conductivities which led to an overall zT of 0.27 at 1000 K for the smallest grain sizes sample.[54]

Sodium cobaltite, Na_xCoO_2, was one of the first oxide thermoelectrics studied.[55] The electrical properties are dependent on the sodium content and a maximum in electrical conductivity has been seen around $x = 0.75$.[56,57] In general the values of zT are higher than with calcium cobaltite.[44] Typical values are in the range 0.1 to 0.75 over a temperature range from 273 K to 973 K as compared to 0.01 to 0.4. One problem that has been noted in the preparation of sodium cobaltites is the volatility of sodium which makes composition control more difficulty as sodium can be lost during sintering.[44]

As with calcium cobaltite the addition of dopants has allowed the improvement of some properties but there has been no significant overall improvement in zT other than with a 10% Ag addition and then only at the highest temperature when a zT of 0.9 was achieved.[58]

The use of boron as a dopant[59] improved the Seebeck coefficient at temperatures below ambient although the overall the zT was reduced.

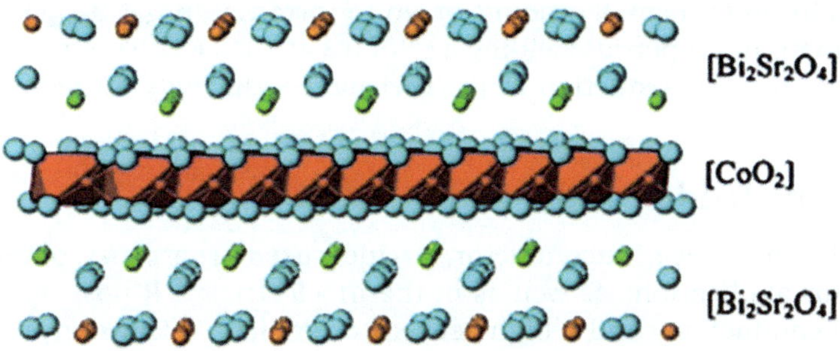

Figure 3.7 Structure of misfit layered $Bi_2Sr_2Co_{1.8}O_{7.7-\delta}$ ceramics.[64] Blue atoms are oxygen, red atoms are bismuth and green atoms strontium. Reprinted from O. Jankovský, D. Sedmidubský, K. Rubešová, Z. Sofer, J. Leitner, K. Ružička and P. Svoboda, Structure, non-stoichiometry and thermodynamic properties of $Bi_{1.85}Sr_2Co_{1.85}O_{7.7-\delta}$ ceramics, *Thermochim. Acta*, **582**, 40–45, Copyright (2014), with permission from Elsevier.

The existence of a correlation between the magnetic and thermoelectric properties was noted which may prove to be important in the future development of these materials.

Bismuth strontium cobalt oxide,[60,61] $Bi_2Sr_2Co_2O_x$, has, like the other cobaltites discussed herein, an incommensurate crystal structure[62,63] that consists of layers of CoO_2 with layers of $Bi_2Sr_2O_4$ between them. Although the bismuth and strontium ions are also present in layers as is shown in Figure 3.7.[64] Single crystal material has been shown to have a high zT of 1.1 at 973 K but on testing bulk ceramics much lower values have been obtained 0.19 at 973 K.[65,66] Although the Bi : Sr : Co ratio is nominally given as 2 : 2 : 2 the actual values used have been varied. The importance of microstructure and density was shown in a study of processing conditions. The combination of grain size and density achieved through partial melting gave enhanced properties as compared to conventional sintering and hot pressing[67] though further optimization was needed.

3.3.3 Calcium Manganate Based Materials

Calcium manganite has an orthorhombic (*Pnma*) structure based on the four unit cells of the simple perovskite structure and can accommodate many dopants on the A-sites and B-sites.[68] Pure calcium manganite has a high Seebeck coefficient ($\sim$250 $\mu V K^{-1}$). However, it has a low electrical conductivity, which contributes to a low power factor, this coupled with a high thermal conductivity gives an overall low zT. Doping on the A-site with lanthanides and on the B-site with the lower members of Groups 5 and 6 has been shown to improve the zT.[69] The effect of these dopants is to introduce some Mn^{3+} on the B site in addition to the Mn^{4+}. In addition the presence of dopants changes the crystal structure with some dopants (Sr-Mo) to tetragonal (*I4/mcm*) symmetry.

The addition of a small amount of Nb to form $CaMn_{0.98}Nb_{0.02}O_3$ gave an improvement in properties yielding a *zT* of 0.3 at 1070 K.[70] This was achieved through a dramatic reduction in the thermal conductivity of the material.

3.3.4 Zinc Oxide

Zinc oxide, an n-type semiconductor, is widely used in varistors, photovoltaic and sensor applications as well as in thermoelectrics.[71] It possesses a high electrical conductivity and a high Seebeck coefficient.[72] However, its simple crystal structure, wurtzite, gives rise to a high thermal conductivity. The emphasis of research therefore has been to reduce the thermal conductivity primarily through the addition of dopants which form nanocomposites within the bulk structure. Aluminium and gallium have been shown to be effective in improving the *zT* either separately or together in ternary doped systems.[72–75] Values of *zT* of 0.65 at 1000 °C have been demonstrated with $Zn_{0.96}Ga_{0.02}Al_{0.02}O$.[75] The structure of these materials has been shown to include distortions in the wurtzite structure,[76] which disrupts the passage of phonons.

The use of nanorods and nanoparticles has also been shown to improve the properties of zinc oxide, although the former gave rise to anisotropic properties.[74] The addition of silver particles to zinc oxide improved the properties by 35 times, yielding a highest *zT* of 0.062 for 7.5 mol% Ag–ZnO nanocomposite at 750 K, but did not reach the levels reported for aluminium and gallium doping.[77]

3.4 Microstructure

Although much work has been carried out on single-crystal specimens of many thermoelectric oxides the commercial development is centred on the use of polycrystalline materials. The reduction in the thermal conductivity of oxide thermoelectrics has been one of the dominant factors in improving the overall properties (*zT*) of materials. The control of microstructure, as with the application all other ceramic materials, is critical to ensure the optimum properties are obtained. The areas of most interest are grain size,[13,54,78,79] the nature of grain boundaries,[80–82] second phases[83,84] and the use of thin films[85] as opposed to bulk ceramics.

The reduction in grain size (and consequent increase in the number of grain boundaries) can affect two of the important thermoelectric properties of oxide ceramics, the first is the electrical conductivity and the second is the lattice thermal conductivity. The former is dependent on the nature of the grain boundaries and will be discussed below. The latter is improved by the increase in phonon scattering that occurs at the grain boundaries. One of the problems with making materials of different grain size is that as the sintering is usually carried out at different temperatures, the density of the ceramic also changes with the porosity, which in turn affects

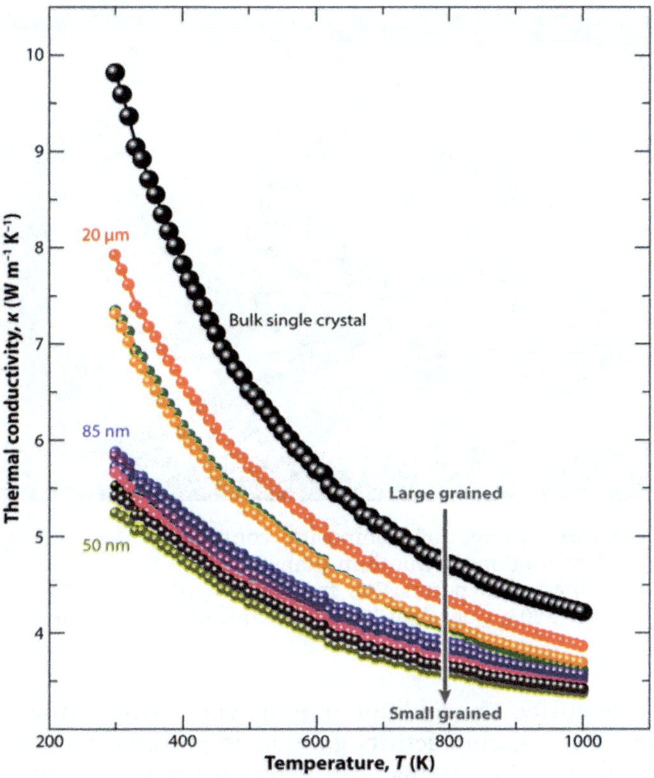

Figure 3.8 Temperature dependence of effective thermal conductivity for strontium titanate polycrystalline ceramics and a bulk single crystal.
Reproduced with permission of *Annu. Rev. Mater. Res.*, © 2010 by Annual Reviews, http://www.annualreviews.org.[14]

properties.[14,54,86,87] The effect of grain size on thermal conductivity after correction for porosity is shown in Figure 3.8.[14]

Changes to the nature of the grain boundaries in praseodymium doped strontium titanate were achieved[82] through the variation in heating rate during the spark plasma sintering process. The grain boundaries at low heating rates contained praseodymium oxide particles whereas at high heating rates a praseodymium rich grain boundary was formed. A 2.4 times increase in power factor was achieved with the praseodymium rich grain boundary phase.

The addition of second phases also serves to alter the heat transfer properties by disrupting phonon movement. The presence of insulating materials, at excessively high concentrations, would also disrupt the electrical conductivity. The addition of yttria zirconia nanoparticles to niobium doped strontium titanate both inter- and intra-granularly[88] led to a reduction in thermal conductivity. The particles acted as phonon scattering centres. Furthermore, the grain size was increased leading to an improvement in

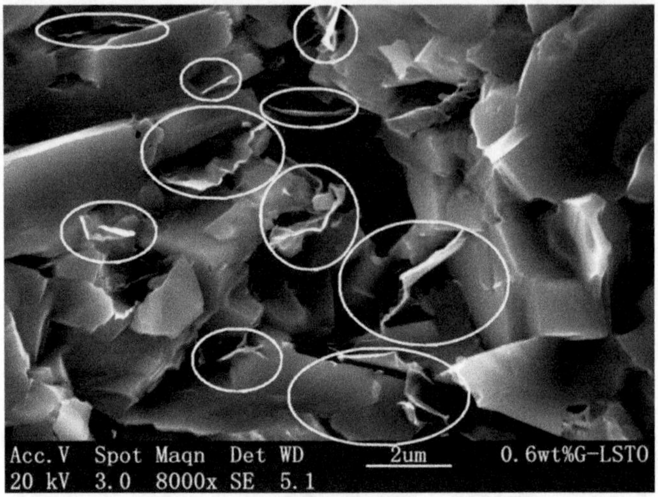

Figure 3.9 Fracture surface of composite comprising 0.6 wt% graphene in lanthanum-doped strontium titanate.
Reprinted with permission from *ACS Appl. Mater. Interfaces*, 2015, 7, 15898–15908. Copyright (2015) American Chemical Society.[84]

electrical conductivity. The addition of graphene,[84] which in itself has high electrical and thermal conductivity gave an improvement in the zT of lanthanum doped strontium titanate. In addition the temperature window of operation was widened which would allow use over a wider range. The presence of the graphene can be seen at the grain boundaries in Figure 3.9.

The use of thin films is especially valuable in the production of sensors.[24] Similarly the use of multilayers can scatter the phonons at each layer without substantially interfering with the electrical conductivity. However, the development of temperature gradients through the film is harder so a relatively higher Seebeck coefficient is required to provide the voltage but will not necessarily give a high power output.

3.5 Module Manufacture

Although there have been many investigations on the development of thermoelectric oxides relatively little has been carried out on the production of complete modules. In addition to optimisation of the individual material properties, a module also needs to be made so that its properties, and most importantly its efficiency, are suitable for the intended application. Most applications use pairs of materials p- and n-type thermoelectric oxides. The use of the pair makes the interconnects simpler as the connection goes from top to top and bottom to bottom of the legs of the ceramic pieces due to the opposite polarity of the two ceramics. The compatibility factor (s)[89] of the two materials is an important consideration. This property must be similar for the two pairs of materials in order for them and the module overall to operate at an optimum efficiency (s), which is given by eqn (3.11)

$$s = \frac{\sqrt{1+zT}-1}{\alpha T} \tag{3.11}$$

where zT is the figure-of-merit and α is the Seebeck coefficient.

The need for the compatibility factor arises from the variation in conversion efficiency of a thermoelectric as a function of the current density. If the two materials are not close to their individual maximum efficiencies then overall the module will operate at a lower overall efficiency.

One of the problems with most thermoelectric materials is that zT exhibits a maximum value with respect to temperature and therefore the effective zT for a device will depend on the actual temperature gradient along it. The best material at 500 °C may not perform very well at 100 °C. There have been a number of ways to overcome this limitation. One is to make the legs of a number of different materials so that each operates at its optimum temperature. This is the segmented approach.[90] A development of this is the cascaded thermoelectric device which consists of a vertical series of mini modules.

Simple and more complex thermoelectric designs all have common issues for effective device manufacture. Firstly, there has to be a good contact with the electrical leads usually achieved by the application of an electrode by sputtering or painting. This material must wet the ceramic and provide an Ohmic low resistance contact otherwise energy is lost within the module. The exact nature of these electrodes will need investigation once the active thermoelectric components are defined.

Long-term phase and compositional stability will also be needed, along with resistance to oxidation and reaction with water.[91] Indeed, any reaction with the electrodes must be minimised.

The applications of thermoelectric devices involve a wide range of environmental conditions. The modules may be subjected to heating and cooling cycles, vibration and impact. All of these will require that the thermal shock resistance, strength and fracture toughness properties are suitable for the intended application.[69,91–98] Within a module there will be a number of different materials present, including the thermoelectric electrodes, interconnects and heat sinks; all will have different coefficients of thermal expansion that will generate *in situ* stresses during operation and in any heating and cooling cycles.

The final consideration for the viable use of thermoelectrics is the system cost but there is no simple ideal $ per watt of power generated parameter, as device size and operating temperature range determine the value.[99]

3.6 Conclusions

Since the initial work on sodium cobaltite the range of oxide materials for thermoelectric applications, especially at higher temperatures, has grown.

The values of zT are approaching 1, which makes commercial viability more likely. However, the complex nature of the materials with variable densities, grain sizes, crystal structures, stoichiometries and oxygen contents means that it is often difficult to compare their properties. The continuing desire for both increased energy efficiency coupled with the need for evermore sensors will continue to provide the incentive for further work in this field.

References

1. A. Date, A. Date, C. Dixon and A. Akbarzadeh, *Renewable Sustainable Energy Rev.*, 2014, **33**, 371–381.
2. M. Dresselhaus, *AIP Conf. Proc.*, 2013, **1519**, 36–39.
3. R. Saidur, M. Rezaei, W. K. Muzammil, M. H. Hassan, S. Paria and M. Hasanuzzaman, *Renewable Sustainable Energy Rev.*, 2012, **16**, 5649–5659.
4. X. F. Zheng, C. X. Liu, Y. Y. Yan and Q. Wang, *Renewable Sustainable Energy Rev.*, 2014, **32**, 486–503.
5. J. E. Köhler, R. Heijl, L. G. H. Staaf, S. Zenkic, E. Svenman, A. Lindblom, A. E. C. Palmqvist and P. Enoksson, *Smart Mater. Struct.*, 2014, **23**, 095042.
6. M. Wahbah, M. Alhawari, B. Mohammad, H. Saleh and M. Ismail, *IEEE J. Emerg. Select. Top. Circuits Syst.*, 2014, **4**, 354–363.
7. I. Chowdhury, R. Prasher, K. Lofgreen, G. Chrysler, S. Narasimhan, R. Mahajan, D. Koester, R. Alley and R. Venkatasubramanian, *Nat. Nanotechnol.*, 2009, **4**, 235–238.
8. Z. B. Liu, L. Zhang, G. C. Gong, H. X. Li and G. F. Tang, *Energy Buildings*, 2015, **102**, 207–216.
9. D. M. Rowe, *CRC Handbook of Thermoelectrics*, CRC Press, 1995.
10. Z. H. Dughaish, *Phys. B*, 2002, **322**, 205–223.
11. G. J. Snyder and E. S. Toberer, *Nat. Mater.*, 2008, **7**, 105–114.
12. E. S. Toberer, A. Zevalkink and G. J. Snyder, *J. Mater. Chem.*, 2011, **21**, 15843.
13. H. Alam and S. Ramakrishna, *Nano Energy*, 2013, **2**, 190–212.
14. K. Koumoto, Y. Wang, R. Zhang, A. Kosuga and R. Funahashi, in *Annual Review of Materials Research*, ed. D. R. Clarke, M. Ruhle and F. Zok, 2010, vol. 40, pp. 363–394.
15. G. Ren, J. Lan, C. Zeng, Y. Liu, B. Zhan, S. Butt, Y.-H. Lin and C.-W. Nan, *J. Miner., Met. Mater. Soc.*, 2014, **67**, 211–221.
16. S. Walia, S. Balendhran, H. Nili, S. Zhuiykov, G. Rosengarten, Q. H. Wang, M. Bhaskaran, S. Sriram, M. S. Strano and K. Kalantar-zadeh, *Prog. Mater. Sci.*, 2013, **58**, 1443–1489.
17. A. Nag and V. Shubha, *J. Electr. Mater.*, 2014, **43**, 962–977.
18. M. Backhaus-Ricoult, J. Rustad, L. Moore, C. Smith and J. Brown, *Appl. Phys. A: Mater. Sci. Process.*, 2014, **116**, 433–470.
19. K. Koumoto, R. Funahashi, E. Guilmeau, Y. Miyazaki, A. Weidenkaff, Y. Wang and C. Wan, *J. Am. Ceram. Soc.*, 2013, **96**, 1–23.

20. A. Kołodziejczak-Radzimska and T. Jesionowski, *Materials*, 2014, **7**, 2833–2881.
21. M. W. Gaultois, T. D. Sparks, C. K. H. Borg, R. Seshadri, W. D. Bonificio and D. R. Clarke, *Chem. Mater.*, 2013, **25**, 2911–2920.
22. M. Ohtaki, *J. Ceram. Soc. Jpn.*, 2011, **119**, 770–775.
23. T. A. Ring, *Fundamentals of Ceramic Powder Processing and Synthesis*, Academic Press, 1996.
24. S. B. K. Moorthy, *Thin Film Structures in Energy Applications*, Springer International Publishing, 2015.
25. A. Sotelo, S. Rasekh, M. A. Torres, P. Bosque, M. A. Madre and J. C. Diez, *J. Solid State Chem.*, 2015, **221**, 247–254.
26. H. Itahara, C. T. Xia, J. Sugiyama and T. Tani, *J. Mater. Chem.*, 2004, **14**, 61–66.
27. H. Itahara, S. Tajima and T. Tani, *J. Ceram. Soc. Jpn.*, 2002, **110**, 1048–1052.
28. W. Zhang, Z. Wang and W. Liu, in *Energy and Environment Materials*, ed. X. F. Tang, Y. Wu, Y. Yao and Z. Z. Zhang, 2013, vol. 743–744, pp. 111–115.
29. F. Deganello, G. Marcì and G. Deganello, *J. Eur. Ceram. Soc.*, 2009, **29**, 439–450.
30. A. K. Tyagi, S. V. Chavan and R. D. Purohit, *Indian J. Pure Appl. Phys.*, 2006, **44**, 113–118.
31. N. Wu, T. C. Holgate, N. Van Nong, N. Pryds and S. Linderoth, *J. Eur. Ceram. Soc.*, 2014, **34**, 925–931.
32. P. P. Hankare, V. T. Vader, U. B. Sankpal, R. P. Patil, A. V. Jadhav and I. S. Mulla, *J. Mater. Sci.: Mater. Electron.*, 2011, **22**, 1109–1115.
33. A. Sotelo, S. Rasekh, M. A. Madre, E. Guilmeau, S. Marinel and J. C. Diez, *J. Eur. Ceram. Soc.*, 2011, **31**, 1763–1769.
34. F. Baumann, J. Fleig, H. Habermeier and J. Maier, *Solid State Ionics*, 2006, **177**, 3187–3191.
35. L. Bi, E. Fabbri and E. Traversa, *Solid State Ionics*, 2012, **214**, 1–5.
36. A. Sotelo, G. Constantinescu, S. Rasekh, M. A. Torres, J. C. Diez and M. A. Madre, *J. Eur. Ceram. Soc.*, 2012, **32**, 2415–2422.
37. M. Suarez, A. Fernandez, J. L. Menendez, R. Torrecillas, H. U. Kessel, J. Hennicke, R. Kirchner and T. Kessel, in *Sintering Applications*, ed. B. Ertug, Intech, 2013.
38. A. S. Bhalla, R. Guo and R. Roy, *Mater. Res. Innovations*, 2000, **4**, 3–26.
39. C. J. Howard, G. R. Lumpkin, R. I. Smith and Z. Zhang, *J. Solid State Chem.*, 2004, **177**, 2726–2732.
40. D. Srivastava, F. Azough, M. Molinari, S. C. Parker and R. Freer, *J. Electr. Mater.*, 2014, **44**, 1803–1808.
41. S. Popuri, A. Scott, R. Downie, M. Hall, E. Suard, R. Decourt, M. Polletc and J.-W. Bos, *RSC Adv.*, 2014, **4**, 33720–33723.
42. K. Park, J. S. Son, S. I. Woo, K. Shin, M.-W. Oh, S.-D. Park and T. Hyeon, *J. Mater. Chem. A*, 2014, **2**, 4217.
43. T. Teranishi, Y. Ishikawa, H. Hayashi, A. Kishimoto, M. Katayama and Y. Inada, *J. Am. Ceram. Soc.*, 2013, **96**, 2852–2856.

44. J. W. Fergus, *J. Eur. Ceram. Soc.*, 2012, **32**, 525–540.
45. S. V. Smaalen, *Incommensurate Crystallography*, Oxford University Press, Oxford, 2007.
46. W. Koshibae, K. Tsutsui and S. Maekawa, *Phys. Rev. B: Condens. Matter Mater. Phys.*, 2000, **62**, 6869–6872.
47. M. Karppinen, H. Fjellvåg, T. Konno, Y. Morita, T. Motohashi and H. Yamauchi, *Chem. Mater.*, 2004, **16**, 2790–2793.
48. S. Li, R. Funahashi, I. Matsubara, K. Ueno, S. Sodeoka and H. Yamada, *Chem. Mater.*, 2000, **12**, 2424–2427.
49. H. Su, Y. Jiang, X. Lan, X. Liu, H. Zhong and D. Yu, *Phys. Status Solidi A*, 2011, **208**, 147–155.
50. Y. Liu, Y. Lin, L. Jiang, C.-W. Nan and Z. Shen, *J. Electroceram.*, 2008, **21**, 748–751.
51. Y. Wang, Y. Sui, J. Cheng, X. Wang and W. Su, *J. Alloys Compds.*, 2009, **477**, 817–821.
52. M. Mikami, N. Ando and R. Funahashi, *J. Solid State Chem.*, 2005, **178**, 2186–2190.
53. L. Zhang, T. T. Tan and S. Li, *RSC Adv.*, 2015, **5**, 28158–28162.
54. M. Gunes and M. Ozenbas, *J. Alloys Compds.*, 2015, **626**, 360–367.
55. I. Terasaki, Y. Sasago and K. Uchinokura, *Phys. Rev. B: Condens. Matter Mater. Phys.*, 1997, **56**, 12685–12687.
56. D. J. Singh and D. Kasinathan, *J. Electr. Mater.*, 2007, **36**, 736–739.
57. P. Liu, G. Chen, Y. Cui, H. Zhang, F. Xiao, L. Wang and H. Nakano, *Solid State Ionics*, 2008, **179**, 2308–2312.
58. M. Ito and D. Furumoto, *J. Alloys Compd.*, 2008, **450**, 517–520.
59. E. Altin, E. Oz, S. Demirel and A. Bayri, *Appl. Phys. A: Mater. Sci. Process.*, 2015, **119**, 1187–1196.
60. R. Funahashi, I. Matsubara and S. Sodeoka, *Appl. Phys. Lett.*, 2000, **76**, 2385–2387.
61. M. Molinari, D. A. Tompsett, S. C. Parker, F. Azough and R. Freer, *J. Mater. Chem. A*, 2014, **2**, 14109–14117.
62. H. Leligny, D. Grebille, O. Perez, A. C. Masset, M. Hervieu and B. Raveau, *Acta Crystallogr., Sect. B: Struct. Sci.*, 2000, **B56**, 173–182.
63. H. Leligny, D. Grebille, O. Pi-Rez, A.-C. Masset, M. Hervieu, C. Michel and B. Raveau, *C. R. Acad. Sci.*, 1999, **Serie IIc**, 409–414.
64. O. Jankovský, D. Sedmidubský, K. Rubešová, Z. Sofer, J. Leitner, K. Ružička and P. Svoboda, *Thermochim. Acta*, 2014, **582**, 40–45.
65. R. Funahashi, I. Matsubara and S. Sodeoka, *Appl. Phys. Lett.*, 2000, **76**, 2385.
66. R. Funahashi and M. Shikano, *Appl. Phys. Lett.*, 2002, **81**, 1459–1461.
67. E. Combe, R. Funahashi, F. Azough and R. Freer, *J. Mater. Res.*, 2014, **29**, 1376–1382.
68. D. Srivastava, F. Azough, R. Freer, E. Combe, R. Funahashi, D. Kepaptsoglou, Q. M. Ramasse, M. Molinari, S. R. Yeandel, J. D. Baran and S. C. Parker, *J. Mater. Chem. C*, 2015, **3**, 12245–12259.

69. P. Thiel, S. Populoh, S. Yoon and A. Weidenkaff, *J. Solid State Chem.*, 2015, **229**, 62–67.
70. L. Bocher, R. Robert, M. H. Aguirre, S. Malo, S. Hebert, A. Maignan and A. Weidenkaff, *Solid State Sci.*, 2008, **10**, 496–501.
71. V. A. Coleman and C. Jagadish, in *Zinc Oxide Bulk, Thin Films and Nanostructures*, ed. C. Jagadish and S. Pearton, Elsevier Science Ltd, Oxford, 2006, pp. 1–20.
72. T. Tsubota, M. Ohtaki, K. Eguchi and H. Arai, *J. Mater. Chem.*, 1997, **7**, 85–90.
73. H. Takemoto, K. Fugane, P. Yan, J. Drennan, M. Saito, T. Mori and H. Yamamura, *RSC Adv.*, 2014, **4**, 2661–2672.
74. L. Han, N. Van Nong, W. Zhang, L. T. Hung, T. Holgate, K. Tashiro, M. Ohtaki, N. Pryds and S. Linderoth, *RSC Adv.*, 2014, **4**, 12353.
75. M. Ohtaki, K. Araki and K. Yamamoto, *J. Electr. Mater.*, 2009, **38**, 1234–1238.
76. Y. Michiue, N. Kimizuka and Y. Kanke, *Acta Crystallogr., Sect. B: Struct. Sci.*, 2008, **64**, 521–526.
77. Z.-H. Wu, H.-Q. Xie and Q.-F. Zeng, *Acta Phys. Sin.*, 2013, **62**, 392–397.
78. Y. Wu, S. W. Finefrock and H. Yang, *Nano Energy*, 2012, **1**, 651–653.
79. M. Backhaus-Ricoult, J. Rustad, D. Vargheese, I. Dutta and K. Work, *J. Electr. Mater.*, 2012, **41**, 1636–1647.
80. J. Wang, X. Ye, X. Yaer, B. Zhang, W. Ma and L. Miao, *Scr. Mater.*, 2015, **99**, 25–28.
81. S. Kim, K. Lee, H. Mun, H. Kim, S. Hwang, J. Roh, D. Yang, W. Shin, X. Li, Y. Lee, J. Snyder and S. Kim, *Science*, 2015, **348**, 109–114.
82. A. Mehdizadeh Dehkordi, S. Bhattacharya, T. Darroudi, J. W. Graff, U. Schwingenschlögl, H. N. Alshareef and T. M. Tritt, *Chem. Mater.*, 2014, **26**, 2478–2485.
83. G. P. Srivastava, *Rep. Prog. Phys.*, 2015, **78**, 026501.
84. Y. Lin, C. Norman, D. Srivastava, F. Azough, L. Wang, M. Robbins, K. Simpson, R. Freer and I. A. Kinloch, *ACS Appl. Mater. Interfaces*, 2015, **7**, 15898–15908.
85. B. Rivas-Murias, J. Manuel Vila-Fungueirino and F. Rivadulla, *Sci. Rep.*, 2015, **5**, 11889.
86. F. Zhang, Q. Lu, T. Li, X. Zhang, J. Zhang and X. Song, *J. Rare Earths*, 2013, **31**, 778–783.
87. L.-L. Li, X.-Y. Qin, Y.-F. Liu and Q.-Z. Liu, *Chin. Phys. B*, 2015, **24**, 067202.
88. N. Wang, H. Chen, H. He, W. Norimatsu, M. Kusunoki and K. Koumoto, *Sci. Rep.*, 2013, **3**, 3449.
89. G. J. Snyder and T. S. Ursell, *Phys. Rev. Lett.*, 2003, **91**, 148301.
90. G. J. Snyder, *Appl. Phys. Lett.*, 2004, **84**, 2436–2438.
91. F. Kahraman, J. C. Diez, S. Rașekh, M. A. Madre, M. A. Torres and A. Sotelo, *Ceram. Int.*, 2015, **41**, 6358–6363.
92. D. Kenfaui, M. Gomina, D. Chateigner and J. G. Noudem, *Ceram. Int.*, 2014, **40**, 10237–10246.

93. F. Kahraman, A. Sotelo, M. A. Madre, S. Rasekh, M. A. Torres and J. C. Diez, *J. Mater. Sci.: Mater. Electr.*, 2015, **26**, 1461–1465.
94. F. Kahraman, M. A. Madre, S. Rasekh, C. Salvador, P. Bosque, M. A. Torres, J. C. Diez and A. Sotelo, *J. Eur. Ceram. Soc.*, 2015, **35**, 3835–3841.
95. D. Kenfaui, D. Chateigner, M. Gomina and J. G. Noudem, *J. Alloys Compd.*, 2010, **490**, 472–479.
96. D. Kenfaui, G. Bonnefont, D. Chateigner, G. Fantozzi, M. Gomina and J. G. Noudem, *Mater. Res. Bull.*, 2010, **45**, 1240–1249.
97. D. Kenfaui, M. Gomina, D. Chateigner and J. G. Noudem, *Ceram. Int.*, 2014, **40**, 10237–10246.
98. S. Pinitsoontorn, N. Lerssongkram, A. Harnwunggmoung, K. Kurosaki and S. Yamanaka, *J. Alloys Compd.*, 2010, **503**, 431–435.
99. S. LeBlanc, S. K. Yee, M. L. Scullin, C. Dames and K. E. Goodson, *Renewable Sustainable Energy Rev.*, 2014, **32**, 313–327.

CHAPTER 4

Nano- and Micro-fabrication Techniques for Improving Thermoelectric Materials and Generators

DOUGLAS J. PAUL

School of Engineering, University of Glasgow, Rankine Building, Oakfield Avenue, Glasgow G12 8LT, UK
Email: Douglas.Paul@glasgow.ac.uk

4.1 Introduction

Thermoelectrics are not a new technology for converting heat into electricity. Thomas Johann Seebeck first demonstrated in 1822 that the application of different temperatures across a material would produce a thermoelectric voltage, now called the Seebeck voltage. Jean Charles Athanase Peltier then demonstrated in 1834 that a current applied to a thermoelectric material could be used to pump heat. It was not until the seminal work of William Thomson, later the first Baron Kelvin in 1852 that the physics behind the Seebeck and Peltier effects were derived. It is this understanding of thermoelectrics that is key to improve the efficiency and output power from thermoelectric generators.

The first practical thermoelectric generators were produced in the 1950s when BiTeSb-alloys were also found to be the highest performance materials for applications close to room temperature. In the 1960s, SiGe alloys were

RSC Energy and Environment Series No. 17
Thermoelectric Materials and Devices
Edited by Iris Nandhakumar, Neil M. White and Stephen Beeby

Published by the Royal Society of Chemistry, www.rsc.org

developed for high temperature thermoelectric generators using radioisotopes as heaters for NASA deep space missions. The resulting radioisotope thermoelectric generators using Pu^{238} as fuel were launched in Voyager 1 in 1977 and Voyager 2 in 1979 and have been operating, for over 38 years at the time of writing this article, without any maintenance. Indeed, this provides a clear demonstration of one of the key benefits of thermoelectric generators: they have no mechanical parts that can easily break and so they are low maintenance, fit-and-forget energy harvesters.

Figure 4.1 provides a view of the key, bulk thermoelectric materials that have been used to fabricate thermoelectric generators and Peltier coolers. The figure-of-merit, $zT = \frac{\alpha^2\sigma}{\kappa}T$ is plotted for both n-type and p-type semiconductor materials where α is the Seebeck coefficient, σ is the electrical conductivity, κ is the thermal conductivity and T is the temperature. There are a number of problems with the materials in Figure 4.1. The first is that the zT values are quite low especially when they are used to calculate the thermodynamic conversion efficiency (see Figure 4.15). The second is that Te is the eighth rarest element on earth and the costs of the material fluctuate as the supply is not sufficiently plentiful. There are therefore concerns that Te supplies could run out in the future. The final issue is that Pb is a toxic material and therefore PbTe cannot be used in consumer products. As PbTe is the highest performing thermoelectric material in the temperature regime for energy harvesting for the automotive sector, new non-toxic thermoelectrics must be found if they are to be used in vehicles.

The research literature for thermoelectrics is full of new thermoelectric materials where nanostructures have been used to provide increased zT performance over the commercially used materials in Figure 4.1. This chapter will derive the physics that determines the electrical and thermal properties that allow new thermoelectric materials to be designed with improved

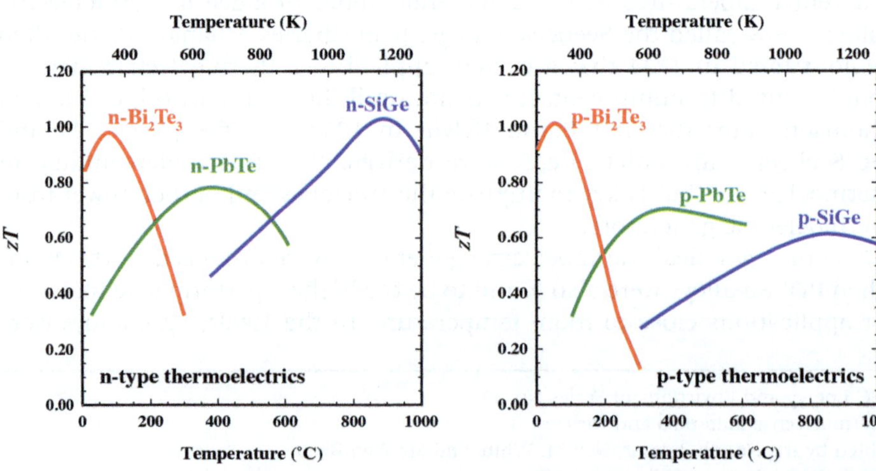

Figure 4.1 The figure-of-merit, zT plotted as a function of temperature for both n-type and p-type thermoelectric materials.
Data from Synder.[19]

figures-of-merit. A review of some of the new low-dimensional materials will be undertaken and the results compared to the bulk materials presented in Figure 4.1. Finally, micro-fabricated thermoelectric generators will be reviewed, looking at the main design parameters that require consideration when such generators are being optimised for specific applications.

4.2 Low-dimensional Electrical Conductivity

The modifications that low-dimensional structures play on the density of states and the electronic transport must be considered before any enhancements from micro- or nano-scale dimensions can be understood. There are many review articles and books on low-dimensional electronic transport[1,2] that the interested reader may consult for more detail. Here, the basics are provided so the reader may understand the key length scales required to perturb and enhance the key parameters for thermoelectric performance.

To work out the dimensionality of the electron or phonon transport in any material, the scattering and characteristic lengths scales defined below must be compared to the important lengths in the sample: the length, L, width, w and the thickness t. Care is always required in this analysis as the electrical and thermal transport may have different dimensionality due to the significant differences in the respective scattering lengths.

The electrical conductivity, σ in the Drude model is defined in terms of the elastic scattering time, τ, the effective mass of the electrons or holes in the material, m^*, and the carrier density, n and the electron charge, q as

$$\sigma = \frac{nq^2\tau}{m^*} \tag{4.1}$$

This is a very simple equation, which disguises the complexity of the transport through all the scattering mechanisms being condensed into the single scattering time τ. In general, the scattering length, l_x for some scattering process x associated with the scattering time, τ_x is linked through the diffusion constant, D as

$$l_x = \sqrt{D\tau_x} \tag{4.2}$$

It is normally the mobility of a semiconductor that is measured and quoted, which is defined as $\mu = \frac{q}{m^*}\tau$. The Einstein relation links the mobility to the diffusion constant through

$$\mu = \frac{qD}{k_B T} \tag{4.3}$$

Only electrons close to the chemical potential (or Fermi energy, E_F) require to be considered in most of the electron transport and the relevant length scale is the Fermi wavelength

$$\lambda_F = \frac{2\pi}{k_F} = \sqrt{\frac{2\pi}{n}} = \frac{h}{\sqrt{2m^* E_F}} \tag{4.4}$$

where h is Planck's constant.

The elastic scattering length of the electrons is frequently called the mean free path, ℓ, although strictly speaking the mean free path is the shortest of all the electron scattering lengths which must include inelastic scattering, phase coherent scattering, electron–electron scattering, impurity scattering and so on. The mean free path, when elastic scattering dominates, is defined generally and for each low dimension as

$$\ell = \upsilon_F \tau = \frac{\hbar k_F}{m^*}\tau \tag{4.5}$$

$$\ell_{3D} = \frac{\hbar}{m^*}\left(3\pi^2\frac{n}{g_v}\right)^{\frac{1}{3}}\frac{\mu m^*}{q} = \frac{\hbar\mu}{q}\left(3\pi^2\frac{n}{g_v}\right)^{\frac{1}{3}} \tag{4.6}$$

$$\ell_{2D} = \frac{\hbar\mu}{q}\sqrt{2\pi\frac{n}{g_v}} \tag{4.7}$$

$$\ell_{1D} = \frac{\hbar\mu}{q}\pi\frac{n}{g_v} \tag{4.8}$$

where g_v is the valley degeneracy of the semiconductor. The above equations have all assumed that the electrons have a spin degeneracy of two, which is only untrue under large magnetic fields sufficient to split the spin degeneracy.

Before calculating the electronic conductivity for some of the nanostructures, we need to first look at the energies of electron states in low-dimensional structures of quantum wells (2D), nanowires (1D) and quantum dots (0D). The simplest 2D system to solve is the quantum well which confines electrons inside dimensions that are smaller than the Fermi wavelength. We require to define a potential in the z-direction where inside the quantum well the electrostatic potential, $V = 0$ and outside the quantum well the potential is infinite ($V = \infty$) as illustrated in Figure 4.2. We can solve for a finite potential outside the quantum well but the infinite potential provides a simpler and easier mathematical solution. To find the wavefunctions, $\psi(z)$ and energy levels, E, inside the quantum well we require to solve the time independent Schrödinger equation which is

$$-\frac{\hbar^2}{2m^*}\frac{d^2\psi(z)}{dz^2} + V(z)\psi(z) = E\psi(z) \tag{4.9}$$

Since outside the quantum well the potential has been set to $V = \infty$, there are no solutions and the electrons cannot occupy any regions outside the quantum well. We have defined the potential inside the quantum well

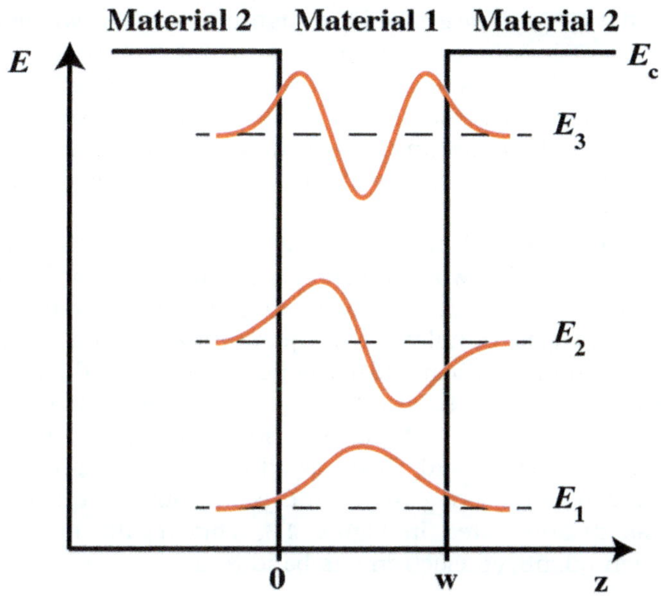

Figure 4.2 The quantised energy states for a quantum well of width, w formed in material 1, which has a lower conduction band edge that material 2.

as $V(z) = 0$ for z between 0 and the quantum well width w and so the Schrödinger equation reduces to the Helmholtz or wave equation as

$$-\frac{\hbar^2}{2m^*}\frac{d^2\psi(z)}{dz^2} = E\psi(z) \tag{4.10}$$

The simplest method to solve this is by using trial solutions. Since this is similar to the wave equation, any sinusoidal vibration that is a travelling wave inside the quantum well could be a solution. Therefore trial solutions could be $\psi(z) = A\sin(kz)$, $\psi(z) = A\cos(kz)$, $\psi(z) = A\exp(ikz)$ and $\psi(z) = A\exp(-ikz)$, where A is the normalised amplitude of the wavefunction and $i = \sqrt{-1}$. Any mixtures of these trial wavefunctions is also a potential solution. The solution must adhere to the boundary conditions which state that $\psi(\pm\infty) = 0$ and $\psi(0) = \psi(z) = 0$ so that the wavefunctions cannot penetrate outside the quantum well. Quantum mechanics also requires that $\psi(z)$ and $\frac{d\psi}{dz} = 0$. When these boundary conditions are applied to all of the potential trial solutions, the only valid solution for the wavefunctions is now

$$\psi(z) = A_n \sin(k_n z) \quad \text{with} \quad k_n = \frac{n\pi}{w} \quad \text{and} \quad n = 1, 2, 3, \ldots. \tag{4.11}$$

The energy of the wavefunctions can be found by substituting eqn (4.11) back into eqn (4.10) which produces

$$E = \frac{\hbar^2 k_n^2}{2m^*} = \frac{\hbar^2 \pi^2 n^2}{2m^* w^2} \quad \text{for } n = 1, 2, 3, \ldots \ldots \tag{4.12}$$

The integers $n = 1, 2, 3, \ldots$ are called the quantum numbers which define the solutions to the Schrödinger equation in terms of quantized energy levels $E_1, E_2, \ldots, E_n$, which are called sub-band states. What is different between the quantum mechanical solutions and the classical solutions to the problem is that in the quantum system the ground state always has finite energy called the zero-point energy of the system whilst the classical ground state has zero energy.

To form a quantum well in semiconductor materials which are the best thermoelectric materials, two semiconductors with different band gaps are required. Provided the conduction band edges are at different energies (or the valence band edges for holes) then a quantum well forms in the material with the lower conduction band edge (highest valence band edge for holes). This is illustrated in Figure 4.2 as a 1D potential through different materials. Real quantum wells are in 3D material so the electrons have parabolic dispersions along the x- and y-axis as illustrated in Figure 4.3. Hence the electrons can move along the x- and y-axes but are quantized in the z-direction as demonstrated in Figure 4.3. This results in the complete energy for each quantized electron sub-band to be

$$E = \frac{\hbar^2 k_{n,x}^2}{2m^*} + \frac{\hbar^2 k_{n,y}^2}{2m^*} + \frac{\hbar^2 \pi^2 n^2}{2m^* w^2} \quad \text{for } n = 1, 2, 3, \ldots\ldots \tag{4.13}$$

One of the parameters, which is key to understanding any potential enhancements to the thermoelectric performance for nanoscale dimensions, is the density of states, which is the number of available states that an electron can occupy in any semiconductor material. The density of states, $g(E)$ is

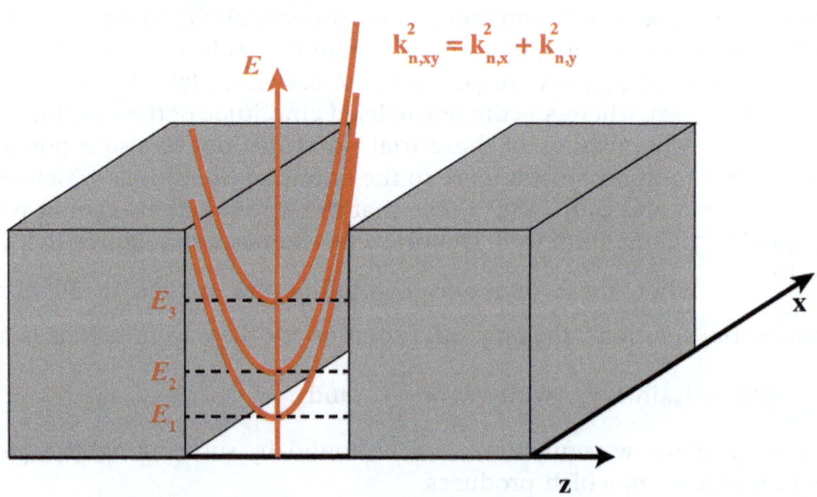

Figure 4.3 The parabolic dispersions in the x-direction for the lowest three sub-band states for a quantum well in the z-direction.

defined as the number of states per unit energy per unit volume of real space where if N is the number of states then

$$g(E)=\frac{\mathrm{d}N}{\mathrm{d}E} \tag{4.14}$$

The density of states is obtained by summing the number of states between E and $E+\mathrm{d}E$ in energy. In **k**-space, the total number of states, N is equal to the volume of the sphere or radius **k** first divided by the volume occupied by one state and then divided by the volume of real space. Therefore for a 3D volume defined by a cube with side length, L then the volume of one state in **k**-space is $\left(\frac{2\pi}{L}\right)^3$. The number of states is therefore given by

$$N=g_\mathrm{v}g_\mathrm{s}\frac{4\pi k^3}{3}\frac{1}{\left(\frac{2\pi}{L}\right)^3}\frac{1}{L^3} \tag{4.15}$$

$$=g_\mathrm{v}g_\mathrm{s}\frac{4\pi k^3}{3(2\pi)^3} \tag{4.16}$$

where g_v is the degeneracy of the valleys and g_s is the spin degeneracy. For an indirect band gap material such as silicon with six valleys at the bottom of the valley band with cubic symmetry, $g_\mathrm{v}=2$ whilst for any direct band gap material, there is a single valley with $g_\mathrm{v}=1$. Bi_2Te_3 is much more complicated and requires relativistic spin orbit corrections[3] to show that the conduction band minima has six valleys with trigonal symmetry so at all temperatures $g_\mathrm{v}=2$ but there are also a number of valleys which are only 30 meV above the conductions band edge and can contribute to the electronic transport at room and higher temperatures. Therefore the number of valleys participating in the electron transport above room temperature many be far higher than the $g_\mathrm{v}=2$ suggests. The spin degeneracy is always two unless it is lifted through the application of a large magnetic field. To work out the density of states we require to split eqn (4.14) into

$$g=\frac{\mathrm{d}N}{\mathrm{d}E}=\frac{\mathrm{d}N}{\mathrm{d}k}\frac{\mathrm{d}k}{\mathrm{d}E} \tag{4.17}$$

Differentiating eqn (4.16) produces

$$\frac{\mathrm{d}N}{\mathrm{d}k}=g_\mathrm{v}g_\mathrm{s}\frac{4\pi k^2}{(2\pi)^3} \tag{4.18}$$

Since the parabolic bands of the effective mass theory provide $E=\frac{\hbar^2k^2}{2m^*}$, which on rearrangement gives

$$k=\sqrt{\frac{2m^*E}{\hbar^2}} \tag{4.19}$$

On taking the derivative with respect to the energy this results in

$$\frac{\mathrm{d}k}{\mathrm{d}E} = \sqrt{\frac{2m^*}{\hbar^2}}\frac{E^{-\frac{1}{2}}}{2} \tag{4.20}$$

Eqn (4.18) and (4.20) can now be combined and using eqn (4.19) to change k into energy results in the 3D density of states

$$g_{3\mathrm{D}}(E) = \frac{g_s g_v}{4\pi^2}\left(\frac{2m^*}{\hbar^2}\right)^{\frac{3}{2}} E^{\frac{1}{2}} \tag{4.21}$$

The same approach can be used to obtain the density of states for 2D systems. The number of states in 2D is given by

$$N_{2\mathrm{D}} = g_v g_s (4\pi k^2) \frac{1}{\left(\frac{2\pi}{L}\right)^2} \frac{1}{L^2} \tag{4.22}$$

$$= g_v g_s \frac{\pi k^2}{(2\pi)^2} \tag{4.23}$$

and repeating the trick of eqn (4.17) produces the 2D density of states as

$$g_{2\mathrm{D}}(E) = g_s g_v \frac{m^*}{2\pi\hbar^2} \tag{4.24}$$

Eqn (4.24) is the result for a single sub-band in the quantum well being populated (see Figure 4.2). Most thermoelectrics are degenerately doped with many sub-bands being populated and so eqn (4.24) requires a summation over all populated sub-band states. Summing over all n populated sub-band states and using the Heaviside step function, Θ results in the following for the 2D density of states

$$g_{2\mathrm{D}}(E) = \sum_{i=1}^{n} g_s g_v \frac{m^*}{2\pi\hbar^2} \Theta(E - E_i) \tag{4.25}$$

The same approach can be followed to produce the 1D and 0D density of states. For all dimensions, the density of states are

$$g_{3\mathrm{D}}(E) = \frac{g_s g_v}{4\pi^2}\left(\frac{2m^*}{\hbar^2}\right)^{\frac{3}{2}} E^{\frac{1}{2}} \tag{4.26}$$

$$g_{2\mathrm{D}}(E) = \sum_{i=1}^{n} g_s g_v \frac{m^*}{2\pi\hbar^2} \Theta(E - E_i) \tag{4.27}$$

$$g_{1D}(E) = \frac{1}{2\pi\hbar}\sum_{i=1}^{n} g_s g_v \sqrt{\frac{m^*}{2}}\pi\hbar^2\Theta(E - E_i) \tag{4.28}$$

$$g_{0D}(E) = \sum_{i=1}^{n} g_s g_v \delta(E - E_i) \tag{4.29}$$

where $\delta(E - E_i)$ is the Dirac delta function. Figure 4.4 provides a plot of the density of states for 3D, 2D, 1D and 0D with examples of the chemical potential (Fermi energy) indicating which states are populated.

To calculate the number of electrons participating in the electronic transport, the Fermi function,

$$f(E) = \frac{1}{\exp\left[\frac{E - E_F}{k_B T}\right] + 1} \tag{4.30}$$

must be multiplied by the density of states and then both integrated over all the energy to give

$$n = \int f(E)g(E)\mathrm{d}E \tag{4.31}$$

For the calculation of the electrical conductivity, the density of states and the derivative of the Fermi function with respect to energy must be integrated over all energies of the electrons participating in the conductivity. The most useful form to understand how nanostructures will affect the electrical conductivity is to follow the approach of Ziman[4] and to ignore energy dependent scattering so that the momentum relaxation time, $\tau(E)$ along with the electron velocity, $v(E)$ define the electrical conductivity as

$$\sigma = \frac{q^2}{3}\int \tau(E)v^2(E)\left[-g(E)\frac{\mathrm{d}f}{\mathrm{d}E}\right]\mathrm{d}E \tag{4.32}$$

Eqn (4.32) provides the mechanisms to calculate σ from only theory but τ is normally obtained experimentally from the mobility and combined with experimental measurements of the carrier density with eqn (4.1) to provide the experimental value of the electrical conductivity.

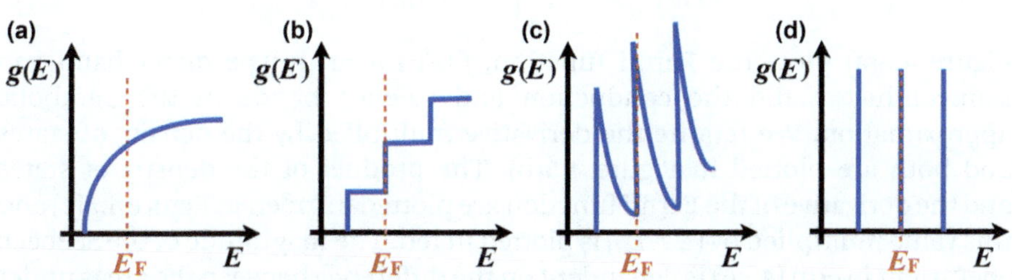

Figure 4.4 The density of states for electrons for (a) 3D, (b) 2D, (c) 1D and (d) 0D low-dimensional systems.

4.3 The Seebeck Coefficient and Low-dimensional Modifications

The Seebeck coefficient can be calculated from knowing the energy dependent electrical conductivity as demonstrated in eqn (4.32) using the Boltzmann transport equation to give

$$\alpha = -\frac{k_B}{q}\int \frac{(E-E_F)}{k_B T}\frac{\sigma(E)}{\sigma}\mathrm{d}E \tag{4.33}$$

where

$$\sigma = \int \sigma(E)\mathrm{d}E = q\int g(E)\mu f(E)(1-f(E))\mathrm{d}E \tag{4.34}$$

Cutler and Mott[5] used this approach in the degenerately doped limit through a Taylor expansion of the electrical conductivity around the chemical potential to demonstrate that

$$\alpha = -\frac{\pi^2}{3q}k_B^2 T\left[\frac{\mathrm{d}\ln(\mu(E)g(E))}{\mathrm{d}E}\right]\Bigg|_{E=E_F} \tag{4.35}$$

This provides a direct route to seeing how to increase the Seebeck coefficient as first suggested by Hicks and Dresselhaus.[6] Figure 4.4 demonstrates how this approach can be used to significantly increase the Seebeck coefficient through the move to 2D, 1D or 0D systems through the use of micro- or nano-structures. It is the derivative of the density of states and mobility that are key at the chemical potential or Fermi energy and as Figure 4.4 demonstrates, the reducing of dimensions results in large discontinuities which enhances the Seebeck coefficient.

Figure 4.5 provides further insight into how this enhancement is obtain and more generally of how to enhance the Seebeck coefficient. If eqn (4.32) is substituted into eqn (4.34) it can be shown that the Seebeck coefficient is given by

$$\alpha = \frac{q^2}{3T\sigma}\int \tau(E)v^2(E)\left[-g(E)\frac{\mathrm{d}f}{\mathrm{d}E}\right](E-E_F)\mathrm{d}E \tag{4.36}$$

Figure 4.5(a) plots the Fermi function, $f(E)$ for an n-type direct band gap semiconductor and the conduction and valence bands in the parabolic approximation. We require the derivative multiplied by the density of states and both are plotted in Figure 4.5(b). The product of the density of states and the derivative of the Fermi function are plotted in green in Figure 4.5(c) and this value multiplied by $(E-E_F)$ is plotted in red. The magnitude of the Seebeck coefficient in eqn (4.36) is dependent on the difference between the areas under the red curves on either side of the chemical potential, E_F in Figure 4.5(c). Therefore any mechanism that increases this asymmetry around the chemical potential will increase the Seebeck coefficient.

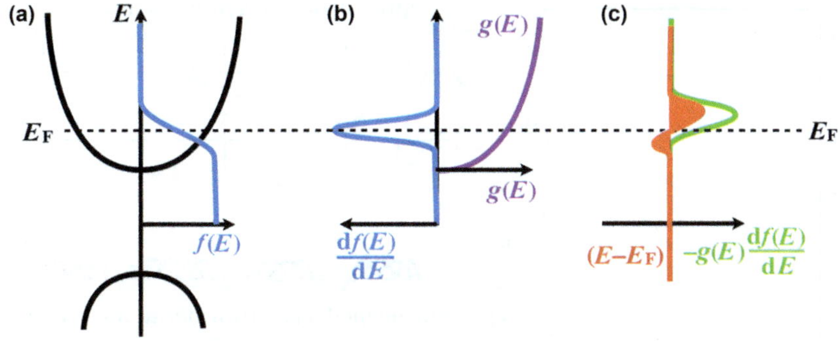

Figure 4.5 (a) The parabolic dispersion of the conduction and valence bands for a degenerately doped semiconductor material. The Fermi function is also plotted to demonstrate the filling of the available states in the conduction band with electrons. (b) The derivative of the Fermi function as a function of energy plotted in (a) along with the 3D density of states. (c) The product of the density of states from (b) and the derivative of the Fermi function from (b) is plotted in green. This product of this function with $(E - E_F)$ is plotted in red. The magnitude of the asymmetry of this function above and below the chemical potential (Fermi energy) determines the magnitude of the Seebeck coefficient.

4.4 Thermal Conductivity

The thermal conductivity, κ is one of the key parameters that many researchers try to reduce to improve the zT figure of merit of thermoelectric materials. Wiedemann and Franz first demonstrated empirically for metals that the thermal conductivity divided by the electrical conductivity times the temperature is a constant. The explanation of the Wiedemann–Franz law was one of the great successes of the Drude theory of electron transport but it has now been shown that this success was fortuitous due to two different factors of 100 cancelling out. Despite this issue, the Wiedemann–Franz rule is still valid and thwarts many attempts to improve the zT of thermoelectric materials.

Quantum mechanics is required to describe accurately the thermal conductivity in a thermoelectric material in the approach now to be described. Thermal conductivity is related to the vibrations of the atoms on the appropriate crystallographic lattice. Phonons are the quantized quasi-particle which describe the collective vibrational excitations of the lattice (modes of vibration). The modern theory of phonons uses a semi-classical approach through considering a spring model connecting the atoms in the lattice together (see Figure 4.6). There are more complete and detailed quantum models of phonons but the simple spring picture provides the same results as the complete theories for the vibrational modes as presented in Figure 4.6.

Two types of vibrational modes are present in any crystalline material. The first are acoustic modes where the neighbouring atoms are displaced

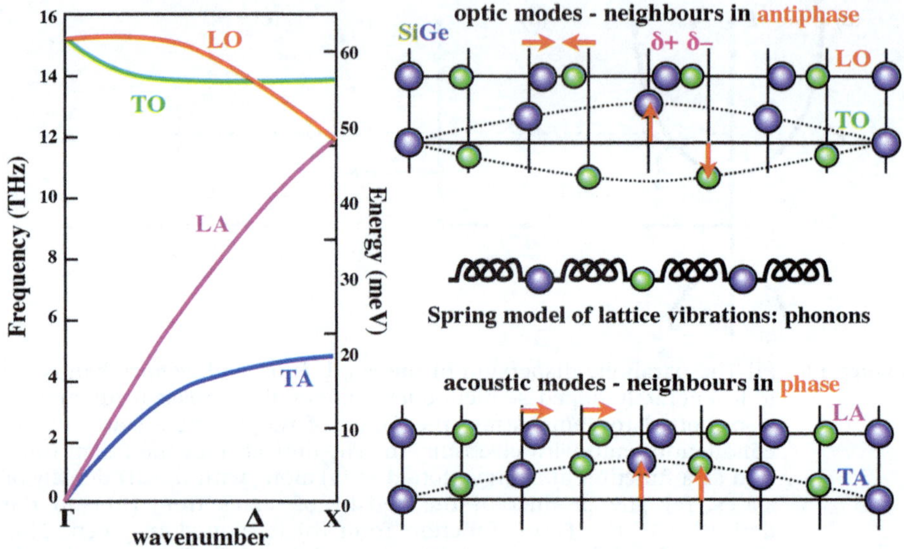

Figure 4.6 Left is the phonon energy dispersion *versus* wavenumber for phonons in Si. Right is a schematic diagram of each of the phonon modes in the dispersion and the basic concept of the modes coming from the model of each atom in the lattice being joined by springs. The acoustic modes have nearest neighbour displacement of atoms in phase whilst the optic modes have the nearest neighbour displacement of atoms in anti-phase.

in phase in either the longitudinal or the transverse direction. The optic modes are higher in energy and correspond to the neighbouring atoms being displaced in anti-phase in either the longitudinal or transverse direction. Figure 4.6 is drawn for the group IV materials Si and Ge in the periodic table but a further polar optic phonon mode also exists for polar semiconductor materials such as Bi_2Te_3, Sb_2Te_3 or any III–V or II–VI materials.

For convenience, the thermal conductivity is divided into the contribution from the lattice vibrations, κ_{ph} and that due to the thermal energy transported with electrons, κ_{el}.

$$\kappa = \kappa_{el} + \kappa_{ph} \tag{4.37}$$

For degenerate semiconductor materials (high carrier densities including metals), $\kappa_{el} \gg \kappa_{ph}$, whilst for non-degenerate semiconductors (low carrier density below the Mott criteria), $\kappa_{ph} \gg \kappa_{el}$. An ideal semiconductor for thermoelectric applications should have both a high carrier density to increase the electrical conductivity and low thermal conductivity so that $\kappa_{ph} \gg \kappa_{el}$ but the Wiedemann–Franz rule prevents this in bulk semiconductors and metals.

In metals or degenerately doped semiconductors that are used for thermoelectrics, κ_{el} dominates the thermal conductivity and the Wiedemann–Franz rule requires

$$\frac{\sigma T}{\kappa} = \frac{3}{\pi^2}\left(\frac{q}{k_B}\right)^2 = \frac{1}{\acute{L}} \tag{4.38}$$

where $\acute{L}$ is the Lorentz number ($= 2.44 \times 10^{-8}$ W Ω K^{-2}). There are a number of regimes where the Wiedemann–Franz rule breaks down, for examples in pure metals at low temperatures. For nanostructures the most interesting regime is in low dimensional structures which potentially allows engineered structures to break the linear connection between the thermal and electrical conductivities. In reality, whilst the linear relationship can be broken, there are still strong links between the electrical and thermal conductivities for many nanostructured materials.

The lattice contribution to the thermal conductivity is in quantum theory the phonon contribution where phonons are the quantised vibrations of the lattice. The phenomenological model using phonon scattering, which is now used to calculate the phonon contribution to the thermal conductivity was first published by Callaway.[7] It uses the Boltzmann transport equation and assumes that the phonon scattering processes can be represented by frequency-dependent relaxation times. The lattice thermal conductivity from the Callaway formula is

$$\kappa_{ph} = \frac{k_B}{2\pi^2}\left(\frac{k_B}{\hbar}\right)^3 \int_0^{\frac{\theta_D}{T}} \frac{\tau_c(x)x^4 e^x}{\upsilon(x)(e^x - 1)^2}\,dx \tag{4.39}$$

where θ_D is the Debye temperature, $x = \frac{\hbar\omega}{k_B T}$, ω is the angular frequency of the phonons, τ_c is the combined phonon scattering time and $\upsilon(x)$ is the phonon velocity. The integral has to include all the phonon modes which contribute to the thermal conductivity but in many nanoscale systems, the length scales forbid a number of the optic phonon modes which can help to reduce the thermal conductivity.

The electrical contribution to the thermal conductivity also is obtained from the Boltzmann transport equation and was first derived by Nag.[8] For a total electron momentum relaxation time of τ for electrons of energy, E, the electron contribution to the thermal conductivity is

$$\kappa_{el} = \frac{\sigma}{q^2 T}\left[\frac{\langle\tau\rangle\langle E^2\tau\rangle - \langle E\tau\rangle^2}{\langle\tau^3\rangle}\right] \tag{4.40}$$

This clearly demonstrates the link between κ_{el} and σ that the Wiedemann–Franz rule describes.

To determine the length scale at which a material or device reduces the number of thermal dimensions, the thermal length scale, L_T corresponding to the length over which thermal smearing and the associated phase randomisation of an electron over the Fermi distribution must be calculated. This can be though of as adding an energy uncertainty of k_BT to the electrons and is given by

$$L_T = \sqrt{\frac{D\hbar}{k_B T}} \tag{4.41}$$

If the phonon wavenumber is defined as q then the group velocity (the velocity of the thermal energy) is defined as $\frac{\partial \omega}{\partial q}$ and represents the instantaneous gradient of the slopes for the dispersion relations in Figure 4.6. The other parameter to determine the number of dimensions in the thermal transport is the phonon mean free path defined as

$$\Lambda_{ph} = \frac{3\kappa_{ph}}{C_v \langle v \rangle \rho} \tag{4.42}$$

where C_v is the specific heat capacity at constant volume, $\langle v \rangle$ is the mean phonon velocity and ρ is the density of phonons. Table 4.1 provides examples of the phonon mean free paths in a range of materials. A number of different theories have been used to provide values. The Debye theory provides a simple cut-off at the Debye temperature for the number of modes being used. The dispersion approach calculates the group velocity taking into account all the modes.[9] Table 4.1 demonstrates the large difference between the models which needs to be considered when calculating the phonon mean free path and considering the dimensionality of the phonon transport and scattering. In all the materials in Table 4.1, the phonon mean free paths are all larger than the equivalent electron mean free paths in the bulk materials. This provides a (small) window where

Table 4.1 The specific heat capacity, group velocity, phonon mean free path and Debye temperature for a range of thermoelectric and semiconductor materials. The data for Si and Ge are from[9] and the Bi_2Te_3 from Jeong,[15] Sb_2Te_3 from Pashinkin[16] and Feng[17] and PbTe from Pashinkin.[18]

Material	Model	Specific heat/ J m^{-3} K^{-3} at 300 K	Group velocity/ m s^{-1}	Phonon mean free path, Λ_{ph}/nm	Debye temperature/K
Si	Debye	1.66×10^6	6400	40.9	645
Si	Dispersion	0.93×10^6	1804	260.4	645
Ge	Debye	1.67×10^6	3900	27.5	360
Ge	Dispersion	0.87×10^6	1042	198.6	360
Bi_2Te_3	Debye	1200	1780	14	165
Sb_2Te_3	Debye	200	1915	0.96	157
PbTe	Debye	156	1730	10	130

nanostructures can be designed to scatter phonons more than electrons to improve the thermoelectric performance.

4.5 Potential Improvements to Thermoelectrics from Nano- and Micro-structures

For all bulk thermoelectric materials, the main way to optimise the thermoelectric performance is to maximise the $zT = \frac{\alpha^2 \sigma}{\kappa} T$ figure of merit. For a given bulk material (see Figure 4.1 for zT values as a function of temperature), this can only be achieved through changing the doping density as is demonstrated for Bi_2Te_3 in Figure 4.7. The electrical conductivity increases with increased doping as does the thermal conductivity due to the Wiedemann–Franz rule. As the doping increases the Seebeck coefficient decreases leading to a maximum in zT at some degenerate doping density, for this figure around 10^{19} cm^{-3}. The power factor, $\alpha^2\sigma$, also has a maximum but this is not necessarily at the same point as zT. For $Bi_{1-x-y}Te_xSb_y$ alloys the alloy concentration can also be changed to provide a maximum value for zT.[10]

The simplest way to produce nanostructure materials which may enhance the thermoelectric properties is to use polycrystalline or amorphous rather than single crystal material. Such approaches have been used particularly with poly-Si and SiGe alloys which are readily available. In particular, for Si, the poly-crystalline form can have zT values greater than the single crystal Si by over a factor of 100. In such materials, grains can be produced from a few nm in diameter up to 100s of nm dependent on annealing temperatures and environments. The grains can aid the scattering of phonons (but also

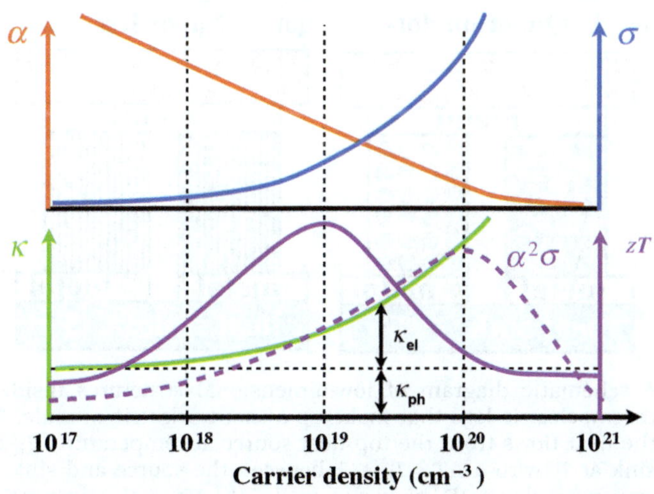

Figure 4.7 The Seebeck coefficient (α in red), electrical conductivity (σ in blue), thermal conductivity (κ in green), zT (in purple) and the power factor ($\alpha^2\sigma$ in dashed purple).

electrons) and simple models can be produced to estimate potential improvements. The phonon scattering time, τ_B by a grain boundary is given by

$$\frac{1}{\tau_B} = \frac{v}{d_g}\left[\frac{1-p(\omega)}{1+p(\omega)}\right] \quad \text{where} \quad p(\omega) = \exp\left[-\pi\left(\frac{2\Delta\omega}{v}\right)^2\right] \tag{4.43}$$

Here, Δ is the surface roughness parameter and d_g is the average size of each grain in the polycrystalline material. Whilst polycrystalline materials normally have far lower thermal conductivity compared to single crystal, they also have lower electrical conductivity so any overall improvements in thermoelectric properties are normally limited for alloy materials. For single elements the enhancements can be significant but this is mainly due to the low zT values for single elements.

Figure 4.8 provides examples of a range of how heterostructures can be used to produce low dimensional enhancements to thermoelectric devices.

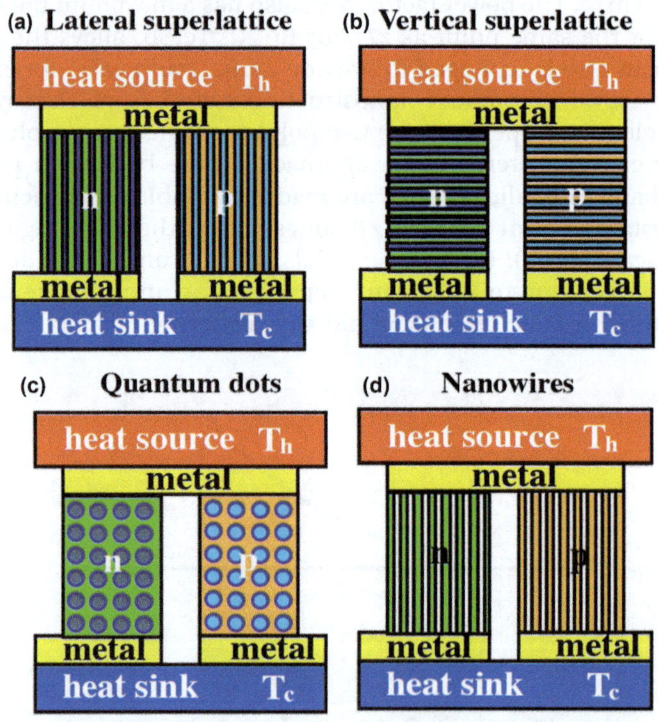

Figure 4.8 A schematic diagram of low-dimensional structures inside a pair of thermoelectric legs that make up a thermoelectric module. In all cases the heat flows from the top heat source at temperature, T_h to the heat sink at T_c with a $\Delta T = T_h - T_c$ between the source and sink. (a) Lateral transport along 2D quantum wells. (b) Vertical transport through a superlattice of quantum wells and potential barriers, (c) Use of 0D quantum dots to scatter phonons (and electrons). (d) A set of 1D nanowires.

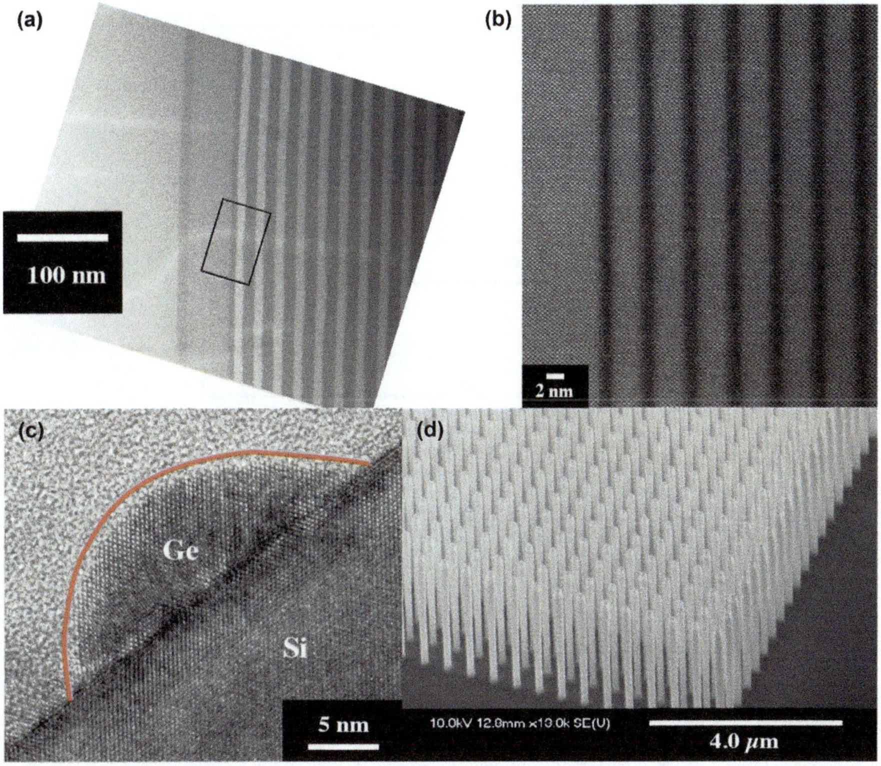

Figure 4.9 Electron microscope images of a range of (a) 2D lateral quantum well, (b) 2D superlattice vertical, (c) 0D quantum dot and (d) 1D nanowires all fabricated from Ge/SiGe hetoerstructure thermoelectric materials.

Figure 4.9 presents transmission electron microscope (TEM) images of a range of Ge quantum wells, which can be used with the electrical and thermal transport either parallel to the quantum wells (Figure 4.9(a)) or perpendicular to the quantum wells (Figure 4.9(b)). Figure 4.9(c) presents an example of a Ge quantum dot grown inside a Si matrix, which can be used to scatter phonons to improve zT whilst Figure 4.9(d) presents 50 nm wide nanowires fabricated from the material presented in Figure 4.9(a).

Figure 4.10(a) compares a set of results from vertical transport superlattices consisting of Ge quantum wells and $Si_{0.5}Ge_{0.5}$ barriers[11] (blue squares) and lateral transport Ge quantum wells with $Si_{0.25}Ge_{0.75}$ barriers[12] (red circles) as a function of quantum well width. Also plotted are the values for bulk p-Ge, p-Si and p-$Si_{0.7}Ge_{0.3}$ all doped at around 10^{19} cm^{-3} to provide a fair comparison. For vertical electron transport through quantum mechanical tunnel barriers, thin barriers and thin quantum wells are required to form minibands to aid the electrical conductivity. It is clear that the lateral transport devices have significantly higher electrical conductivity. This is because the electron transport is 2D along the modulation doped quantum

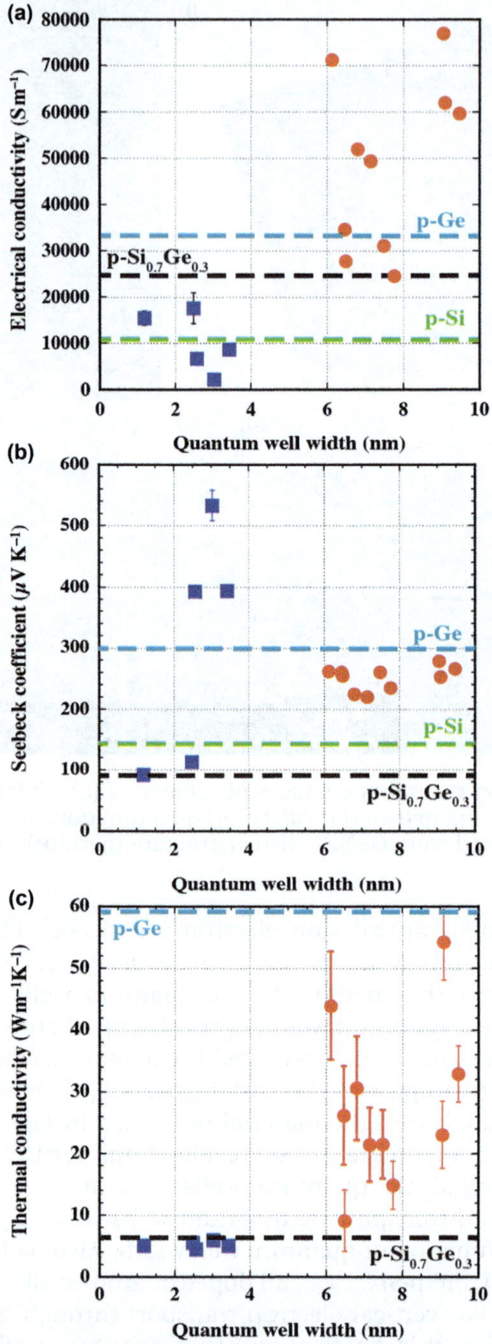

Figure 4.10 (a) The electrical conductivity, (b) the Seebeck coefficient and (c) the thermal conductivity for 2D vertical superlattice thermoelectric material of p-type Ge quantum wells and $Si_{0.5}Ge_{0.5}$ barriers[11] (blue squares) and 2D lateral p-type Ge quantum wells with $Si_{0.25}Ge_{0.75}$ barriers[12] (red circles) as a function of quantum well width. Also plotted are data for 3D p-Si, 3D p-Ge and 3D p-$Si_{0.7}Ge_{0.3}$ all at 300 K.

wells which have low interface roughness and low Coulombic scattering with the wavefunction predominantly in the Ge quantum well. For the vertical structures the wavefunctions have to overlap with the whole quantum mechanical barrier which results in far stronger interface roughness scattering and lower mobility. Most importantly is that the lateral quantum well material can have electrical conductivity over twice the value of bulk material due to the confinement of the electrons into quantum wells and the reduction of Coulombic scattering through modulation doping.

This is not the whole story. It is far easier through eqn (4.35) to enhance the Seebeck coefficient with thin quantum wells. This results for the present samples in the vertical superlattices having higher Seebeck coefficients compared to the lateral structures (Figure 4.10(b)). Bulk p-Ge already has a very high Seebeck coefficient compared to many materials but using thin quantum wells allows significantly higher Seebeck coefficients in the 2D samples. Figure 4.10(c) demonstrates the Wiedemann–Franz rule link to the electrical conductivity in Figure 4.10(a) with the samples with the lowest electrical conductivity also having the lowest thermal conductivity. The vertical structures produce thermal conductivity values below all $Si_{1-x}Ge_x$ bulk alloy values but the lateral structures all have thermal conductivities significantly above $Si_{1-x}Ge_x$ bulk alloys. These results suggest further optimisation is required to produce the best quantum well widths for the best thermoelectric figures of merit.

zT and the power factor for these Ge quantum well materials are presented in Figure 4.11. Whilst both the vertical and lateral quantum well materials can both get *zT* values significantly higher than bulk p-Si, p-Ge or p-$Si_{0.7}Ge_{0.3}$ (Figure 4.11(a)), only the lateral transport quantum well devices have higher power factors (Figure 4.11(b)). These results demonstrate some of the issues in optimising the thermoelectric performance and demonstrate that having

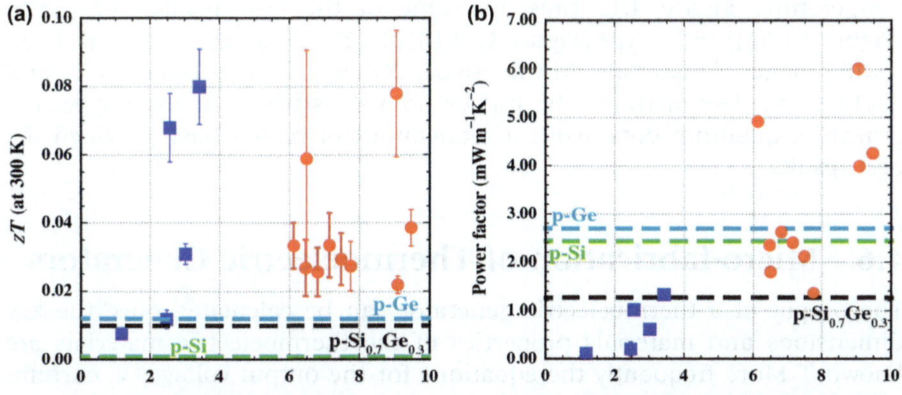

Figure 4.11 (a) The *zT* and (b) the power factor for the materials in Figure 4.9 at 300 K as a function of quantum well width. Also plotted are data for 3D p-Si, 3D p-Ge and 3D p-$Si_{0.7}Ge_{0.3}$ all at 300 K.

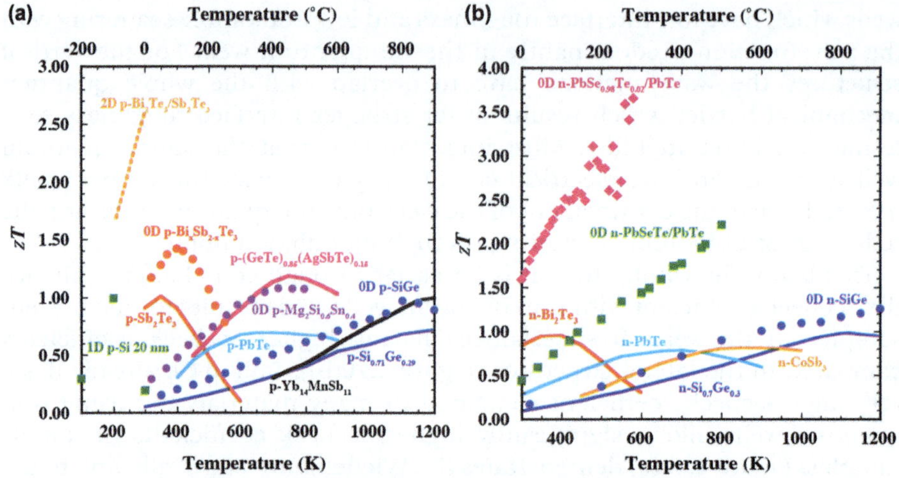

Figure 4.12 A comparison of the zT values for p-type materials (left) and n-type materials (right) as a function of temperature. The references are as follows: 3D p-Sb_2Te_3,[19] p-PbTe,[15] p-$CeFe_4Sb_{12}$,[15] p-$Yb_{14}MnSb_{11}$,[15] p-$Si_{0.71}Ge_{0.29}$,[20] 2D p-Bi_2Te_3/Sb_2Te_3,[21] 1D Si,[22] 0D p-SiGe,[23] p-$(GeTe)_{0.85}(AgSbTe)_{0.15}$,[24] 0D p-$Bi_xSb_{2-x}Te_3$,[25] 0D $Mg_2Si_{0.4}Sn_{0.6}$,[26] 3D n-Bi_2Te_{315}, n-PbTe,[15] n-$CoSb_3$,[15] n-$Si_{0.7}Ge_{0.3}$,[16] 0D PbSeTe,[27] 0D n-SiGe,[28] 0D n-$PbSe_{0.98}Te_{0.02}$/PbTe.[29]

a high electrical conductivity to allow electrical impedance matching to a load is much more important for a given zT than the lowest possible thermal conductivity.

The results presented above for SiGe low dimensional materials provide a good insight into the compromises in the physics to optimise zT and power factors but SiGe alloys are not the best thermoelectric materials at room temperature. Figure 4.12 presents some of the best results for p-type (Figure 4.12(a)) and n-type (Figure 4.12(b)) 3D, 2D, 1D and 0D materials from a wide range of materials. In the p-type, 2D superlattices from Bi_2Te_3 and Sb_2Te_3 have demonstrated the highest zT values whilst in the n-type, the growth of quantum dots inside a semiconductor matrix have provided the best results.

4.6 Micro-fabrication of Thermoelectric Generators

The output of a thermoelectric generator can be calculated once the key dimensions and materials properties of the thermoelectric materials are known.[13] More frequently the equations for the output voltage, V, current, I and power, P are used to optimise the design for a given set of thermoelectric p- and n-type materials. As the Seebeck coefficient of most thermoelectric materials are $\ll 1$ mV K^{-1}, a large number of pairs of n- and p-type thermoelectric legs are always required electrically in series

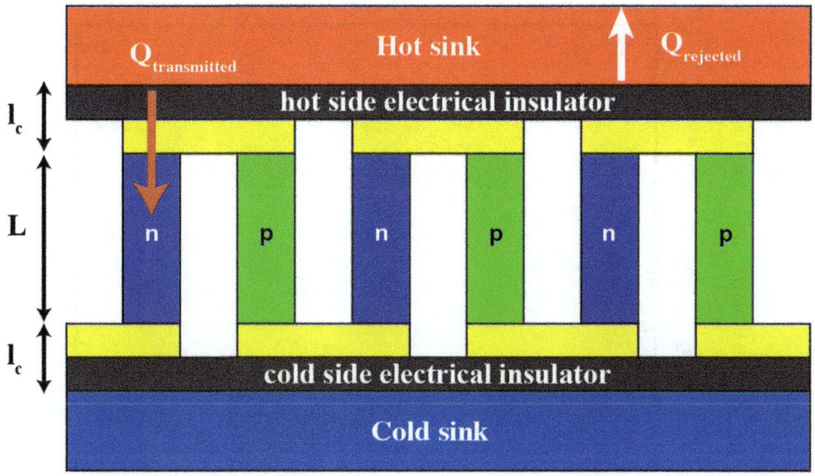

Figure 4.13 A schematic diagram of a complete module with leg length, L and leg contact length, l_c.

to generate a useful voltage >1 V. Figure 4.13 provides a schematic diagram of a few sets of thermoelectric legs of length, L inside a generator and connected together by metal contacts with a length, l_c. If there are N legs, of area A where κ is the thermal conductivity of the thermoelectric material, κ_c is the thermal conductivity of the metal contacts, ρ_c is the resistivity of the metal contacts, α is the Seebeck coefficient, σ is the electrical conductivity, then the voltage, current and power are

$$V = \frac{\alpha N \Delta T}{1 + 2\dfrac{\kappa l_c}{\kappa_c L}} \tag{4.43}$$

$$I = \frac{\alpha \sigma A \Delta T}{2(\rho_c \sigma + L)\left(1 + 2\dfrac{\kappa l_c}{\kappa_c L}\right)} \tag{4.44}$$

$$P = \frac{\alpha^2 \sigma A N \Delta T^2}{2(\rho_c \sigma + L)\left(1 + 2\dfrac{\kappa l_c}{\kappa_c L}\right)^2} \tag{4.45}$$

The power equation, eqn (4.45), illuminates a number of issues. The first is that the output power is dominated by the figure of merit, the power factor, $\alpha^2\sigma$. The second is that the power is proportional to the number of legs and the area of the legs. The power is also proportional to ΔT^2. Also eqn (4.45) demonstrates that shorter legs produce higher power but only until the contact resistance to the Ohmic contacts becomes significant compared to the thermoelectric legs. This is illustrated in Figure 4.14 where Bi_2Te_3 and Ge/$Si_{0.25}Ge_{0.75}$ materials are compared with different Ohmic contact

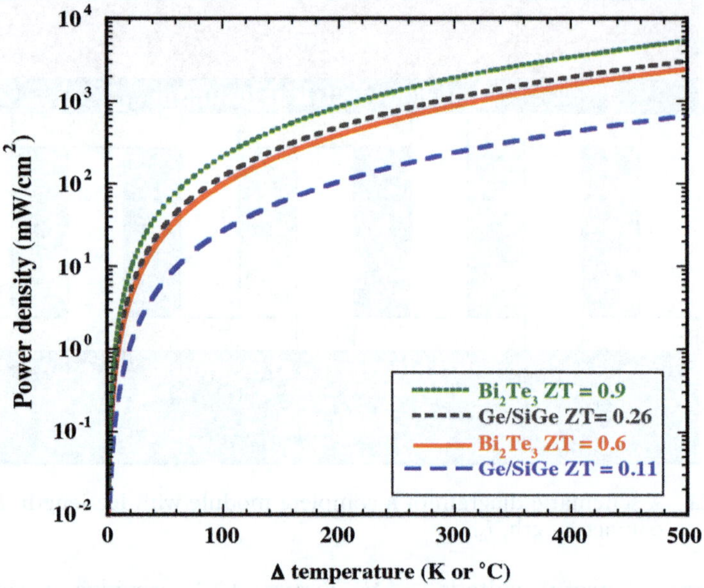

Figure 4.14 The power density *vs.* ΔT for two materials with different parameters to illustrate how short legs result in the contact resistance dominating the performance of the thermoelectric generator using eqn (4.45). The thermoelectric generator had 525 legs of area 500 μm × 50 μm with a leg length, $L = 20$ μm and $l_c = 10$ nm.

Table 4.2 The parameters used to calculate the power density from thermoelectric generators in Figure 4.13.

Material units	$\alpha_p - \alpha_n$/μV K^{-1}	σ/S m^{-1}	κ/W m^{-1} K^{-1}	zT at 300 K	ρ_c/Ω cm^2
Bi_2Te_3	500	71 468	2.0	0.67	10^{-7}
Bi_2Te_3	500	71 468	2.0	0.67	10^{-7}
Ge/$Si_{0.25}Ge_{0.75}$	600	70 000	7.3	0.26	10^{-8}
Ge/$Si_{0.25}Ge_{0.75}$	500	27 900	4.8	0.22	10^{-8}

resistivity properties (see Table 4.2 for parameters). Whilst Bi_2Te_3 is a far superior thermoelectric material in terms of zT, if the Ohmic contacts are not high performance then the contact losses can reduce the generated power significantly. This highlights the great importance of high performance Ohmic contacts with low resistance. This higher power for shorter leg length is one of the reasons microfabricated modules are starting to dominate especially for low ΔT applications (<50 °C) as they can produce far higher power density for less thermoelectric material and therefore are cheaper to manufacture in volume.

We have assumed that both the n- and p-type thermoelectric materials have similar performance whilst this is seldom true. For thermal impedance matching, the following is required for the leg lengths, areas, electrical conductivities and thermal conductivities for the n-type and p-type thermoelectric materials:

$$\frac{L_n A_p}{L_p A_n} = \sqrt{\frac{\sigma_n \kappa_n}{\sigma_p \kappa_p}} \tag{4.46}$$

The mean value of zT can also be calculated as

$$zT = \frac{(\alpha_p - \alpha_n)^2}{\left[\sqrt{\frac{\kappa_p}{\sigma_p}} + \sqrt{\frac{\kappa_n}{\sigma_n}}\right]} T \tag{4.47}$$

Therefore the following is required in eqn (4.43) to (4.45)

$$\alpha = \alpha_p - \alpha_n \tag{4.48}$$

$$\frac{\kappa}{\sigma} = \sqrt{\frac{\kappa_p}{\sigma_p}} + \sqrt{\frac{\kappa_n}{\sigma_n}} \tag{4.49}$$

The amount of heat and physical electrical transport are determined by the conductances rather than the conductivities. The thermal conductance, K through the generator is given by

$$K = \frac{\kappa_p A_p}{L_p} + \frac{\kappa_n A_n}{L_n} \tag{4.50}$$

whilst the electrical resistance is

$$R = \frac{L_p}{\sigma_p A_p} + \frac{L_n}{\sigma_n A_n} \tag{4.51}$$

From these equations, the thermoelectric power generation efficiency can be derived using all the parameters, which can be easily measured on the real generator to be

$$\eta = \frac{I\left[(\alpha_p - \alpha_n)\Delta T - IR\right]}{K\Delta T + (\alpha_p - \alpha_n)IT_h - \frac{1}{2}I^2 R} \tag{4.52}$$

Figure 4.15 provides a comparison of the thermodynamic conversion efficiency as a function of zT for present $zT = 0.7$ materials against thermal engines (see Figure 4.1 for the bulk material zT values as a function of temperature). This $zT = 0.7$ is the typical effective zT, which is delivered from BiSbTe-alloys once an average of the zT throughout the temperature range of operation is considered along with defects and fabrication losses. It is clear to see that the performance of present bulk thermoelectrics is still far below the Rankine and Stirling engines using in many different applications. This is only true for large generation systems where the power being

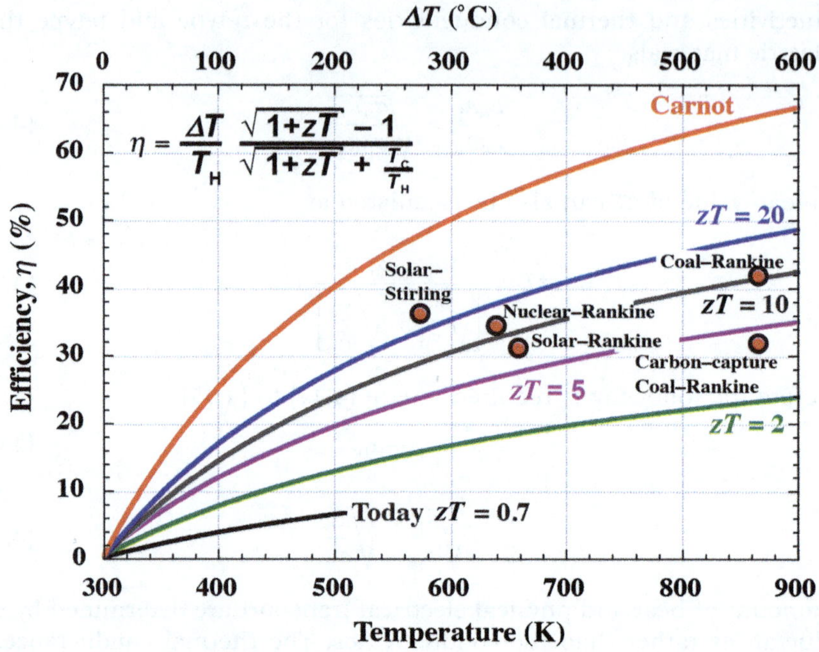

Figure 4.15 The maximum thermoelectric generation efficiency, η as a function of ΔT for a range of different zT parameters. The performance is compared with the Carnot efficiency and a range of Rankine and Stirling engine systems.

considered in >1 W. For small systems and especially microfabricated systems of only a few mm^3 in volume, the changes to the Reynolds number for fluid flow results in thermoelectric generators being more efficient than the thermal engines. What is clear from Figure 4.15 is that present thermoelectrics have efficiencies significantly below the thermal cycles and especially the Carnot limit. It is clear that higher zT materials are required to increase this efficiency which is the main reason there are so many nanostructured thermoelectric results being published in the literature.

At present thermoelectric markets for Peltier coolers are larger than for thermoelectric generation. For Peltier cooler applications, the coefficient of performance is defined as

$$\phi = \frac{(\alpha_p - \alpha_n)IT_c - K\Delta T - \frac{1}{2}I^2R}{I\left[(\alpha_p - \alpha_n)\Delta T - IR\right]} \tag{4.53}$$

Whilst the direct connection to zT is not provided by this equation, it is clear that low R (so high σ), high α and low K (so also low κ) are all key for high coefficients of performance of Peltier coolers and therefore high zT materials are also required.

Eqn (4.43) to (4.45) do provide a good analytical approach to understand the consequences of changing key parameters such as the leg length or area. More frequently numerical solvers such as HFSS or Comsol are used to calculate the thermal and electrical transport through the whole structure to provide more accurate results.[14] These results can be fed through to the later equations with optimisation routines to aim the design of high performance generators and Peltier coolers.

In operation, the heat supplied to a thermoelectric module may not be constant and therefore the output current, voltage and power are not constant. Therefore, some form of power conditioning is required to electrically impedance match the generator to any battery, capacitor or electrical load to the generator. Power conditioning circuits such as maximum power point tracking systems are therefore essential to optimise the useful output power from the generator.

4.7 Conclusions

The microfabrication of thermoelectric generators is now a mature technology with benefits of higher power density than pick-and-place or hand built generators along with shorter thermoelectric legs resulting in far less thermoelectric materials required for the generator. These generators are therefore smaller and cheaper than many of the original hand built generators and there is a strong market push for these devices for the internet of things especially for thermostats powered by thermal energy harvesters for smart building controls. Nanostructures materials have demonstrated improved thermoelectric parameters over their bulk material counterparts. Whilst many impressive zT values have been reported in the literature, there have been extremely few demonstrations of these nanostructured materials providing improved performance in thermoelectric generators. Until this can be achieved, these materials will be confined to academic studies rather than full scale use in thermal energy harvesting and Peltier coolers.

References

1. T. Ihn, *Semiconductor Nanostructures: Quantum States and Electronic Transport*, Oxford University Press, 2010.
2. J. H. Davies, *The Physics of Low Dimensional Semiconductors*, Cambridge University Press, 1997.
3. G. S. Nolan, J. Sharp and J. Goldsmid, *Thermoelectrics: Basic Principles and New Materials Developments*, Springer, 2001.
4. J. M. Ziman, *Electrons and Phonons*, Oxford University Press, 1960.
5. M. Cutler and N. F. Mott, *Phys. Rev.*, 1969, **181**, 1336.
6. L. D. Hicks and M. S. Dresselhaus, *Phys. Rev. B*, 1993, **47**(19), 12727.
7. J. Callaway, *Phys. Rev.*, 1959, **113**, 1046.

8. B. R. Nag, *Electron Transport in Compound Semiconductors*, *Solid State Sciences*, Springer, 1980, vol. 11.
9. G. Chen, *Phys. Rev. B*, 1998, **57**(23), 14958.
10. H. J. Goldsmid, *Materials*, 2014, **7**, 2577.
11. L. Ferre Llin, A. Samarelli, S. Cecchi, T. Etzelstorfer, E. Müller Gubler, D. Chrastina, G. Isella, J. Stangl, J. M. R. Weaver, P. S. Dobson and D. J. Paul, *Appl. Phys. Lett.*, 2013, **103**, 143507.
12. A. Samarelli, L. Ferre Llin, S. Cecchi, J. Frigerio, T. Etzelstorfer, E. Müller, Y. Zhang, J. R. Watling, D. Chrastina, G. Isella, J. Stangl, J. P. Hague, J. M. R. Weaver, P. Dobson and D. J. Paul, *J. Appl. Phys.*, 2013, **113**, 233704.
13. D. W. Rowe and G. Min, *IEE Proc.: Sci., Meas. Technol.*, 1996, **143**(6), 351.
14. W. Li, M. C. Paul, N. Sellami, X.-L. Meng, T. K. Mallick, E. Fernandez, A. R. Knox, A. Montecucco, J. Siviter, P. Mullen, A. Ashraf, A. Samarelli, L. Ferre Llin, D. J. Paul, D. H. Gregory, M. Gao, T. Sweet, F. Azough, R. Lowndes and R. Freer, *Energy Procedia*, 2015, **75**, 325.
15. C. Jeong, S. Datta and M. Lundstron, *J. Appl. Phys.*, 2011, **109**, 073718.
16. A. S. Pashinkin, A. S. Malkova and M. S. Mikhailova, *Russian J. Phys. Chem.*, 2008, **82**, 878.
17. T. Feng and X. Ruan, *J. Nanomat.*, 2014, **2014**, 206370.
18. A. S. Pashinkin, V. Zlomanov and A. S. Malkova, *Inorg. Mater.*, 1994, **30**, 1118.
19. G. J. Synder and E. S. Toberer, *Nat. Mater.*, 2008, **7**, 105.
20. J. P. Dismukes, E. Ekstrom, D. S. Beers, E. F. Steigmeier and I. Kudman, *J. Appl. Phys.*, 1964, **35**(10), 2899.
21. R. Venkatasubramanian, E. Sivola, T. Colpitts and B. O'Quinn, *Nature*, 2001, **413**, 597.
22. A. I. Boukai, Y. Bunimovich, J. Tahir-Kheli, J. K. Yu, W. A. Goddard and J. R. Heath, *Nature*, 2008, **451**, 597.
23. G. Joshi, H. Lee, Y. C. Lan, X. W. Wang, G. H. Zhu, D. Z. Wang, R. W. Gould, D. C. Cuff, M. Y. Tang, M. S. Dresselhaus, G. Chen and Z. F. Ren, *Nano Lett.*, 2008, **8**, 4670.
24. *Thermoelectrics Handbook: Micro to Nano*, ed. D. M. Rowe, CRC Press, Taylor and Francis, 2006.
25. Y. Ma, Q. Hao, B. Poudel, Y. Lan, B. Yu, D. Wang, G. Chen and Z. Ren, *Nano Lett.*, 2008, **8**, 2580.
26. Q. Zhang, J. He, T. J. Zhu, S. N. Zhang, X. B. Zhao and T. M. Tritt, *Appl. Phys. Lett.*, 2008, **93**(10), 102109.
27. T. C. Harman, P. J. Taylor, M. P. Walsh and B. E. LaForge, *Science*, 2002, **297**, 2229.
28. X. W. Wang, H. Lee, Y. C. Lan, G. H. Zhu, G. Joshi, D. Z. Wang, J. Yang, A. J. Muto, M. Y. Tang, J. Klatsky, S. Song, M. S. Dresselhaus, G. Chen and Z. F. Ren, *Appl. Phys. Lett.*, 2008, **93**(19), 193121.
29. T. C. Harman, M. P. Walsh, B. E. Laforge and G. W. Turner, *J. Electron. Mater.*, 2005, **34**, L19.

CHAPTER 5

Review of the Methods for Thermal Conductivity Measurements Most Appropriate for Thermoelectric Materials

EKATERINA SELEZNEVA, CLARK STACEY, PABLO DÍAZ-CHAO, ANDRES MUÑIZ-PINIELLA AND ALEXANDRE CUENAT*

Materials Division, National Physical Laboratory, Teddington, Middlesex TW11 0LW, UK
*Email: alexandre.cuenat@npl.co.uk

5.1 Introduction

5.1.1 Thermoelectric Challenges

Thermoelectric (TE) materials and modules have been of increasing interest since the early 1990s, where a significant improvement in their efficiency was theoretically predicted by exploiting nanoscale effects.[1] The subsequent improvements in TE materials, together with the well-known success of TE generators in space applications, have attracted the attention of several industries, such as automotive companies and foundries that investigate the implementation of TE modules in their production lines.[2,3]

RSC Energy and Environment Series No. 17
Thermoelectric Materials and Devices
Edited by Iris Nandhakumar, Neil M. White and Stephen Beeby

Published by the Royal Society of Chemistry, www.rsc.org

The performance of a thermoelectric material is conventionally evaluated in terms of its thermoelectric figure-of-merit $zT = S^2\sigma T/\kappa$, where S, σ, and κ are the Seebeck coefficient, electrical and thermal conductivities, respectively, and T is the average temperature between hot and cold sides. The enhancement of zT is very challenging due to interconnection of the transport coefficients, such that optimization of one of them has negative effect on the others.[4] Apart from the individual cases of reported high zT values, which have not been independently confirmed,[5] the relative increase of zT is often within 15–20%. Accurate knowledge of all the transport coefficients is thus essential to set directions for future thermoelectric research.

Of all the thermoelectric coefficients, thermal conductivity is one of the most difficult to measure. Different from electrical conduction, heat conduction is virtually impossible to isolate solely using insulation. In a typical thermal property measurement system there will always be parasitic heat transfer present, such as conduction along sensor and heater leads or radiative (infrared) heat exchange between instrument components and the environment. Therefore, the exact heat flux passing through the specimen becomes difficult to determine.

There are many well-established methods for thermal conductivity measurements. However, the thermoelectric community does not seem to make use of all the knowledge available. It is often the case in thermoelectric papers that the thermal conductivity measurement set-ups are over-simplified and the uncertainties in the measurements underestimated.

5.1.2 Thermal Conductivity Measurements

In a real system a thermoelectric material is always found in contact with a heat exchanger. Hence, it is pragmatic to consider the interface thermal resistance as an inherent property of the material. Therefore, what is actually of interest for a real life application is not the thermal conductivity of a material, but rather the effective thermal resistance, which is a characteristic of both the material intrinsic property and the interface. Measuring the effective thermal resistance also enables a more accurate comparison of materials within the same lab. Ensuring that all the specimens are of the same size and shape, the errors associated with thermal interface resistance and uncertainty in determination of the specimen dimensions[6] are eliminated. However, in order to link this property with theoretical research, it is desirable to obtain the intrinsic property of the material, the thermal conductivity.

Thermal conductivity, κ, is the property of a material that quantifies its ability to transport heat when subject to a temperature gradient. It can be determined according to the commonly used Fourier heat conduction equation by measuring a temperature gradient across a specimen in response to a known heat flux passing through it:

$$\frac{\dot{Q}}{A} = -\kappa\frac{\Delta T}{\Delta l} \tag{5.1}$$

where $\dot{Q}$ is the heat flux [W] passing through the specimen, A and Δl are the specimen cross-section [m^2] and length [m], respectively, and ΔT is the temperature gradient [K].

The Fourier equation in this form implies that the thermal conductivity is constant along the direction of heat flux. However, thermal conductivity often being a function of temperature, may vary significantly along this direction. Generally, a measured value for thermal conductivity is assumed to relate to the mean specimen temperature. However, this assumption can only be used if the thermal conductivity has a linear temperature dependence or when a sufficiently small temperature gradient is applied in order to approximate a linear temperature dependence.

It is important to understand that there is no universal method for determining the thermal conductivity covering the whole range of temperatures and thermal conductivities (five orders of magnitude), as each method has its own specific problems and limitations. There are various technical issues that can be encountered depending on the thermal conductivity range of the material of interest. For low thermal conductivity materials ($\kappa = 0.005$–0.1 Wm^{-1} K^{-1}), the heat losses are a big concern, as any insulation will have similar conductivity to the test specimen and only a small quantity of heat is required to create a reasonable temperature difference. High thermal conductivity materials ($\kappa > 5$ Wm^{-1} K^{-1}) require high heat fluxes in order to establish a sufficient thermal gradient, and thus it is difficult to make a heater that can easily be thermally guarded and still provide enough heat flux density. Additionally, high heat fluxes can make it difficult to position temperature sensors in order to make a representative temperature measurement and not alter the intended heat flow. For medium thermal conductivity materials ($\kappa = 0.1$–5 Wm^{-1} K^{-1}), which is the range in which most thermoelectrics belong, the major source of error is associated with thermal contact resistances affecting both the measurement of heat flux and the temperature gradient.

Depending on the thermal conductivity range, different specimen geometries are used. For low thermal conductivity materials, flat specimens (plates or disks) help to minimize the lateral heat losses. For high thermal conductivity materials, long narrow specimens (cylinders) enable an increased thermal gradient along the specimen. The heat losses through the side walls in the latter case are of lesser concern, as the conduction through the specimen will be the dominant heat conduction path. The measurements become particularly challenging at high temperatures when the thermal losses due to radiant heat transfer and convection are extremely difficult to control.

The thermal conductivity measurement methods can be divided into two main categories: (1) steady-state and (2) transient. In steady-state methods the measurement is made once a stable heat transfer equilibrium has been established within the specimen and instrument. Whereas in transient measurement, the thermal conductivity of the material is obtained by fitting the time dependence of temperature rise within the specimen to a heat conduction model with a number of parameters. Transient techniques tend to have simpler experimental set-ups and require less time to take a measurement. However, the initial heating is normally applied to the system in an equilibrium starting state, so times between repeat measurements can still be long. Steady-state instruments will often have larger and more complex experimental set-ups, including multiple sensors and active thermal guarding that includes both multiple heaters and heat extractions. The larger instrument components and the requirement to reach and maintain a stable heat transfer equilibrium means that steady-state measurements can take longer (hours or days) than transient measurement. Nevertheless, it is far more straightforward to evaluate the data and the propagation of uncertainties into the overall uncertainty for the measured thermal conductivity value in steady-state methods.[7]

The steady-state methods can be subdivided into absolute and comparative methods. In absolute methods, the heat flux through a specimen is measured directly by measuring the electrical power supplied to the heater ($\dot{Q} = IV$) converted into heat. In comparative methods the heat flux is found by measuring the output of the heat flux transducer that either has known thermal conductivity or is calibrated using reference specimens with known thermal conductivities. The uncertainty of the comparative measurements is normally higher than that of equivalent absolute methods, since the uncertainty in the measured value of thermal conductivity will always include the uncertainty in the thermal conductivity of the reference material.

Finally, depending on whether the material property is measured directly or derived from another property, both steady-state and transient methods can also be classified as either direct or indirect. In particular, many transient methods (*e.g.* laser flash, time-domain thermoreflectance) measure thermal diffusivity and the thermal conductivity of a specimen is then derived using additional measurements of heat capacity and density. The uncertainty on the thermal conductivity measured indirectly is likely to be higher since it will combine the uncertainties from the measurements of several other physical quantities.

In the following sections we review the main techniques that can be adopted to measure the thermal conductivity of TE materials.

5.2 Steady-state Methods

5.2.1 Guarded Hot Plate

The Guarded Hot Plate method is the primary (*i.e.* the operation can be completely described and understood[8]) absolute standard for measuring the thermal conduction properties of a material with low thermal conductivity[9] and is described in ISO 8302[10] and ASTM-C177.[11] The overall measurement uncertainty of this method for insulation materials at room temperature can be $\pm 1\%$ and about $\pm 5\%$ over the full thermal conductivity and temperatures range of -170 to 800 °C.[12] These values only apply for primary instruments developed in the national standards laboratories and are likely to be larger for other systems. Many groups have used the thermal guarding concept for measuring thermal conductivity, as well as other thermal-related properties.[13–23]

5.2.1.1 Description of the Technique

The Guarded Hot Plate is an absolute measurement technique based on steady-state unidirectional heat flow across a specimen sandwiched between the hot plate and the heat sink. The main components of the Guarded Hot Plate system are schematically represented in Figure 5.1. There are two identical specimens located on either side of the hot plate and two isothermal cold surfaces working as heat sinks on the outer sides of the two specimens. In the absence of heat losses, the heat flux is equally divided between the two specimens and is determined by measuring the amount of electrical power supplied to the heater. The electrical power is maintained at a fixed value to obtain a steady-state temperature gradient across the specimens, typically in the range of 20 to 50 K. Measuring the temperatures on either side of the specimen enables the apparent thermal conductivity κ to be calculated according to Fourier's heat conduction equation:

$$-\kappa\frac{\Delta T}{\Delta l}=\frac{IV}{2A} \tag{5.2}$$

The factor of two in the denominator indicates that the heat flux is equally divided between the two specimens.

Fourier's law assumes that the heat flux through the specimen is one-dimensional, *i.e.* there are no heat losses to the environment at the edges of the hot plate or the specimens. This is achieved by means of the primary lateral guard and secondary edge guards which are kept at same temperatures as the corresponding parts of the assembly and separated from them by a small gap, which is often filled with insulation to reduce convection and radiant heat transfer. The use of the primary lateral guard is an integral part of the Guarded Hot Plate technique, while a secondary guard around the

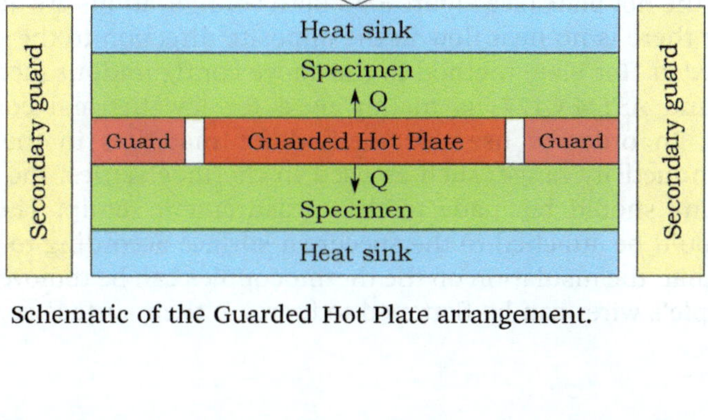

Figure 5.1 Schematic of the Guarded Hot Plate arrangement.

edge is typically used when the measurement temperatures are more than ± 10 K from ambient. A reproducible compressive load is applied to the stack in order to ensure a reproducible thermal contact resistance between the components. The metering area through which the heat passes through the specimen is defined as being halfway across the gap between the central heater plate and the lateral guard.

In spite of the seeming simplicity of the method, realization of all the necessary conditions is a significant engineering and metrology challenge. For an accurate measurement it is necessary for the lateral heat losses to be maintained below 0.5%, as the power supplied to the heater is only used to provide the temperature gradient across the specimen. This condition includes a requirement that the temperature difference between the central plate and the guard surface is much less than 0.2 K. The lateral guard width needs to be about one-half of the linear dimension of the metering area and the maximum specimen thickness—one-third of the metering area unless the edge guards with linear temperature gradient are used. These requirements are extremely difficult to fulfil for small samples ($\sim$1 cm and below) such as typically produced in thermoelectric R&D.

Minimizing the net heat flow across the gap requires far more consideration than may initially be thought. In order to increase the sensitivity, a large number of thermocouples are connected in series with junctions located alternatively in grooves in the central plate and the lateral guard plate. The temperature measurement is affected by the thermocouple's effective insertion depth, so the grooves are run parallel to the gap for as long as practical before crossing the gap. Another reason to use a large number of thermocouples is to reduce the systematic error connected with non-uniformity of the temperatures across the gap.

To further reduce heat transfer at the edges of the specimen the surroundings are usually kept at the mean specimen temperature, *i.e.* the average of the hot and cold plate temperatures.[10,11]

The apparatus can be also operated with a single specimen.[24] Two configurations are possible: (1) the two cold plates are controlled independently and one of them is maintained at the hot plate temperature and (2) one of the specimens is replaced by an auxiliary heater plate sandwiched between two layers of insulation. The cold plate on that side enables smooth temperature control of the auxiliary heater plate maintained at the temperature of the hot plate (see Figure 5.2). Both configurations are designed to ensure that there is no heat flow in the opposite direction to the specimen.

The Guarded Hot Plate method in the above configurations (according to ISO 8302 and ASTM-C177) is mainly used for low thermal conductivity insulations. In order to use this method for materials in the medium thermal conductivity range, such as used in thermoelectrics, the following modifications should be made to the measurement set-up. The thermocouples should be attached to the specimen surface according to ISO 8302. Further to that, the insulation on the thermocouples can be removed and the thermocouple's wires can be flattened to increase their surface contact with

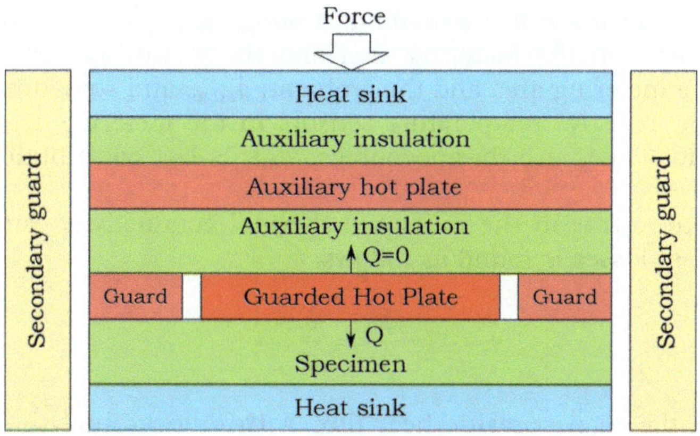

Figure 5.2 Schematic of the Guarded Hot Plate arrangement in the single-sided mode.

the specimen. Thermal contact sheets, such as foamed silicone rubber, should be placed between the specimen and the plates to eliminate air gaps associated with the specimen bowing due to differential expansion and/or due to the surface thermocouples. A compressive load greater than 10 kPa is recommended to ensure the thermal contact sheets conform to the specimen and fill any air gaps. When measuring materials with $\kappa > 0.5$ Wm^{-1} K^{-1}, the thermocouples should be mounted in small grooves machined in the specimen surface. The recommended techniques for materials with thermal conductivity higher than 2 Wm^{-1} K^{-1} are described in the following sections.

5.2.1.2 Method Limitations

The biggest challenges for measuring TE materials using a Guarded Hot Plate are the relatively large plate and specimen dimensions required for practical implementation of accurate temperature balancing and thermal contact resistances making it difficult to measuring surface temperatures in regions of high heat flux. The specimens' high area to thickness ratio results in increased contribution of the interface thermal resistance that creates an upper limit for the specimens' thermal conductivity of 2 Wm^{-1} K^{-1}. The measurements can also be time consuming as the time to achieve heat transfer equilibrium increases in an apparatus with a large thermal mass.

5.2.1.3 Sources of Error

The purpose of the Guarded Hot Plate apparatus is to establish a heat flow within the specimen that is normal to the central heater and passes through a known metering area to the cold plates, with no net heat flowing laterally. That is, that the heat flow through the specimen is one-dimensional. All deviations from the idealized configuration create parasitic heat flows and decrease the accuracy of the method. Among such parasitic flows are the heat flow between the metering area and the guard, $\dot{q}_{gap}$, the heat flow between the metering area and the ambient, $\dot{q}_{edge}$, and sometimes the heat produced by resistive temperature sensors in the metering area, $\dot{q}_r$ (most Guarded Hot Plates use thermocouples, so this last contribution is rarely included).

The relative error in the measured thermal conductivity caused by the parasitic heat losses is found as follows:

$$\frac{\Delta\kappa}{\kappa} = \frac{\dot{q}}{\dot{Q}} \tag{5.3}$$

where $\dot{q}$ is the total parasitic heat flux. A theoretical analysis on how to determine these heat losses can be found in ref. 25–28.

The most significant source of error when measuring medium thermal conductivity materials is likely to be in the measurements of the temperature difference across the specimen. This can be minimized by increasing the temperature difference, but at some point this may introduce other errors due to the increased heat flux density, increased differential expansion, and variation in the thermal conductivity with temperature, meaning the temperature gradient becomes less linear.

It is very important to use temperature sensors that are well matched in term of their temperature calibration offsets. Great care should be taken to mount them, so they are at the temperature of the specimen surface and with sufficient insertion depth so that their leads are in an isothermal region for as greater distance as can be practical achieved (see Figure 5.3).

The other uncertainties of the method are associated with measurements of specimen thickness while in the apparatus, electrical power supplied to the central heater, absolute mean specimen temperature, and the size of the metering area through which the heat travels. A detailed account of these sources of error can be found in the literature.[17,29]

5.2.2 High-temperature Measurements

Guarded Hot Plate measurements at high temperatures (nominally up to 800 °C) mean larger differences between the temperature of the specimen and the ambient temperature surrounding the apparatus. So parasitic heat

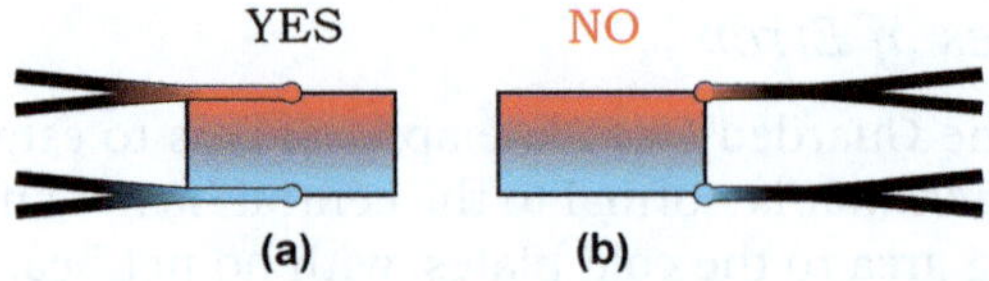

Figure 5.3 Schematic of adequate (having sufficient isothermal contact with the specimen surface) (a), and insufficient (b) thermocouple attachments.

loss, due to radiant heat exchange, becomes far more significant. However, the most significant difficulties at high temperatures are due to changes in the materials used to make the apparatus components.

Temperature measurement tends to be an even greater challenge at higher temperatures, as differences between individual thermocouples become more significant. Fine wire thermocouples, which are better for minimising heat loss from the plates, can suffer from calibration drift even when mineral insulated and protected by metal sheaths. There can also be issues with the materials used to make the apparatus plates, which need to remain flat after being thermally cycled to high temperature during each test.

The methods for reducing thermal contact resistance (described at the end of Section 5.2.1.2) become extremely difficult to implement at higher temperatures. Although the thermal contact resistance is reduced at high temperatures, due to the increased radiative heat exchange across any air gaps, measuring the specimen surface temperatures in regions of high heat flux and high temperature still introduces very large measurement errors that can make this method impractical for measuring the medium thermal conductivity materials found in thermoelectric applications.

5.2.3 Radial Heat Flow Method

The radial heat flow method is another absolute steady-state method and is perhaps one of the most reliable methods for thermal conductivity measurements, especially at high temperatures, as its geometry makes it far easier to minimize the effect of parasitic heat losses. This tends to be the primary method used at the temperatures above 1000–1200 °C.[13]

5.2.3.1 Description of the Technique

In this method the heat is generated by a heater (usually a wire) mounted along the central longitudinal axis of a cylindrically shaped specimen. The temperature gradient is thus established through the specimen radius and measured by two thermocouples located at distances R_1 and R_2 from the axis (see Figure 5.4). The thermal conductivity κ is found as:[30]

$$\kappa = \frac{\dot{Q}\ln\left(\dfrac{R_2}{R_1}\right)}{2\pi L(T_1 - T_2)} \tag{5.4}$$

where $\dot{Q}$ is the heat rate [W] and L is the specimen length [m].

Mounting the heater and the thermocouples is a complex task, which can be simplified by slicing the specimen in two halves along the axis. The flat surface of each half is then polished to remove the cutting roughness. The two halves are kept together by applying an external radial compression so the parasitic heat losses through the cut are minimized.[30]

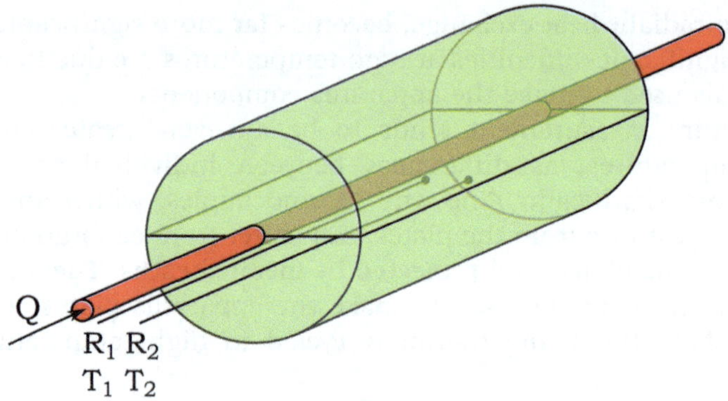

Figure 5.4 Schematic of the radial heat flow method.

5.2.3.2 Method Limitations and Sources of Error

The main limitation of this technique is the size and isotropy of the specimen. The specimen diameter should be sufficient to accommodate two thermocouples across the radius, while the heat path being in all radial directions means that the method is only appropriate for materials isotropic in the plane perpendicular to the heater wire.

The major contribution to the measurement uncertainty in this method comes from determination of the precise location of the thermocouples. The uncertainty in this term is often 10–20%.[30]

The accuracy in the determination of the temperature sensor position is limited by the sensor diameter ($\pm 0.5\times$(sensor diameter)) in each measurement point. The finite size of temperature sensors also limits the accuracy of the temperature readings, as the measured temperature will be within a thermal gradient across the sensor diameter. The lower the thermal conductivity of the specimen, the higher will be the corresponding thermal gradient, and thus higher the uncertainty.

The accuracy improves for larger specimens, as sensor sizes can more easily be small relative to the specimen dimensions. It is also recommended that the specimen length to the diameter ratio is >4 in order to keep the temperature field distortion caused by the ends of the specimen within 2–3%. Additional thermal guarding that is maintained at the same temperature as the heater can be used at the ends of the specimen to minimise heat losses down the heater and sensor leads.

5.2.4 High-temperature Measurements

Measurements at high temperatures will require large thermal gradients along the specimens' radius. This means that the thermal conductivity, being a function of temperature, may vary significantly in this direction. Unless the thermal conductivity of a material has a linear temperature dependence, eqn (5.4) is no longer valid. Additionally, large thermal gradients,

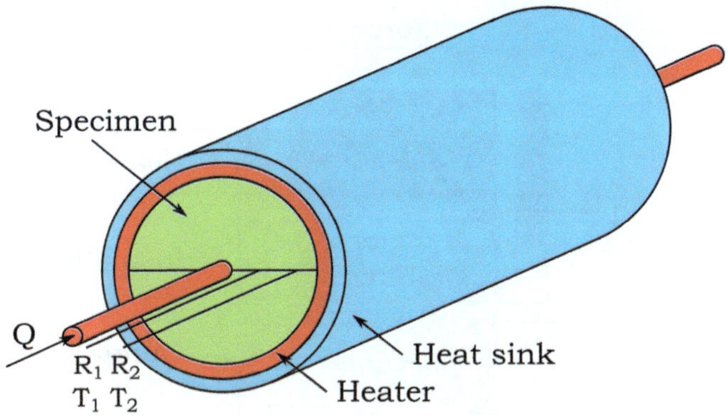

Figure 5.5 Schematic of the radial heat flow method configured for high-temperature measurements.

will result in higher uncertainty in the temperature measurements, as it was explained in Section 5.2.3.2.

These issues can be resolved by adding an external cylindrical heater as illustrated in Figure 5.5. The heat sink is used to enable a smooth temperature control of the heater. This design would allow a smaller temperature gradient to be maintained across the specimen at high temperatures.

5.2.5 Longitudinal Heat Flow

The longitudinal heat flow method belongs to the family of comparative methods based on the ASTM-E1225[31] and ASTM-D5470 standards.[32]

Being a comparative method, it has a higher measurement uncertainty than that of an equivalent absolute method, as the uncertainty in the thermal conductivity value of the reference specimen (determined by an absolute method) will need to be included with other uncertainties associated with the method. For instance, the data obtained during a round-robin testing had up to 18% variation from the mean value.[32] However, carefully calibrated apparatus can have reproducibility equal or better than that of the Guarded Hot Plate.[33] The combination of lower complexity and higher versatility of the experimental set-ups based on this method makes them widely used for medium to high thermal conductivity materials. The examples of the apparatus can be found in ref. 22, 34–41. There are also examples of this type of method converted into making absolute measurements by providing thermal guarding of the main heater.[13,42]

5.2.5.1 Description of the Technique

The main components of the comparative longitudinal heat flow apparatus, based on ASTM-E1225 standard, are schematically represented in Figure 5.6. A specimen with unknown thermal conductivity is sandwiched between two reference materials, also called meter bars. The meter bars are made from

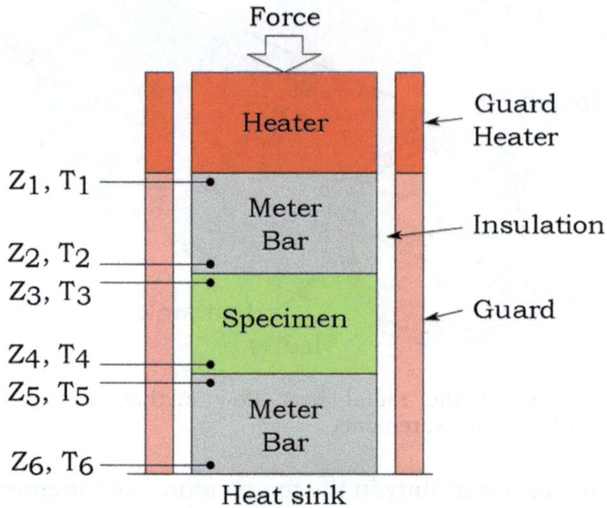

Figure 5.6 Schematic of the comparative-longitudinal heat flow method.

high thermal conductivity materials. The recommended thermal conductivity of the bar material is $> 50\ \mathrm{Wm^{-1}\,K^{-1}}$. There is a heat source on top of the structure and a heat sink on the bottom. A force is applied to the column to ensure a good thermal contact between the specimen and meter bars. When the meter bars and the specimens have a cylindrical shape, the technique is often referred to as the cut-bar method. When thin specimens and meter bars with large cross-sectional areas are used, the technique is referred to as the flat-slab-comparative method. In principle, any shape can be used as long as the cross-sections of the meter bars and the specimen coincide. The stack is surrounded by a thermal insulation material. The complete assembly is enclosed in an active guard shield to further minimize the lateral heat losses. The guard can either be kept at the mean temperature of the column or, in case of better design, have a temperature gradient with the guard matching corresponding positions of the column.

The principle of this method is in passing a heat flux through the meter bars—specimen column and comparing the respective temperature gradients, which will be inversely proportional to the materials' thermal conductivities. The value of the thermal conductivity κ_s of the specimen can be determined by the following equation:

$$\kappa_s = \frac{Z_3 - Z_4}{T_3 - T_4} \cdot \frac{\kappa_M}{2} \cdot \left(\frac{T_1 - T_2}{Z_1 - Z_2} + \frac{T_5 - T_6}{Z_5 - Z_6} \right) \tag{5.5}$$

where κ_M is the thermal conductivity of the meter bars.[31]

It is preferable to use internationally certified reference materials for the meter bars. Other reference materials can be used as long as there is general acceptance of their thermal conductivity values.

When the specimens are either too thin or not sufficiently stable mechanically to allow placement of thermocouples the ASTM-D5470 standard may be more applicable, as the thermal gradient is measured only in the meter bars and no temperature sensors are mounted in the specimen. The thermal conductivity κ_s of the specimen is then found as:[32]

$$\kappa_s = \frac{\dot{Q}_M \cdot d_s}{A_s \cdot (T_{hot} - T_{cold})} \tag{5.6}$$

where d_s and A_s is the specimen thickness and cross-section area [m^2], T_{hot} and T_{cold} are the temperatures at the surfaces of the hot and cold meter bars in contact with the specimen [K], and $\dot{Q}_M$ is the average heat flux through the specimen [W] derived as:

$$\dot{Q}_M = \frac{\dot{Q}_{12} + \dot{Q}_{56}}{2} \tag{5.7a}$$

$$\dot{Q}_{12} = \frac{\kappa_M \cdot A_M}{Z_1 - Z_2} \times (T_1 - T_2) \tag{5.7b}$$

$$\dot{Q}_{56} = \frac{\kappa_M \cdot A_M}{Z_5 - Z_6} \times (T_5 - T_6) \tag{5.7c}$$

where $\dot{Q}_{12}$ and $\dot{Q}_{56}$ are the heat fluxes in the top and bottom meter bars and A_M is the cross-section area [m^2].

In order to minimize the effect of the interfaces' thermal resistances on the temperature measurement, the temperature sensors are located not directly at the interfaces, but within a distance from them. However, the interfaces' thermal resistances still represent a significant measurement issue when no thermocouples are mounted in the specimen.

5.2.5.2 Method Limitations

Both the specimen and the meter bars should have higher thermal conductivities relative to that of the insulation layer. If it is not the case, corrections would be necessary to compensate for heat flows through the insulation. The deviation from uniform heat flow is $<$10% when κ_M/κ_I and κ_s/κ_I are greater than 30, where κ_I is the thermal conductivity of the insulation.[31]

As already mentioned in Section 5.2.5.1, there is no particular limitation on the specimen geometry. However, it is generally recommended to use flat geometry (plates or disks) for the medium thermal conductivity, and long specimens (cylinders) when the thermal conductivity is high. This is to minimize the lateral heat losses for the medium thermal conductivity specimens and to increase the temperature gradient through those with high thermal conductivity. It is also recommended that the cross-section of the specimen and the meter bars coincide. A detailed analysis of the geometry effects on the measurement uncertainty can be found in ref. 43.

Perhaps the most crucial limitation of the method is connected with the lack of standard materials for the reference bars.

5.2.5.3 Sources of Error

The uncertainties associated with the measurements based on comparative methods are constituted of two parts: (1) uncertainty associated with thermal conductivity value of the reference materials and (2) uncertainty related to the precision and bias of the apparatus.

If the meter bar material is chosen from internationally certified reference materials, the uncertainty value would be provided by the supplier or national standards laboratories involved in characterizing the reference value for thermal conductivity. It is a common misconception to take the value of reference materials as 'error-free'.

In Table 5.1 are summarized the available reference materials with the corresponding measurement uncertainty.[44,45]

The sources of error that contribute to the total apparatus error are discussed below.

(1) *Temperature and position accuracy of the temperature sensors*
Similarly to the radial flow method, a significant source of error in this method comes from determination of the positions of the thermocouples and the temperatures (see Section 5.2.3.2). Knowing the precise position of the thermocouples is even more critical when the thermocouples are not placed radially, as the distances enter in the eqn (5.5) without the logarithmic term.[30] The size and aspect ratio of the specimen can also make it extremely difficult to achieve sufficient insertion depth of the temperature sensor to match its calibration or make representative temperature measurements.

(2) *Other sources of error*
There are several sources of errors for which the contribution to uncertainty can be determined only approximately. Most of them are connected with the parasitic lateral heat flows from the heater, the specimen, and the reference bars edges.

If the thermal conductivity of the insulation is not sufficiently low and the thermal conductivity of the specimen is different (higher or lower) from that

Table 5.1 Selection of reference materials and the associated measurement uncertainty.

Material	Thermal conductivity/Wm^{-1} K^{-1}	Estimated uncertainty
Perspex	0.188 (20 °C)	±2.5%
Pyroceram (BCR-724)	4.09 (20 °C)	±6.5%
Inconel 600	13.9 (100 °C)	±4.8%
Stainless steel 304	15.7 (100 °C)	±4.8%
Pure iron	63.8 (100 °C)	±6.1%

of the meter bars, the heat flux will pass through the insulation, avoiding the part of the stack with lower thermal conductivity. The associated uncertainty can be estimated as follows:

$$\Delta\kappa = \kappa_S - \kappa_S' \approx m\kappa_I\left(\frac{\kappa_S'}{\kappa_M} - 1\right) \quad (5.8)$$

where κ_S' is the measured specimen thermal conductivity and m is the geometry factor in the order of two.[46]

An additional measurement uncertainty rises when the thermal gradient through the part of the stack with higher thermal conductivity would be too low for a representative temperature measurement.

5.2.6 High-temperature Measurements

At high temperatures the longitudinal heat flow method has similar issues to those of the guarded hot plate (Section 5.2.2) and radial flow methods (Section 5.2.4). Similarly, the issue of large temperature gradients can be resolved by incorporating an additional heater at the cold side in order to be able to establish small temperature gradients through the measurement stack.

5.2.7 Heat-flow Meter

In this section on heat-flow meters, we review the two apparatus for thermal conductivity measurements conforming to standards ASTM-C518[47] and ASTM-E1530.[48] Note that, even though less common, the term 'heat-flow meter' can also refer to the longitudinal heat flow method (Section 5.2.5), as well as to a transducer that outputs electrical signal proportional to heat flux applied to its surface and also an apparatus for measuring radiant flux during fire testing.

The standards ASTM-C518 and ASTM-E1530 are used in many commercial systems, such as LaserComp FOX series,[49] Linseis HFM Lambda,[50] Netzsch HFM Lambda,[51] TA Instruments DTC series,[52] Ulvac-Riko GH series.[53]

5.2.7.1 Description of the Technique

The design of apparatus conforming to ASTM-C518 is usually intended for large area (305 mm × 305 mm) low density thermal insulation materials. The principle of these apparatus in many ways resembles the guarded hot plate and thus can also be adopted for thermoelectric materials if surface thermocouples, contact sheets and high compressive loads are similarly employed. However, very few commercial apparatus are equipped with either surface mountable thermocouples or the minimum compressive load of 10 kPa and may not be suitable for some TE materials.

The guarded heat flow meter conforming to ASTM-E1530 may seem similar to the ASTM-C518 heat flow meter, but in fact it is designed for

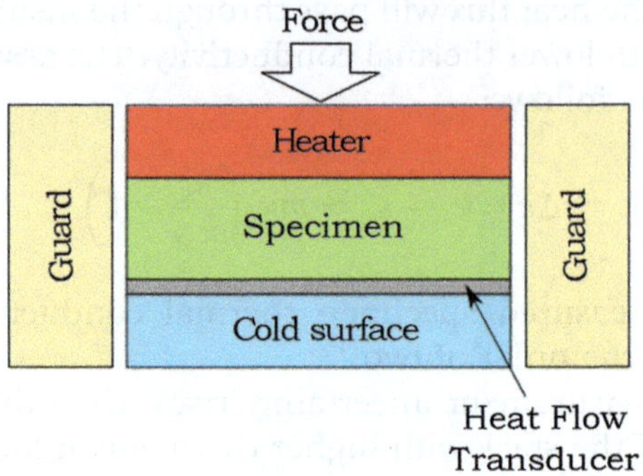

Figure 5.7 Schematic of the guarded heat flow meter apparatus.

smaller area, higher thermal conductivity specimens, thus it is more appropriate for thermoelectric specimens.

The main components of the guarded heat flow meter set-up are schematically presented in Figure 5.7. A specimen and a heat flow transducer are sandwiched between a heater and a cold plate. A compressive load is applied to the stack to ensure a consistent thermal contact between the layers. A guard surrounding the stack is kept at a uniform mean temperature of the heater and the cold plate. The electrical output of the heat flow transducer is proportional to the heat flow passing through the specimen.

The calibration of the heat flow transducer is performed either using specimen with known thermal conductivity (previously determined by an absolute method) or by placing the transducer with a dummy specimen between the plates of the one-sided guarded hot plate apparatus.[54]

It is recommended to calibrate the heat flow transducer using several standard materials covering the range of specimen thermal resistances and thicknesses. As long as the calibration conditions are reproduced during the measurement the heat losses can be considered eliminated. When this is the case, an accuracy of $\pm 5\%$ and a reproducibility of 1% or better can be obtained at room temperature.[9]

5.2.7.2 Method Limitations and Sources of Error

The specimen geometry and thermal resistance should be in the range of those of the standard materials used for the calibration of the heat flow transducer. Deviations may introduce additional errors, for instance, due to lateral heat losses or gains caused by changes in the specimen lateral surface area.

There is a limited choice of standard reference materials already for low thermal conductivity values and practically none for the medium range. The latter is particularly relevant for the ASTM-E1530, since it limits the calibration possibilities of the heat flux transducer.

The uncertainty of the method will consist of the uncertainty associated with the materials used to calibrate the heat flow transducer and the uncertainty of the apparatus itself. The method assumes that the calibration constant of the heat flow transducer does not change. In reality it may vary, depending on the thermal properties of the specimen and test conditions. A complex error verification is required to establish the effect of various test conditions on the uncertainty of the results.[55]

5.3 Transient Methods

Transient methods have become increasingly popular in the last decades. Among them there are widely-used techniques, such as laser flash[56] and 3-omega,[57] as well as many others, such as hot strip technique,[58] hot wire,[59] thermal probe,[60] photoacoustic,[61] time-,[62] and frequency-domain thermoreflectance,[63] thermoreflectance imaging,[64] and Raman.[65]

In these methods heat is applied to the specimen and time dependence of the temperature rise is recorded. The thermal conductivity of a specimen is obtained by fitting these data to a heat conduction model with a number of parameters.

In many of the transient methods the heat source is periodically modulated. When this is the case, only a portion of a specimen near the surface is heated up, as the thermal penetration depth is inversely proportional to the modulation frequency of the heat source. The radiative heat losses from the surface is thus minimized compared to that in the steady state methods. In the case of the 3-omega method, the calculated error due to radiative losses is less than 2% even at 1000 K.[66]

However, it is a mistake to think that a small portion of a specimen is representative of its thermal properties. Due to the nature of the manufacturing processes, many thermoelectric specimens will have variations in density over a short spatial distances. This also makes the heat conduction models developed for homogeneous isotropic material strictly speaking not applicable.

Furthermore, as it was pointed out in the introduction, in a final application, the thermal properties of a complete system, including its interface thermal resistance with a heat exchanger define the performance of a thermoelectric material.

Undoubtedly, for some specimen geometries, such as thin-films, use of the transient methods is the most common and practical approach. There are, however, several issues associated with these methods which often seem to be underestimated.

It is often the case that the uncertainties associated with the fitting the materials' temporal response to a heat conduction model are either given approximately or even completely neglected. The materials data themselves may also have large or even unknown uncertainty.

Many transient methods are indirect methods and derive the thermal conductivity from the measurements of the thermal diffusivity according to the following equation:

$$\kappa = \alpha \rho C_{\mathrm{p}} \tag{5.9}$$

where α is the thermal diffusivity, ρ is the density, and C_p is the heat capacity. The uncertainty on the thermal conductivity value will thus combine the uncertainties from the measurements of all these physical quantities. It is worth pointing out that there can be significant uncertainties in determination of both the specimen density and the heat capacity. For instance, accurate heat capacity measurements are still very challenging: more than 15% of data scatter was observed in the international round-robin study with all the laboratories following specific guide lines.[67]

To summarize, the current state-of-the-art of the transient methods does not allow the uncertainty evaluation to the same extent of the steady-state methods. Therefore, we encourage researchers in the thermoelectric field to be prudent reporting the thermal conductivity results obtained by transient methods. It is good practice to implement and compare the results of several transient methods as well as to carry out inter-laboratory round-robins.

5.3.1 Metrological Approach

As we pointed out in the previous sections, measurement of the thermal conductivity is particularly challenging due to the difficulties controlling heat flux in the system and measuring temperatures in regions of high heat flux. The main consequence of such difficulties is the existence of heat losses and measurement errors which ultimately results in an increase of the uncertainty in the measured value of the thermal conductivity, and consequently, in a loss of accuracy.

To evaluate the uncertainty in the measurement of thermal conductivity with any measurement system, whether it is a bespoke design or commercial, it is good practice to start doing a series of performance checks. The objective of the performance checks is to evaluate the influence that any variable or parameter used during the preparation of the experiment or in the measurement protocol may have on the measured value of thermal conductivity. To do this, an offset is intentionally introduced in such variables or parameters and their influence on the result is evaluated. For example, in the case of the guarded hot plate method the temperature of the primary lateral guard should match the temperature of the central heater (see Section 5.2.1.1). To know how much a temperature offset in the primary lateral guard would affect the resulting thermal conductivity, an intentional offset can be introduced. The comparison between the obtained values of thermal conductivity with offset values of, for instance, +2 K and −2 K will give information on such an influence of the primary lateral guard on the measurement result. This extreme case can then be used in evaluating the effect of any uncertainties introduced by the sensors and control systems used to balance the temperature of the primary lateral guard during a thermal conductivity measurement. The performance checks give valuable information on the variables that most affect the measurement uncertainty of thermal conductivity. As a result of these checks, the influences of such variables are quantified and the uncertainty budget can then be obtained.

The evaluation of uncertainty is done following internationally agreed guidelines.[68–70] In these guidelines, two types of errors are defined: systematic and random errors. The systematic errors (also known as bias) are those that remain constant in replicate measurements, whereas the random errors (affecting precision) are those that vary unpredictably in replicate measurements. Due to such a stochastic nature, the random errors are responsible for the dispersion of results when replicate measurements are taken. Since the systematic errors cannot be identified by the dispersion of results in replicate measurements, they are more difficult to spot and quantify. The systematic errors, such as calibration of instruments (or lack of), insufficient immersion depth of temperature sensors or heat losses due to poor (or non-existing) thermal guarding, are the most common sources of uncertainty in thermal conductivity measurement apparatus.

When the overall uncertainty is being calculated both types of errors are included, but systematic errors are treated differently. Whenever possible the systematic errors should be compensated and a correction factor applied before evaluating the measurement uncertainty. However, it is not possible to compensate for many systematic errors, such as uncertainty in calibration of thermocouples and resolution of digital multimeters. It should be noted that any correction factor used has its own uncertainty that has to be taken into account in the evaluation of uncertainties of the measurement result. To help identify the systematic error sources and quantify such correction factors, reference standard specimens can be used. In these cases, the uncertainty of the specimens has to be considered as well. It is a common misconception to take the value of reference values or calibration specimens as 'error-free'.

Note that in the mentioned guidelines on evaluation of the measurement uncertainty, the concepts of precision and accuracy are defined. The measurement precision is defined as 'closeness of agreement between indications or measured quantity values obtained by replicate measurements on the same or similar objects under specified conditions' whereas accuracy is defined as 'closeness of agreement between a measured quantity value and a true quantity value'. In other words, the measurement precision gives information on the dispersion of measurement results. Therefore, reducing the contribution of the random errors increases the precision of the measurement, but not necessarily its accuracy. To increase the accuracy as well, the systematic errors have to be minimised.[71]

Usually, once the performance checks are done, the systematic errors are corrected, and the uncertainty budget obtained, the measurements of standard reference specimens are used to validate the system. If the obtained results do not agree within their uncertainties, the uncertainty calculations and measurements (performance checks) should be reviewed and, eventually, redone.

The interested reader can find further information on references in 69 and 70. Both documents are based on the GUM[68] but simplified for beginners.

5.4 Conclusions

Accurate measurements are essential for the future of any research and are particularly crucial in thermoelectric field where relative improvement in the efficiency is often comparable to the uncertainty of the measurement.

In spite of seeming simplicity of the phenomenon, making an accurate measurement of thermal conductivity is a difficult task. The highest accuracy of the order 1–5% is achieved in the steady-state absolute guarded hot plate method, which itself involves significant metrology and engineering challenges. Additionally there is an upper limit for the specimens' thermal conductivity of 2 $Wm^{-1}K^{-1}$, which can be measured by the guarded hot plate method.

The family of steady-state comparative methods employ less complex and more versatile experimental set-ups with some of them being also appropriate for medium to high thermal conductivity materials, where most thermoelectrics belong. Even though the accuracy of these methods is lower than that of the absolute methods, carefully calibrated apparatuses can have reproducibility equal or better than that of the guarded hot plate. This makes the comparative methods a practical solution for the applications such as quality control and verification.

The current state-of-the-art of the transient techniques do not allow a rigorous estimation of the uncertainty budget. We thus encourage researchers to perform comparisons of the results obtained by different techniques as well to carry out inter-laboratory round-robins.

Whether it is a bespoke design or commercial measurement system, it is a good practice to perform a series of performance checks followed by calculation of uncertainty budget of the system according to internationally agreed guidelines (GUM). This enables correct comparisons of the measurement results between systems or laboratories.

Acknowledgements

This work is funded by the UK National Measurement Office Materials Programme and through EMRP Project ENG 51-SolCell. The EMRP is jointly funded by the EMRP participating countries within EURAMET and the European Union.

References

1. L. D. Hicks and M. S. Dresselhaus, Effect of quantum-well structures on the thermomagnetic figure of merit, *Phys. Rev. B: Condens. Matter Mater. Phys.*, 1993, **47**, 12727–12731.
2. M. Zebarjadi *et al.*, Perspectives on thermoelectrics: from fundamentals to device applications, *Energy Environ. Sci.*, 2012, **5**, 5147–5162.
3. H. Kawamoto, R&D Trends in High Efficiency Thermoelectric Conversion Materials for Waste Heat Recovery, *Sci. Technol. Trends*, 2009, **30**, 54–69.

4. D. M. Rowe, *Thermoelectric Handbook: Macro to Nano*, 2006.
5. R. Venkatasubramanian *et al.*, Thin-film thermoelectric devices with high room-temperature figures of merit, *Nature*, 2001, **413**, 597–602.
6. D. Flack, *Good Practice Guide No. 40: Calipers and Micrometers*, National Physical Laboratory, 2014.
7. U. Hammerschmidt *et al.*, Critical Review of Industrial Techniques for Thermal-Conductivity Measurements of Thermal Insulation Materials, *Int. J. Thermophys.*, 2015, **36**, 1530–1544.
8. P. Taylor, H. Kipphardt and P. De Bièvre, The definition of primary method of measurement (PMM) of the "highest metrological quality": a challenge in understanding and communication, *Accredit. Qual. Assur.*, 2001, **6**, 103–106.
9. D. Salmon and R. Tye, *An Inter-comparison of Guarded Hot Plate and Heat Flow Meter Apparatus Within the United Kingdom and Ireland*, 2001.
10. ISO-8302, *Thermal Insulation - Determination of Steady-state Thermal Resistance and Related Properties - Guarded Hot Plate Apparatus*, 2013.
11. ASTM-C177, *Standard Test Method for Steady-State Heat Flux Measurements and Thermal Transmission Properties by Means of the Guarded-Hot-Plate*, 2013.
12. B. Hay *et al.*, Analysis of Thermal-Conductivity Measurement Data from International Comparison of National Laboratories, *Int. J. Thermophys.*, 2012, **34**, 737–762.
13. A. V. Petrov, *Metodika izmereniya teploprovodnosti poluprovodnikov pri vysokih temperaturah [Technique for Measuring the Thermal Conductivity at High Temperatures] (in Russian)*, Akademia Nauk Publ., 1963, pp. 27–35.
14. W. Hemminger and R. Jugel, A Guarded Hot-Plate Apparatus for Thermal Conductivity Measurements over the Temperature Range - 75 to 200 °C, *Int. J. Thermophys.*, 1985, **6**, 483–498.
15. D. Salmon, Thermal conductivity of insulations using guarded hot plates, including recent developments and sources of reference materials, *Meas. Sci. Technol.*, 2001, **12**, R89–R98.
16. L. Rauscher *et al.*, Efficiency determination and general characterization of thermoelectric generators using an absolute measurement of the heat flow, *Meas. Sci. Technol.*, 2005, **16**, 1054–1060.
17. J. Wu and R. Morrell, Corrections for thermal expansion in thermal conductivity measurement of insulations using the high-temperature guarded hot-plate method, *Int. J. Thermophys.*, 2012, **33**, 330–341.
18. S. Kwon *et al.*, *Measurement of the Figure-of-Merit of Thermoelectric Devices*. XX IMEKO World Congress Metrology for Green Growth, 2012.
19. M. Li, H. Zhang and Y. Ju, Design and construction of a guarded hot plate apparatus operating down to liquid nitrogen temperature, *Rev. Sci. Instrum.*, 2012, **83**, 075106.
20. S. B. Mahajan, *A Test Setup for Characterizing High-Temperature Thermoelectric Modules*, 2013.
21. P. Kolodner, High-precision thermal and electrical characterization of thermoelectric modules, *Rev. Sci. Instrum.*, 2014, **85**, 054901.

22. H. Wang *et al.*, Determination of thermoelectric module efficiency: A survey, *J. Electron. Mater.*, 2014, **43**, 2274–2286.
23. V. Scoarnec, J. Hameury and B. Hay, A New Guarded Hot Plate Designed for Thermal-Conductivity Measurements at High Temperature, *Int. J. Thermophys.*, 2014, **36**, 540–556.
24. ASTM-C1044, *Standard Practice for Using a Guarded-Hot-Plate Apparatus or Thin-Heater Apparatus in the Single-Sided Mode*, 2012.
25. W. Woodside, Deviations from one-dimensional heat flow in guarded hot-plate measurements, *Rev. Sci. Instrum.*, 1957, **28**, 1033–1037.
26. W. Woodside and A. G. Wilson, *Unbalance Errors in Guarded Hot Plate Measurements*, ASTM STP 217, 1957, pp. 32–48.
27. W. Woodside, *Analysis of Errors due to Edge Heat Loss in Guarded Hot Plates*, ASTM STP 217, 1957, pp. 49–64.
28. U. Hammerschmidt, Guarded Hot-Plate (GHP) Method: Uncertainty Assessment, *Int. J. Thermophys.*, 2002, **23**, 1551–1570.
29. B. Rennex, *Summary of Error Analysis for the National Bureau of Standards 1016-mm Guarded Hot Plate and Considerations Regarding Systematic Error For the Heat Flow Meter Apparatus*, ASTM STP 879 "Guarded Hot Plate and Heat Flow Meter Methodology", 1985, pp. 69–85.
30. G. A. Slack and C. Glassbrenner, Thermal conductivity of germanium from 3 °K to 1020 °K, *Phys. Rev.*, 1960, **120**, 782–789.
31. ASTM-E1225, *Standard Test Method for Thermal Conductivity of Solids by means of the Guarded-Comparative-Longitudinal Heat Flow Technique*, 2009.
32. ASTM-D5470, *Standard Test Method for Thermal Transmission Properties of Thermally Conductive Electrical Insulation Materials*, 2012.
33. F. De Ponte, *Present and Future Research on Guarded Hot Plate and Heat Flow Meter Apparatus*, 1985, ASTM STP 879 "Guarded Hot Plate and Heat Flow Meter Methodology", pp. 101–120.
34. J. Francl and W. D. Kingery, Apparatus for Determining Thermal Conductivity by a Comparative Method, *J. Am. Ceram. Soc.*, 1954, **37**, 80–84.
35. S. S. Ballard, K. A. McCarthy and W. C. Davis, A Method for measuring the thermal conductivity of small samples of poorly conducting materials such as optical crystals, *Rev. Sci. Instrum.*, 1950, **21**, 905–907.
36. R. G. Morris and J. G. Hust, Thermal Conductivity Measurements of Silicon from 30 °C to 425 °C, *Phys. Rev.*, 1961, **124**, 1426–1430.
37. A. J. Slifka, Thermal-conductivity apparatus for steady-state, comparative measurement of ceramic coatings, *J. Res. Natl. Inst. Stand. Technol.*, 2000, **105**, 591–605.
38. K. Lofgreen, *A Metrology for Comprehensive Thermoelectric Device Characterization*, 2011.
39. C. Jensen *et al.*, Design and validation of a high-temperature comparative thermal-conductivity measurement system, *Int. J. Thermophys.*, 2012, **33**, 311–329.

40. R. L. Kallaher, C. A. Latham and F. Sharifi, An apparatus for concurrent measurement of thermoelectric material parameters, *Rev. Sci. Instrum.*, 2013, **84**, 013907.
41. T. Sakamoto *et al.*, Selection and Evaluation of Thermal Interface Materials for Reduction of the Thermal Contact Resistance of Thermoelectric Generators, *J. Electron. Mater.*, 2014, **43**, 3792–3800.
42. J. Wu *et al.*, Design and Performance Checks of the NPL Axial Heat Flow Apparatus, *Int. J. Thermophys.*, 2015, **36**, 529–539.
43. C. Xing *et al.*, Uncertainty analysis on the design of thermal conductivity measurement by a guarded cut-bar technique, *Measur. Sci. Technol.*, 2011, **22**, 075702.
44. *NPL Reference Materials*. Available from: http://www.npl.co.uk/science-technology/thermal-performance/areas/thermal-conductivity/reference-materials.
45. D. Salmon *et al.*, *Certification of Thermal Conductivity and Thermal Diffusivity up to 1025 K of a Glass-ceramic Reference Material BCR-724*, 2007.
46. R. R. Heikes and R. W. Ure, *Thermoelectricity: Science and Engineering*, Interscience Publishers, New York - London, 1961.
47. ASTM-C518, *Standard Test Method for Steady State Thermal Transmission Properties by Means of the Heat Flow Meter Apparatus*, 2010.
48. ASTM-E1530, *Evaluating the Resistance to Thermal Transmission of Materials by the Guarded Heat Flow Meter Technique*, 2011.
49. *LaserComp Inc*. Available from: http://www.lasercomp.com.
50. *Linseis Messgeräte GmbH*. Available from: http://www.linseis.com.
51. *Netzsch-Gerätebau GmbH*. Available from: https://www.netzsch-thermal-analysis.com.
52. *TA Instruments*. Available from: http://www.thermophysical.tainstruments.com.
53. *Ulvac-Riko*. Available from: http://www.advance-riko.com/en.
54. M. Bomberg and R. Solvason, *Discussion of Heat Flow Meter Apparatus and Transfer Standards Used for Error Analysis*. ASTM STP 879 "Guarded Hot Plate and Heat Flow Meter Methodology", pp. 140–153.
55. J. J. W. Tzeng, T. W. Weber, and D. W. Krassowski, *Technical Review on Thermal Conductivity Measurement Techniques for Thin Thermal Interfaces*, 16th IEEE SEMI-THERM Symposium, 2000, pp. 174–181.
56. ASTM-E1461, Standard Test Method for Thermal Diffusivity by the Flash Method, 2013.
57. D. G. Cahill, Thermal conductivity measurement from 30 to 750 K: The 3ω method, *Rev. Sci. Instrum.*, 1990, **61**, 802–808.
58. M. Gustavsson and L. Hälldahl, Thermal conductivity measurements of thin insulating layers deposited on high-conducting sheets, *Int. J. Thermophys.*, 2006, **27**, 195–208.
59. ASTM-C1113, *Standard Test Method for Thermal Conductivity of Refractories by Hot Wire (Platinum Resistance Thermometer Technique)*, 2013.
60. ASTM-D5334, *Standard Test Method for Determination of Thermal Conductivity of Soil and Soft Rock by Thermal Needle Probe Procedure*, 2014.

61. B. Bonno, J. L. Laporte and R. T. D'León, Determination of thermal parameters of PVDF using a photoacoustic technique, *Measur. Sci. Technol.*, 2001, **12**, 671–675.
62. C. A. Paddock and G. L. Eesley, Transient thermoreflectance from thin metal films, *J. Appl. Phys.*, 1986, **60**, 285–290.
63. A. J. Schmidt, R. Cheaito and M. Chiesa, A frequency-domain thermoreflectance method for the characterization of thermal properties, *Rev. Sci. Instrum.*, 2009, **80**, 094901.
64. Y. Zhang *et al.*, Characterization of Heat Transfer Along a Silicon Nanowire Using Thermorefectance Technique, *IEEE Trans. Nanotechnol.*, 2006, **5**, 67–74.
65. S. Perichon *et al.*, Measurement of porous silicon thermal conductivity by micro-Raman scattering, *J. Appl. Phys.*, 1999, **86**, 4700–4702.
66. D. G. Cahill and R. O. Pohl, Thermal conductivity of amorphous solids above the plateau, *Phys. Rev. B: Condens. Matter Mater. Phys.*, 1987, **35**, 4067–4073.
67. H. Wang *et al.*, Transport properties of bulk thermoelectrics: An international round-robin study, part II: Thermal diffusivity, specific heat, and thermal conductivity, *J. Electron. Mater.*, 2013, **42**, 1073–1084.
68. Joint Committee for guides in metrology (JCGM), *Evaluation of Measurement Data - Guide to the Expression of Uncertainty in Measurement (GUM)*, International Organization for Standardization, Geneva, 2008, ISBN.
69. United Kingdom Accreditation Service (UKAS), *The Expression of Uncertainty and Confidence in Measurement*, 2012.
70. S. Bell, *Good Practice Guide No. 11 "A Beginner's Guide to Uncertainty of Measurement"*, 1999, National Physical Laboratory.
71. Joint Committee for guides in metrology (JCGM), *International Vocabulary of Metrology - Basic and General Concepts and Associated Terms (VIM)*, International Organization for Standardization, Geneva, 2012, ISBN.

CHAPTER 6

High-throughput Thermoelectric Measurement Techniques

JORGE GARCÍA-CAÑADAS[a,b] AND GAO MIN*[a]

[a] Cardiff University, School of Engineering, The Parade, Cardiff CF24 3AA, UK; [b] Universitat Jaume I, Department of Industrial Systems Engineering and Design, Campus del Riu Sec, Castellón 12071, Spain
*Email: Min@cardiff.ac.uk

6.1 Introduction

Progress in the field of thermoelectrics has always been strongly associated with the discovery of suitable thermoelectric materials. Over previous decades, the search for improved thermoelectric materials has remained predominantly experimental, which usually involves extensive preparation of materials of various compositions/structures and in-depth characterisation of their thermoelectric properties. Recently, progress in material preparation technologies, such as direct laser deposition, is showing great promise of high-throughput synthesis of new compounds/structures. This paves the way for the possibility of accelerated discovery of new thermoelectric materials through systematic investigation of alloy systems, particularly in ternary or quaternary systems where there are still huge number of compounds unexplored. For this purpose, the capability of high-speed characterization of thermoelectric properties is required.

Characterization of thermoelectric materials usually requires determination of the Seebeck coefficient S, electrical resistivity ρ, and thermal

RSC Energy and Environment Series No. 17
Thermoelectric Materials and Devices
Edited by Iris Nandhakumar, Neil M. White and Stephen Beeby

Published by the Royal Society of Chemistry, www.rsc.org

conductivity λ, or direct measurement of thermoelectric figure-of-merit, $zT = S^2T/\rho\lambda$. In addition, thermoelectric properties are functions of temperature and measurements over a wide temperature range are also needed. Currently, the state-of-the-art facilities for thermoelectric measurement are designed mainly for accuracy and in-depth characterization at expense of the speed. These techniques usually require lengthy procedures, such as preparation of stringently required sample geometry, delicate operation of mounting the sample and making the satisfactory contacts and conducting measurement in a vacuum or inert atmosphere. It is unlikely that these techniques are able to provide sufficient speed required by high-throughput characterization.

In order to achieve high-throughput characterization, suitable measurement techniques, together with appropriate strategies, are required. In general, all electrically conductive materials exhibit a thermoelectric effect to some extent, but only very few show moderate thermoelectric figure-of-merit that is several orders of magnitude higher. The same is true for the power factor (defined as S^2/ρ). Clearly, an effective strategy is to follow a two-stage approach involving: (1) carrying out screening test by measuring the power factor at room temperature to identify the promising materials and (2) further in-depth characterization of all three thermoelectric quantities over a wide temperature range. The second stage involves only a small number of identified materials that can be evaluated effectively using the state-of-the-art commercial equipment. However, the first stage involves huge number of samples and a rapid measurement system needs to be developed. In this chapter, we describe a recently developed thermoelectric characterization system of multifunctional probes for high-throughput applications. In addition, fast screening of thermoelectric properties may be achieved by direct measurement of zT using impedance spectroscopy technique. The fundamentals and the applications of impedance spectroscopy for rapid characterisation of thermoelectric properties are also discussed.

6.2 Multifunctional Probes

6.2.1 Measuring Principles

The Seebeck coefficient is defined as the open-circuit voltage ΔV that is produced across a thermoelectric material subjected to a small temperature difference ΔT

$$S \equiv \lim_{\Delta T \to 0} \frac{\Delta V}{\Delta T}. \tag{6.1}$$

Several considerations have to be taken into account to properly determine S experimentally. For example, it is important to perform the voltage measurement at the same points (or as close as possible) where the temperature difference is measured. Moreover, since two wires (usually Cu) from the voltmeter have to be used to obtain ΔV, one or both of them also present a temperature difference across, which introduces a Seebeck voltage in the ΔV measurement that has to be taken into account and properly subtracted.

The electrical resistivity is determined from the ratio between the voltage applied to a material and the current I that flows through it, normalized by the material dimensions. The measurements should be carried out under isothermal conditions to avoid the errors due to the Seebeck effect. For a material with homogeneous cross-sectional area A and length L, the electrical resistivity is given by

$$\rho = \frac{\Delta V}{I}\frac{A}{L} \tag{6.2}$$

where, $R_\Omega = \Delta V/I$ is the ohmic resistance. It is very important when measuring ρ to ensure that the voltage drop (ΔV) measured completely falls across the material and any voltage drops that can be originated at the contacts (contact resistances) are properly eliminated. This is especially important in semiconductor materials, where contact resistance could lie in a similar range to the material resistance and hence become no longer negligible. For this reason, electrical resistivity is typically measured using a four-probe technique, where two of the probes are used to supply the electric current while the other two are employed for measuring the voltage. It is important to ensure that the voltage is measured without current flow (open-circuit) so that the contribution due to the contact resistance is eliminated.

The thermal conductivity is by far the most difficult thermoelectric property to measure accurately. It is the proportionality constant of Fourier's law, which under a constant T gradient takes the form

$$Q = -\lambda \frac{\Delta T}{L}, \tag{6.3}$$

where Q is the heat flux uniformly flowing through the material. The main problem in the determination of λ lies in the fact that the heat flow cannot be completely confined within the material. Unlike electrical current that only flows through conductive materials, heat can be transferred to the ambient (by convection and radiation) and/or to the bodies in contact with the sample (by conduction). Losses by conduction can occur from the sample to the measuring equipment, such as thermocouples. They are typically made up of metals that are good thermal conductors and thus easily remove heat from the sample when they are brought into contact, which can reduce the temperature at the measuring point. This is called the cold finger effect and thermocouples of thin wires are used to minimize this effect.

6.2.2 Design and Fabrication of Multifunctional Probes

In general, thermoelectric measurements require either electrical or thermal stimulus applied to a sample and the detection of responses through electrical or thermal contacts. Conventionally, this is achieved by bringing individual components into contact with the sample. In this section, we describe a novel design of a multifunctional probe that integrates three functionalities: electrical contact, thermocouple, and heater, into one single probe. This considerably simplifies the experimental set-up and measurement procedures for thermoelectric measurements. It is a crucial component that enables the development of a high-throughput thermoelectric test system described in the following section.

Multifunctional probes can be fabricated[1] using a Cu tube (*e.g.*, 30 mm long and 1.6 mm in diameter), which by itself serves as electrical probe to apply current or measure voltage. In order to measure temperature, the tube can be transformed into a thermocouple by welding a constantan wire at the tip of the tube (Figure 6.1(a)). The steps to achieve this are shown in Figure 6.1(b). The tube tip was swaged into a pencil point with a small hole in the tip (step 1). Then, a constantan wire with the insulation removed from its end was threaded from the tip until the insulation bottomed out at the entrance to the hole (step 2). Finally, the constantan wire and copper sleeve were welded at the tip (step 3) and sharpened to a point (step 4) forming a T-type thermocouple. To provide the heating functionality, an electrical insulated heater coil (*e.g.* made of constantan wire) is inserted inside the Cu tube as shown in Figure 6.1(a).

If the Cu tube employed was not made with Cu material of same purity as required for the standard used in thermocouple construction, this could introduce differences in the value of the tabulated Seebeck coefficient of the copper-constantan couple (40.85 $\mu V\ K^{-1}$ at 300 K).[2] In such cases, the probe

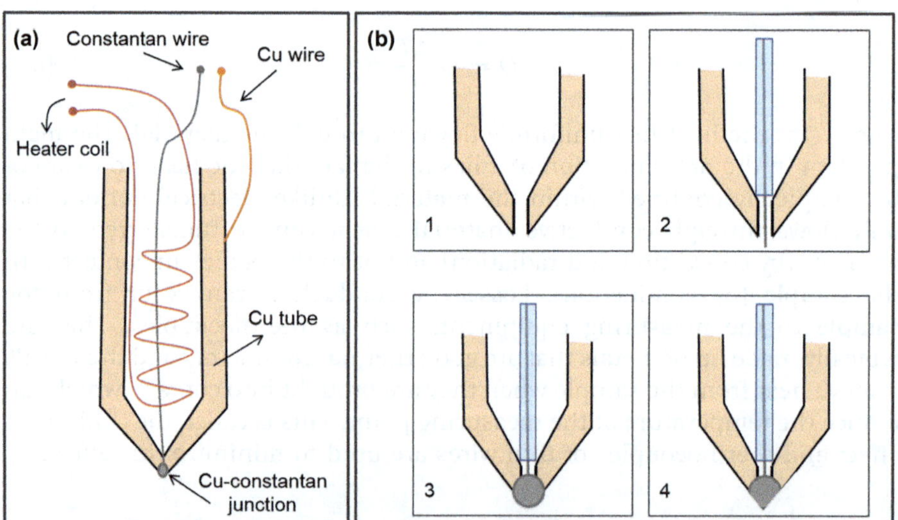

Figure 6.1 (a) Schematic diagram of a multifunctional probe[14] and (b) its fabrication procedure. Figure 6.1(a) reproduced with permission from J. García-Cañadas and G. Min, *Rev. Sci. Instrum.*, 2014, **85**, 043906. Copyright 2014, AIP Publishing.

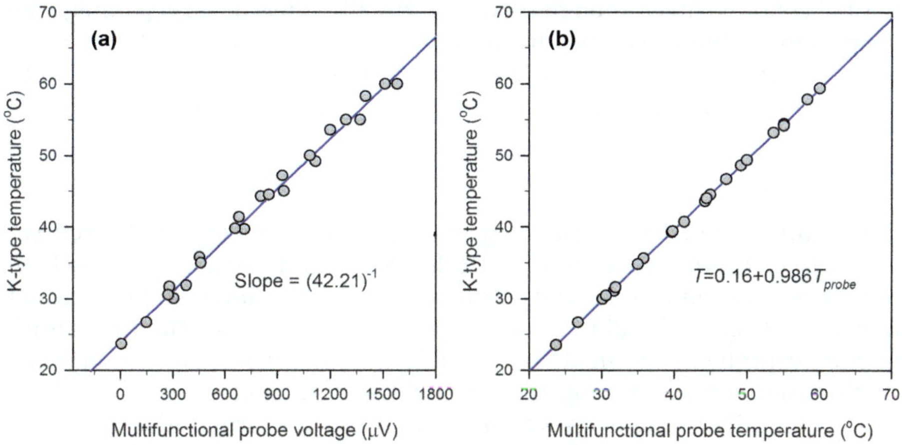

Figure 6.2 (a) Temperature from a commercial K-type thermocouple *vs.* voltage output of the multifunctional probe. The slope obtained (42.21 μV K^{-1}) was used for the calculation of the multifunctional probe temperature in (b). Lines show the linear fit to the results.
Reproduced with permission from J. García-Cañadas and G. Min, *Rev. Sci. Instrum.*, 2014, **85**, 043906. Copyright 2014, AIP Publishing.[14]

needs to be calibrated. This can be done by immersing the probe tip in stirred water heated at different temperatures by a hot plate and calibrated against a commercial thermocouple (K-type in this work). Figure 6.2(a) shows the temperature reading from the K-type thermocouple *versus* the voltage output of the probe (the Seebeck voltage by copper-constantan thermocouple). The slope of the linear fit provides the actual Seebeck coefficient for the copper-constantan couple $S_{\text{Cu-Con}} = 42.21$ μV K^{-1} at ≈297 K, which is slightly different (1.36 μV K^{-1}) from the standard reference value.

In Figure 6.2(b) the temperature obtained from the multifunctional probe, which was calculated using the voltage output (in μV), the actual $S_{\text{Cu-Con}}$ and the room temperature ($T_{\text{probe}} = T_{\text{room}} + \Delta V/42.21$), is plotted *vs.* the temperature of the commercial thermocouple. The linear fit of Figure 6.2(b) provides origin and slope values very close to 0 and 1, respectively, proving a reliable measurement of the temperature. This calibration was performed for all probes used in the equipment and the equations resulting from the linear regression (Figure 6.2(b)) were used for the final tip temperature determination in order to have a very precise match with commercial thermocouples.

6.2.3 Measurement Systems and Procedures

Two set-ups were developed for high-throughput characterization of thermoelectric materials using the above-mentioned multifunctional probes. The first one aims at measuring bot the Seebeck coefficient and electrical resistivity, while the second one was designed to determine the thermal conductivity. Although they were initially constructed separately, the integration of both set-ups into a single system is feasible. In order to facilitate rapid measurement, the set-ups described below were designed to operate at around room temperatures and without using vacuum.

6.2.3.1 Seebeck Coefficient and Electrical Resistivity

The set-up designed for measuring the Seebeck coefficient and electrical resistivity is shown in Figure 6.3. It follows a Van der Pauw layout[3] where four probes are contacted at the perimeter of a plate-shape sample (it should have at least one of its dimensions larger than the others). The four multi-functional probes were held by four micropositioners with custom-made plastic probe arms replacing the original metallic ones to minimize heat losses. The micropositioners were fixed to a top platform and arranged to approach the sample from four different directions. Its use facilitates the movement of the probes in all directions and allows the adaptation of the apparatus to different sample geometries.

The sample to be measured is placed on a plastic sample holder that provides a rapid sample positioning. The sample holder was fixed to a motorised stage for quick loading/uploading which was screwed to a bottom platform and can be controlled by a computer *via* USB. In order to provide electrical current to several elements (probes and heater) simultaneously,

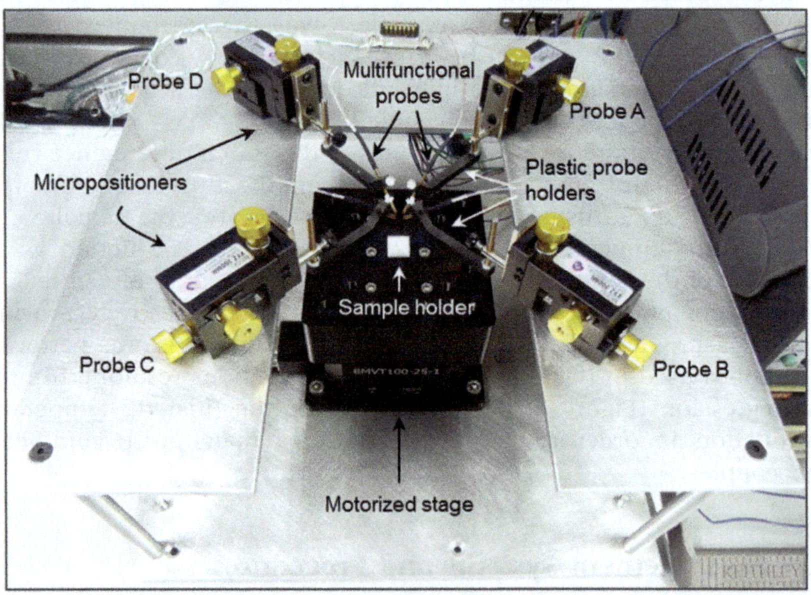

Figure 6.3 Four multifunctional probe set-up for measurement of the Seebeck coefficient and the electrical resistivity.
Reproduced with permission from J. García-Cañadas and G. Min, *Rev. Sci. Instrum.*, 2014, **85**, 043906. Copyright 2014, AIP Publishing.[14]

a triple channel DC power supply was employed, which was controlled by a computer *via* USB port. A multimeter was used to measure resistances and voltages, which was also connected to the computer by a GPIB to USB adaptor. For the rapid switching of the current and voltage, a 16-channel relay module with remote control *via* USB was used. All the equipment and measurement procedures were controlled using Labview 2011 in a PC. This allows a fully automated system. Details of the equipment and suppliers can be found in ref. 1.

The Seebeck coefficient measurement was achieved by contacting two of the multifunctional probes (probe A and probe D) to the sample perimeter. A current was supplied to the heater coil of probe A (no heater was required in the other probes) in order to achieve a temperature of ~3 K higher than that of probe D, the temperature of which usually remained close to room temperature. Once the ΔT was established, it was measured automatically as described above. The open-circuit voltage ΔV between the two probes was measured by the multimeter and finally the Seebeck coefficient of the sample S_s was calculated using,

$$S_s = \frac{\Delta V}{\Delta T} + S_{Cu}, \tag{6.4}$$

where, $S_{Cu} = 1.83\ \mu V\ K^{-1}$ is the Seebeck coefficient of pure Cu.[4] However, the Cu tube employed differs in purity from pure Cu and an appropriate value for eqn (6.4) is $S_{Cu} = 3.2\ \mu V\ K^{-1}$, which was determined by calibration from the experimental results of Figure 6.2(a).

The electrical resistivity was measured using the Van der Pauw method,[3] where four multifunctional probes (probes A, B, C and D) were contacted at four different points on the perimeter of the sample. This technique does not require a clearly defined geometry of the sample, apart from the plate-shape with homogeneous thickness. A good electrical contact between the sample and the probes was ensured by monitoring the resistance between the probes until the saturation is observed with increasing pressure.

A constant DC current was applied to two adjacent probes and the voltage induced at the other two probes was measured. For example, a current I_{AB} entered the sample through probe A and left from probe B, generating a voltage difference $\Delta V_{CD} = V_C - V_D$ between probes C and D. A resistance $R_{AB,CD} = \Delta V_{CD}/I_{AB}$ was then obtained. In order to minimise the possible errors raising from thermoelectric effects, the direction of the current was changed and $R_{BA,DC} = \Delta V_{DC}/I_{BA}$ was calculated. Since the thermoelectric voltages can be cancelled out when averaging, probe A (with a heater coil inside) can remain hot during the electrical resistivity measurement. This avoids the need to wait for the probe to cool down and/or heat up between the measurements of the Seebeck coefficient and electrical conductivity. This feature turns out to be extremely useful for high-speed measurements changing between the Seebeck coefficient and electrical resistivity.

By switching the current applied to probes C and D and measuring the voltage in probes A and B, we obtain in the same way $R_{CD,AB}$ and $R_{DC,BA}$. Averaging these four resistances gives $R_A = (R_{AB,CD} + R_{BA,DC} + R_{CD,AB} + R_{DC,BA})/4$. Similarly, $R_B = (R_{BC,AD} + R_{CB,DA} + R_{AD,BC} + R_{DA,CB})/4$ was also calculated. Both R_A and R_B are required to determine the sheet resistance R_S by

$$\exp(-\pi R_A/R_S) + \exp(-\pi R_A/R_S) = 1. \tag{6.5}$$

Finally, the electrical resistivity is given by

$$\rho = R_s d \tag{6.6}$$

where d is the sample thickness. Eqn (6.5) can only be solved analytically when $R_A = R_B$. For the rest of the cases, it has to be solved numerically. This was achieved by implementing the iteration algorithm given by NIST[5] in LabView.

6.2.3.2 *Thermal Conductivity*

The determination of the thermal conductivity is achieved based on the analysis of temperature decay of the tip of a heated probe when it is brought into contact with the sample to be measured. The experimental set-up is shown in Figure 6.4, which employs a heated multifunctional probe (described above) to capture the temperature transition from one steady state to another. In the apparatus the probe is inserted into a plastic

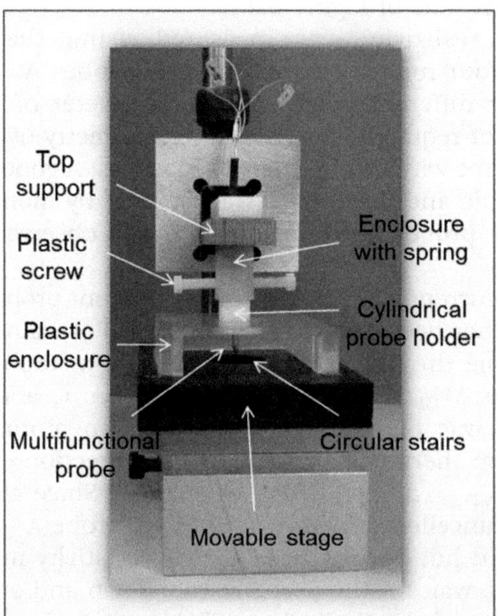

Figure 6.4 Thermal transient setup for the determination of the thermal conductivity.

cylindrical holder and fixed by two plastic screws. The cylinder is supported by a plastic enclosure screwed onto the top support with a spring placed inside the enclosure, which allows the displacement of the probe when pressed against the sample. At the base of the set-up, a movable stage is located. The stage was made of plastic and contains concentric circular stairs to locate sample discs of different diameters (5–15 mm), similarly to the sample holder described in the previous section (Figure 6.3). The height of each step of the stairs is 0.5 mm. In this way, the probe always contacts the disc samples at their centre and the samples have an air gap underneath.

During the measurement, the probe was initially heated up to reach a steady-state temperature without being in contact with the sample. The contact to the sample was then made by lifting the stage and monitoring the temperature change of the probe. The moment when the temperature of the probe starts to decrease, indicates that initial contact has been made. To ensure a good and repeatable contact, the stage is lifted slightly further to a preset position. The zone around the sample was protected by a plastic enclosure and covered with isolating wool (not shown in Figure 6.4) to minimise the heat loss from the probe due to convection. The same power supply and multimeter for the Seebeck coefficient and electrical resistivity measurements (see previous section) are employed for powering the heater coil and measuring the voltage output of the probe, respectively. All the equipment and measurements were controlled and recorded using Labview 2011 by a computer.

The thermal conductivity is determined by identifying the correlation between the temperature change of the probe and the thermal conductivity of the sample. For this, calibration was performed using three reference materials with disc shape of dimensions close to 13 mm diameter and 2 mm thickness. For each calibration, the probe was heated up with a constant current of 82 mA applied to its heater coil until a steady state temperature (usually between 68 to 74 °C) was reached. Then, the sample was brought into contact with the hot probe by lifting the stage, which resulted in a decrease of the tip temperature of the probe until a new steady state was reached (usually after 800–1000 s). The temperature difference between the initial and final steady state, ΔT, can be obtained experimentally from the recorded temperature plots as shown in Figure 6.5(a). The calibrations were performed using 3 reference samples with different thermal conductivity values: a commercial PTFE disc (0.25 $\mathrm{W\,m^{-1}\,K^{-1}}$), a BCR-724A (4.06 $\mathrm{W\,m^{-1}\,K^{-1}}$) standard reference material (SRM) from LGC Standards (UK) and a Stainless Steel 1461 SRM (14.30 $\mathrm{W\,m^{-1}\,K^{-1}}$) from the National Bureau of Standards (USA).

Figure 6.5(a) shows the thermal transients recorded as a function of time t for the above-mentioned three reference materials. An initial steady-state temperature is reached ($t < 30$ s) after applying the current to the heater coil when the input Joule heating is balanced by the heat losses from the probe. The slightly different initial temperatures between the different experiments (see values for $t < 30$ s) are due to fluctuations in the ambient conditions.

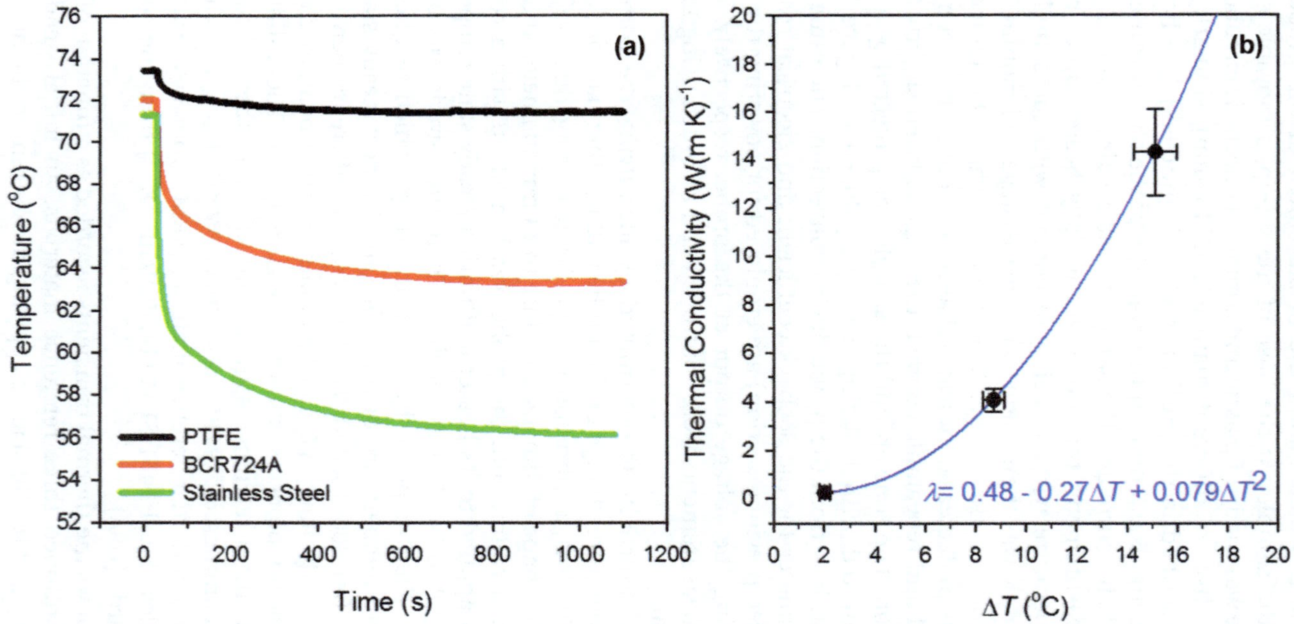

Figure 6.5 (a) Thermal transient experiments performed using PTFE (black), BCR724A (red) and Stainless Steel (green) reference materials. Samples were contacted with the probe at $t \approx 30$ s. (b) Temperature differences between the initial and final steady state (ΔT) *vs.* their thermal conductivity (circles) for three reference samples. The parabolic fit to the experimental points (line) and the resulting calibration equation are also shown.

When the probe is brought into contact with the sample ($t > 30$ s), heat diffuses from the probe into the sample and a new steady state temperature is achieved ($t \gtrsim 900$ s) when the total heat fluxes are again balanced. It is evident that a larger ΔT corresponds to a higher thermal conductivity of the sample because of its larger capability to remove heat from the probe.

The correlation between the thermal conductivity of the samples and their corresponding temperature change, ΔT, is shown in Figure 6.5(b). The curve fitting of experimental data gives a parabolic relationship for the range investigated

$$\lambda = 0.48 - 0.27\ \Delta T + 0.079 \Delta T^2. \tag{6.7}$$

Based on this equation, the thermal conductivity of a sample can be calculated using experimentally determined ΔT. It should be noted that the validity of this relationship has only been demonstrated over a limited range (approximately 0.2 to 15 $\mathrm{W\,m^{-1}\,K^{-1}}$) due to the availability of only three reference materials.

Since this set-up uses the same components as that for the Seebeck coefficient and electrical resistivity measurement (*e.g.*, multifunctional probe, sample stage, sample holder, power supply and multimeter), it is feasible to integrate both set-ups into one system for high-throughput measurements of all three important thermoelectric quantities.

6.2.4 Precision, Accuracy and Rapidness

The precision (deviation from the mean value of repeated measurements) of the developed techniques was evaluated through reparability tests and the accuracy (deviation from the true value) was assessed by comparing with the reference materials and well-established techniques. In addition, the speed of measurements will be evaluated to highlight the unique advantage of these developed techniques.

The Seebeck coefficient measurements were compared with a well-established hot probe apparatus and the Bi_2Te_3 standard reference material (SRM 3451) from NIST (USA). Three samples with different Seebeck coefficients were employed, which include a p-type Bi_4SnTe_7 alloy, an n-type $GeBi_4Te_7$ and an n-type Bi_2Te_3 from NIST. The values measured using both techniques are shown in Table 6.1, together with the standard deviations and the systematic errors determined from 10 repeated measurements. The results show small standard deviations for all three samples, indicating a good repeatability of the developed technique. Systematic errors of less than 2% demonstrate a good agreement between the developed apparatus and the well-established hot probe method. A more accurate evaluation of the accuracy can be achieved by directly comparing the measured value ($-220.02\ \mu\mathrm{V\,K^{-1}}$) of the SRM with the reference value ($-230.69\ \mu\mathrm{V\,K^{-1}}$) provided by NIST. A percentage error of 4.4% is obtained in this case, which provides sufficient accuracy for the purpose of high-throughput characterization of thermoelectric materials.

Table 6.1 Comparison of the Seebeck coefficients obtained by the multifunctional probe technique and the hot-probe technique. Small systematic and random errors indicate good precision and accuracy of the developed technique.

Sample	Hot-probe value/ $\mu V\ K^{-1} \pm$ st. dev.	4-multifunctional probe value/ $\mu V\ K^{-1} \pm$ st. dev.	Random error (%)	Systematic error (%)
p-type Bi_4SnTe_7	105.85 ± 0.76	106.21 ± 0.82	0.77	0.34
n-type $GeBi_4Te_7$	-139.43 ± 4.46	-138.15 ± 1.29	0.93	0.92
n-type Bi_2Te_3 SRM	-223.77 ± 0.99	-220.02 ± 1.15	0.52	1.68

Table 6.2 Comparison of electrical resistivity determined by the multifunctional probe technique and the reference value of the stainless steel SRM and the measured value using four probe technique for the n-type $GeBi_4Te_7$.

Sample	Reference value/ Ω cm $\pm$ st. dev.	Four multifunctional probe value/Ω cm $\pm$ st. dev.	Random error (%)	Systematic error (%)
Stainless Steel (SRM)	8.09×10^{-5}	$7.90 \times 10^{-5} \pm 0.031 \times 10^{-5}$	0.39	2.42
n-type $GeBi_4Te_7$	$2.30 \times 10^{-3} \pm 0.025 \times 10^{-3}$	$2.39 \times 10^{-3} \pm 0.0084 \times 10^{-3}$	0.35	3.96

For the electrical resistivity measurements, a disc-shaped sample prepared using austenitic stainless steel (SRM 1461) from NIST (USA) and the same n-type $GeBi_4Te_7$ disc sample used above for the Seebeck coefficient evaluation (with ρ value typical of thermoelectric samples) were employed. All the measurements were carried out with probe A kept hot to simulate the practical requirement for 'simultaneous' measurements of the Seebeck coefficient and electrical resistivity. The values obtained were compared with the reference value from NIST and those from a commercial four probe apparatus, respectively. Table 6.2 shows the standard deviations from a set of 10 measurements and the systematic errors against the reference values.

The results shown in Tables 6.1 and 6.2 demonstrate the capability of the developed set-up for measuring the Seebeck coefficient and electrical resistivity with reasonably good precision and accuracy. Furthermore, a unique advantage of this facility lies in the rapidness of the measurements. Once the satisfactory contacts with the sample are made, it only takes around 20 s to determine both the Seebeck coefficient and electrical resistivity. If the samples have the same dimensions, there is no need to reposition the probes (which can take 1 or 2 min). The sample change and making new contacts can be performed very quickly by placing the sample in the sample holder and elevating it against the probes on a motorised stage. Additionally, the use of the plastic arms of micro-positioners with an appropriate contacting angle facilitates fast and secured pressure contacts, which eliminates the need for using the solders or conductive paints to achieve satisfactory contacts with the sample.

Table 6.3 Error analysis of the thermal transient measurement. Random error was determined from the average thermal conductivity ($\lambda_{average}$) and the standard deviations. Systematic error was determined by comparing the average value with true values (λ_{true}), provided by the reference sample supplier.

Sample	λ_{true}/W m^{-1} K^{-1}	$\lambda_{average}$/W m^{-1} K^{-1} ± st. dev.	Random error (%)	Systematic error (%)
PTFE	0.25	0.26 ± 0.009	3.5	4.0
BCR-724A	4.06	4.14 ± 0.46	11.1	2.0
Stainless Steel	14.30	14.51 ± 1.80	12.4	1.5
Pyrex 7740	1.10	1.12 ± 0.037	3.3	1.8

The repeatability of the set-up for thermal conductivity measurement was evaluated by performing a total of 21 tests on all three samples used in the calibration curve. Table 6.3 shows the mean values and standard deviations with the corresponding random errors, indicating the repeatability of the technique. The random errors for BCR-724A and stainless steel are 11.1% and 12.4%, respectively. They are significantly higher than PTFE (3.5%) and Pyrex 7740 (3.3%). Such variations may be due to the differences in thermal contact area. The surfaces of solids are not microscopically flat, which can result in randomness in the actual contact area at the solid/solid interface.[6] The amount of heat diffusing into the sample is expected to significantly depend on the contact area, which could be difficult to reproduce even for the same sample. Moreover, the reproducibility of the same contact area may differ for different materials. This explains the observed deviation in the repeatability for different materials.

The accuracy of the technique for thermal conductivity measurement was evaluated using Pyrex 7740 (Netzsch). The sample has the same dimensions as the other reference samples. The measurements were repeated seven times and the results are shown in Table 6.3. The mean value ($\lambda_{average}$) obtained from the experiments shows a good agreement with the published reference value (λ_{true}), which gives a systematic error of 1.8%. For all reference samples employed, the systematic errors can also be calculated using their respective mean values and the reference values listed in Table 6.3, which is less than 4.0%. The results demonstrate a good accuracy with reasonable repeatability, providing a new method for rapid measurement of thermal conductivity. In addition, this method is simple and inexpensive, which makes it a suitable technique for high-throughput screening of the thermal conductivity in materials research. It is to be noted that the heat conduction takes place in the radial directions during measurements using this technique. A careful interpretation of the results is needed for anisotropic materials such as Bi_2Te_3. Also, the precision and accuracy shown in Table 6.3 can only be guaranteed for a specific sample shape (disc) and dimension (13 mm in diameter and 2 mm thickness). For samples with different shapes and sizes, new calibration is required for achieving the stated precision and accuracy.

6.3 Impedance Spectroscopy

Impedance spectroscopy is a characterization technique widely used in many fields of research, such as photovoltaics,[7,8] fuel cells,[9] supercapacitors[10] and corrosion.[11] The main advantage of this method over other characterization techniques resides in the ability to detect and separate the electronic related processes that govern the performance of the device. This is achieved by measuring the impedance response of the system under the excitation of a small ac signal (either voltage or current) with or without a dc bias. The response is monitored as a function of frequency and hence, rapid processes show up at high frequencies while slow processes only appear at low frequencies.

In the field of thermoelectrics, impedance spectroscopy has not been extensively used but some work has been carried out to demonstrate its feasibility for thermoelectric characterization.[12–14] Recently, it has been shown that the impedance spectroscopy is a promising technique for complete evaluation of thermoelectric materials and devices under suspended conditions.[15] In this section we present the fundamentals of the method, the theoretical framework (or equivalent circuits) that underlies the thermoelectric response, the experimental set-up developed for thermoelectric characterization, and its applications in materials screening and module evaluation.

6.3.1 Fundamentals of Impedance Spectroscopy

In an impedance experiment a small amplitude sinusoidal current of certain frequency f, oscillating around a constant value I_{dc}, is applied to the system. This induces a sinusoidal variation of the potential difference with an amplitude ΔV_{ac}, which is proportional to the current I_{ac} and oscillates around a dc value ΔV_{dc}. Figure 6.6(a) shows schematically the excitation and response waveforms, together with their corresponding equations. The response signal could be shifted in time (phase delay) respect to the excitation wave, which is quantified by the phase angle ϕ. If the current is ahead of the voltage as shown in Figure 6.6(a), the system is said to exhibit a capacitive-like response with ϕ being negative. In the opposite case, it shows an inductive-like behaviour and ϕ is positive. The impedance Z is defined as the ratio of the voltage vector to the current vector, which can be described in a complex plane (or Nyquist diagram) by a point represented by the magnitude $|Z| = \Delta V_{ac}/I_{ac}$ and the phase angle ϕ, as shown in Figure 6.6(b). If the impedance is measured at different frequencies, typically around 40 to 50 values over 1 MHz to 10 mHz, an impedance spectrum is obtained. Apart from the complex plane representation, the data can also be presented using the Bode plots, where the change in magnitude and phase angle as a function of frequency is explicitly presented.

Experimentally, a current waveform can be used as the excitation signal to produce a voltage response (galavanostatic mode), and *vice versa*, a voltage

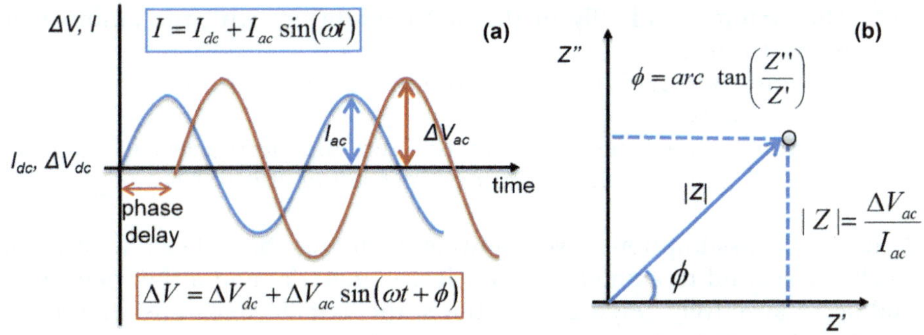

Figure 6.6 (a) Schematic diagram of the current and voltage waveforms employed in an impedance experiment. $\omega = 2\pi f$ is the angular frequency. (b) Representation of a point of the impedance spectrum by a vector of magnitude $|Z|$ and the phase angle Π on the complex plane (Nyquist diagram).

signal can be employed to generate a current response (potentiostatic mode). In general, both procedures lead to the same impedance spectrum because of linear relationship between the excitation and response signals.

Typically, the understanding of the impedance spectrum requires identification of its corresponding equivalent circuit, which usually consists of a network of resistors, capacitors, inductors or other type of elements. Once the impedance spectrum is recorded experimentally, the equivalent circuits that produce good fitting to the recorded impedance spectrum will be identified. Frequently, more than one equivalent circuit could match the experimental impedance response, but there is only one that properly describes the physics of the system, which needs to be determined with the help of theoretical modelling of the physical processes. Once the most appropriate equivalent circuit is identified and fitting is performed properly, the parameters of the different circuit elements are quantified and the system characterized.

6.3.2 Theoretical Framework

The theoretical model and equivalent circuit required to interpret the impedance response of both thermoelectric materials and devices have been recently reported.[15] They are obtained based on the following assumptions:

(a) Thermoelectric element has a uniform cross-sectional area, *A*, and a length, *L*, with very thin metallic layer (*e.g.*, silver paint) coated on the contact surfaces.
(b) The system is tested under adiabatic (suspended) conditions—no heat exchange with surroundings.
(c) All the thermal and thermoelectric parameters are independent of temperature.

(d) The system is initially in thermal equilibrium with the ambient of temperature T_i.
(e) The Joule heating is neglected both in the bulk of the materials and at the junctions.
(f) Thermal contact resistance at the ceramic/thermoelectric material interface and thermal spreading-constriction effects[16] are neglected.

Under these assumptions two equivalent circuits have been identified, which correspond to a single thermoelectric material and a thermoelectric module, respectively. Figure 6.7 shows the circuit diagrams and their corresponding impedance responses.

The circuit elements appeared in Figure 6.7 are summarised in Table 6.4. R is a pure resistor, equivalent to the ohmic resistance of the system, R_Ω. It should be noted that when determined experimentally R_Ω contains contributions from both parasitic R_p (wires, electrodes and contact resistances) and resistance of thermoelectric materials. Z_{WCT} and Z_{Wa} are known as the constant temperature (WCT) and adiabatic (Wa) Warburg impedances respectively,[15] and their response is defined by

$$Z_{WCT} = R_{TE}\left(\frac{j\omega}{\omega_{TE}}\right)^{-0.5}\tanh\left\{\left(\frac{j\omega}{\omega_{TE}}\right)^{0.5}\right\} \tag{6.8}$$

$$Z_{Wa} = R_C\left(\frac{j\omega}{\omega_C}\right)^{-0.5}\coth\left\{\left(\frac{j\omega}{\omega_C}\right)^{0.5}\right\}, \tag{6.9}$$

where, $j=\sqrt{-1}$, $\omega=2\pi f$, R_{TE} and R_M are thermoelectric resistances[17] of the thermoelement and the ceramic layer of the module, respectively. ω_{TE} and ω_C are the corresponding characteristic angular frequencies.

λ_{TE}, α_{TE} and $C_{p,TE}$ are the thermal conductivity, thermal diffusivity and specific heat capacity of thermoelectric materials, respectively. λ_C and α_C are the thermal conductivity and thermal diffusivity of ceramic layer in a thermoelectric module. L_C is the thickness of the ceramic layers. It can be seen from Table 6.4 that for a single thermoelement ($Z=R_\Omega+Z_{WCT}$) all its thermal constants (λ_{TE}, α_{TE} and $C_{p,TE}$) can be determined from its impedance response if the Seebeck coefficient is known. For a thermoelectric device ($Z=R_\Omega+1/(Z^{-1}{}_{WCT} + Z^{-1}{}_{Wa})$), it is possible to determine all thermoelectric parameters of a module if the thermal conductivity of the ceramic layer (usually $\lambda_C\approx 30\ \mathrm{W\,m^{-1}\,K^{-1}}$ for alumina) is known. The Seebeck coefficient can be extracted from R_C. Then, λ_{TE} can be obtained from R_{TE} and the total ohmic resistance from R. Moreover, the thermoelectric figure-of merit of the module, $(zT)_{module}$, can be determined using

$$(zT)_{module} = \frac{R_{TE}}{R_\Omega} = \frac{2NS^2T_iL}{\left(R_p + \dfrac{2N\rho L}{A}\right)\lambda_{TE}A} \tag{6.10}$$

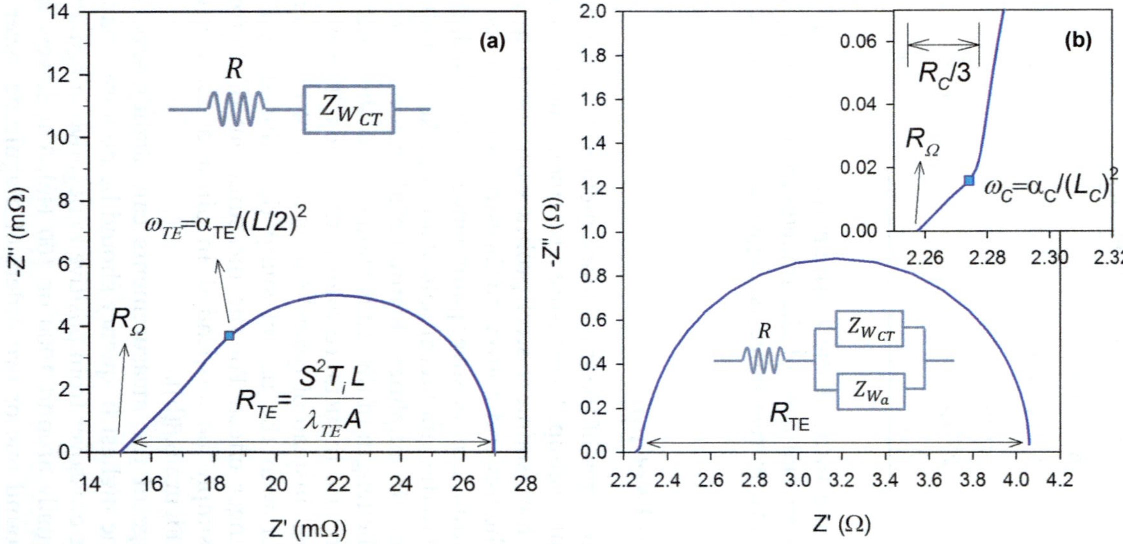

Figure 6.7 Simulated impedance spectrum of (a) a single thermoelectric element and (b) a thermoelectric module of 127 couples. Typical dimensions and thermoelectric properties of commercial Bi_2Te_3 material were used. The corresponding equivalent circuits are represented in the insets. Square points represent characteristic angular frequencies corresponding to $2\pi\omega_{TE}$ and $2\pi\omega_C$ as indicated, which allow the estimation of ω_{TE} and ω_C. The inset in (b) is a magnification to the high frequency part. Reproduced from J. García-Cañadas and G. Min, *J. Appl. Phys.*, 2014, **116**, 174510. Copyright 2014, AIP Publsihing LLC.[15]

Table 6.4 Equivalent circuit elements and the equations for thermoelement and thermoelectric module, respectively. λ_{TE} and α_{TE} are the thermal conductivity and thermal diffusivity of the thermoelement, respectively. λ_C, α_C and L_C are the thermal conductivity, thermal diffusivity and thickness of the ceramic layer in a thermoelectric module of N thermocouples, respectively.

Element	Parameters	Equation for single element	Equation for device
R	R_Ω	$R_p + \rho L/A$	$R_p + 2N\rho L/A$
Z_{WCT}	R_{TE}	$S^2 T_i L/(\lambda_{TE} A)$	$2NS^2 T_i L/(\lambda_{TE} A)$
	ω_{TE}	$\alpha_{TE}/(L/2)^2$	$\alpha_{TE}/(L/2)^2$
Z_{Wa}	R_C	$2S^2 T_i L_C/(\lambda_C A)$	$4NS^2 T_i L_C/(\lambda_C A)$
	ω_C	α_C/L_C^2	α_C/L_C^2

Eqn (6.10) has the same form of the Harman relationship.[18] In fact, the Harman method is a special case of the impedance technique. It can be shown that if the parasitic resistance is neglected, $(zT)_{module}$ equals the material's zT.

6.3.3 Experimental Set-up

One of the main advantages of the impedance method is that it requires a very simple experimental set-up. To measure a thermoelement, the sample is sandwiched by two short stainless steel probes with sharp tips. Prior to contact the sample, the top and bottom surfaces of the thermoelement are usually coated with a thin layer of silver paint. Once sandwiched, the probes are directly connected to the input terminals of an impedance analyzer (such as a potentiostat or lock-in amplifier). Figure 6.8(a) shows a schematic diagram of the set-up. The length of the leads should be as shorter as possible to avoid possible parasitic inductance due to the leads. The probes were made of stainless steel, instead of typical Cu, because of its lower thermal conductivity. Together with the use of sharp tips, this design helps to minimize the cold finger effect. For the measurement of thermoelectric modules, they can simply be suspended in air and connected to the impedance analyzer (Figure 6.8(b)).

The frequency range of the measurements can slightly vary from one system to another. The highest frequency should be no lower than the first frequency value of the crossover from positive to negative imaginary part on the complex plot (usually around 1000 or 100 Hz). The lowest frequency should be at least around one or two orders of magnitude lower than the frequency at which the imaginary part of the impedance reaches the maximum (usually between 50 to 1 mHz). The amplitude of the sinusoidal signal can also vary, but usually lies around 20 mA in galvanostatic mode and 10 mV in potentiostatic mode. For a better estimation, the best value is normally the lowest at which the impedance response is clearly observed

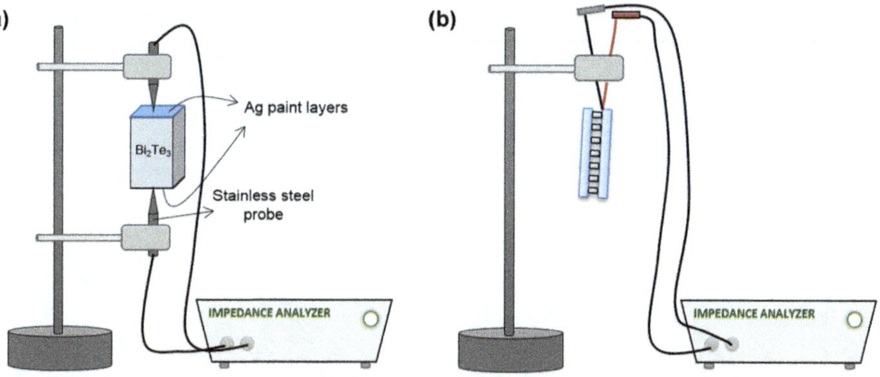

Figure 6.8 Schematic diagrams of the experimental set-ups employed for impedance measurements of (a) a thermoelement and (b) a thermoelectric module under suspended conditions. Not to scale.

without the presence of significant noises. The dc level must be set at 0 A or 0 V, respectively, in the galvanostatic or potentiostatic mode. If the dc level is not set at zero, a temperature gradient will be developed in the sample and the required initial condition of homogeneous temperature cannot be met.

The fitting to the impedance spectrum for identification of equivalent circuits could be performed using the software supplied with the equipment by most of the manufacturers, or the dedicated programs such as *Zview*. They all offer the possibility to create and edit equivalent circuits and other additional options. Alternatively, software packages such as *Labview* or *Matlab*, which are able to perform non-linear least square fittings, could be also used. Once the fitting is performed, it is important to evaluate the relative errors of the parameters of interest. Sometimes, the fitting curve may look very similar to the experimental results but large errors might appear, indicating that the selected circuit elements are not suitable or optimized.

6.3.4 Measurement Analysis and Applications

The validity of the equivalent circuits of Figure 6.7 for the thermoelectric impedance response has been investigated experimentally using a thermoelement and a commercial thermoelectric device. Figure 6.9(a) shows the experimental response (circles) of a p-type Bi_2Te_3 thermoelement with a cross-sectional area of 1.4×1.4 mm^2 and a length of 1.6 mm. Figure 6.9(b) shows the response of a commercial Bi_2Te_3-based thermoelectric module, consisting of 252 thermoelements of 1 mm^2 cross-sectional area and 1.5 mm length. The thickness of the ceramic plates is 0.8 mm.

The experimental results of both samples show an excellent fit to the theoretical responses predicted by the corresponding equivalent circuits (lines). Table 6.5 presents the thermoelectric properties obtained from the measured impedance response. Typical values for commercial Bi_2Te_3 are

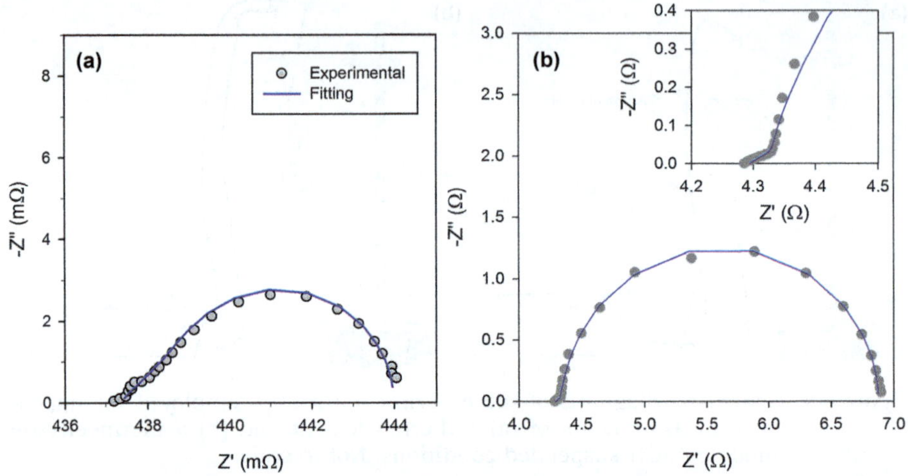

Figure 6.9 Experimental impedance response (circles) of (a) a *p*-type Bi_2Te_3 thermoelement and (b) a commercial bismuth telluride-based thermoelectric module (from European Thermodynamics). The inset in (b) represents a magnification of the high frequency part. Lines indicate the fittings to the equivalent circuits.[15] Figure 6.9(b) reproduced from J. García-Cañadas and G. Min, *J. Appl. Phys.*, 2014, **116**, 174510. Copyright 2014, AIP Publsihing LLC.

also included for comparison (data in brackets). A good agreement is evident for the thermal properties of the thermoelement, which confirms the suitability of the theoretical framework. It should be noted that the R_{TE} value for a thermoelement is in the order of few mΩ, which is close to the limitation of the equipment, indicating the limitation of this technique for accurate measurement of low zT materials ($zT < 0.1$).

In the case of the thermoelectric device, it can be observed from Table 6.5 that α_{TE} and $C_{p,TE}$ deviate significantly from the expected value. This is due to the fact that the heat produced in the system by the Peltier effect, not only diffuses in the thermoelements, but it enters into the ceramic layers,[15] which results in a lower value for α_{TE}. The useful information that can be extracted from this impedance analysis comes from the three resistances R_Ω, R_C and R_{TE}. R_Ω, given by the high-frequency intercept with the real axis, directly provides the total ohmic resistance (frequently called the internal resistance). R_C can be determined from the impedance response as shown in the inset of Figure 6.7(b) and consequently the Seebeck coefficient of the module can be calculated using the corresponding expression listed in Table 6.4 for a given λ_C (≈ 30 W m^{-1} K^{-1} for Al_2O_3). Once S is known, the thermal conductivity of the thermoelement can be calculated using the expression of R_{TE} in Table 6.4. Finally, the module zT can be calculated from the ratio of R_{TE}/R_Ω (eqn (6.10)). It can be seen from Table 6.5 that all the quantities obtained from the measurement show a good agreement with the expected values.

Table 6.5 Fitting parameters and calculated thermoelectric properties from impedance measurements of a thermoelement and a commercial thermoelectric module, respectively. The thermal parameters of the thermoelement were calculated using the Seebeck coefficient value (195 μV K^{-1}) measured by a hot-probe apparatus. The values between brackets, when provided, are of typical commercial bismuth telluride modules provided by the manufacturers (Custom Thermoelectrics USA in this case). The density of the material taken was 7.53 g cm^{-3} and $T_i = 300$ K.

Sample	R_Ω/Ω	R_{TE}/Ω	ω_{TE}/rad s^{-1}	λ_{TE}/W m^{-1} K^{-1}	α_{TE}/cm^2 s^{-1}	$C_{p,TE}$/J g^{-1} K^{-1}	S/μV K^{-1}	zT	R_c/Ω	ω_c/rad s^{-1}
Element	0.437	0.00668	1.78	1.37 (1.50)	0.0036 (0.0037)	0.45 (0.54)	—	—	—	—
Module	4.292	2.585	0.24	1.62	0.0013	1.65	193.5 (200)	0.60	0.149	6.08

It is worth noting that all the module properties shown in Table 6.5 were determined from a single run of impedance measurement, which takes only few minutes (typically 5 to 10 min). In addition, the measurements involve only the electrical connections between the suspended device and impedance analyser, without the need to attach thermocouples for temperature measurement. Such features make this method ideal for the high-throughput evaluation of thermoelectric properties. In addition, the impedance technique can be employed to identify promising thermoelectric materials by simply inspecting the Nyquist plot for appearance of a semicircle.

6.4 Conclusions

The high-throughput evaluation of thermoelectric materials and devices requires the rapid determination of the three main thermoelectric properties (Seebeck coefficient, electrical resistivity and thermal conductivity) and the corresponding figure of merit (zT). In order to meet such requirement, two techniques have been developed and presented in this chapter, which are based on the use of multifunctional probes and the frequency domain analysis (impedance spectroscopy), respectively.

The multifunctional probe is the fundamental part of the first technique. It serves as electrical contact, thermocouple and heater. This novel design enables the integration of the Van der Pauw technique (for electrical resistivity measurement) with the hot-probe technique (for the Seebeck coefficient measurement), which has the advantages of simplicity and rapidness. The system developed using the multifunctional probes demonstrated the ability of rapid measurement of both the Seebeck coefficient and electrical resistivity in less than 20 s, with the precision and accuracy comparable to the well-established techniques. The measurement of thermal conductivity using a multifunctional probe has also proved to be feasible. The prototype system developed shows that the systematic error is less than 5% while the random error is over 10% for thermal conductivity measurement of homogenous materials. The measurement in this case takes about 20 min to complete.

The impedance technique was also investigated for rapid measurement of thermoelectric properties of both thermoelements and thermoelectric modules. The equivalent circuits required for interpretation of the impedance response were determined and validated by experiments. In the case of thermoelements, the thermal conductivity can be determined if the Seebeck coefficient is known. However, the signal produced by a thermoelement is usually very small—close to the limitation of the equipment, and consequently, is expected to be less suitable for characterisation of low-zT materials. In the case of thermoelectric modules, it proved to be a more suitable method for rapid evaluation of thermoelectric devices because several key thermoelectric parameters can be determined from a single run of experiment, which takes a few minutes. For fast screening purposes, identifying promising thermoelectric materials can be achieved by simply checking if a semicircle has appeared on the Nyquist plot.

Acknowledgements

The authors wish to acknowledge financial support from the Accelerated Metallurgy Project, which is co-funded by the European Commission in the 7th Framework Programme (contract NMP4-LA-2011-263206), by the European Space Agency and by the individual partner organizations. We thank European Thermodynamics Ltd. for providing thermoelectric modules and samples, the technicians at the mechanical and electrical workshop of the Cardiff School of Engineering and Lourdes Márquez-García for her support to the experimental work.

References

1. J. García-Cañadas and G. Min, *Rev. Sci. Instrum.*, 2014, **85**, 043906.
2. NIST ITS-90 Thermocouple Database, http://srdata.nist.gov/its90/main/.
3. L. J. Van der Pauw, *Philips Res. Rep.*, 1958, **13**, 1–9.
4. D. M. Rowe, *CRC Handbook of Thermoelectrics*, CRC Press, London, 1995.
5. NIST Algorithm for sheet resistance calculation, http://www.nist.gov/pml/div683/hall_algorithm.cfm.
6. F. P. Incropera and D. P. DeWitt, *Fundamentals of Heat and Mass Transfer*, John Wiley & Sons, USA, 2002.
7. I. Mora-Sero, G. A. Garcia-Belmonte, P. P. Boix, M. A. Vazquez and J. Bisquert, *Energy Environ. Sci.*, 2009, **2**, 678–686.
8. F. Fabregat-Santiago, G. Garcia-Belmonte, I. Mora-Sero and J. Bisquert, *Phys. Chem. Chem. Phys.*, 2011, **13**, 9083–9118.
9. X. Z. Yuan, H. J. Wang, J. C. Sun and J. J. Zhang, *Int. J. Hydrogen Energy*, 2007, **32**, 4365–4380.
10. R. Kotz, M. Hahn and R. Gallay, *J. Power Sources*, 2006, **154**, 550–555.
11. G. W. Walter, *Corros. Sci.*, 1986, **26**, 681–703.
12. A. D. Downey, T. P. Hogan and B. Cook, *Rev. Sci. Instrum.*, 2007, **78**, 93904.
13. A. De Marchi and V. Giaretto, *Rev. Sci. Instrum.*, 2011, **82**, 104904.
14. A. De Marchi and V. Giaretto, *Rev. Sci. Instrum.*, 2014, **85**, 043906.
15. J. García-Cañadas and G. Min, *J. Appl. Phys.*, 2014, **116**, 174510.
16. F. Casalegno, A. De Marchi and V. Giaretto, *Rev. Sci. Instrum.*, 2013, **84**, 024901.
17. J. Garcia-Canadas and G. Min, *J. Electron. Mater.*, 2014, **43**, 2411–2414.
18. T. C. Harman, *J. Appl. Phys.*, 1958, **29**, 1373–1374.

CHAPTER 7

System Design Considerations for Thermoelectric Energy Recovery

RICHARD STOBART,* ZHIJIA YANG AND SONG LAN

Department of Aeronautical and Automotive Engineering, Loughborough University, Loughborough LE11 3TU, UK
*Email: r.k.stobart@lboro.ac.uk

7.1 Introduction

7.1.1 The Potential for Waste Heat Recovery

In a typical internal combustion engine (ICE) as little as one third of the fuel energy content is converted to useful shaft work.[1] This is a partly due to the mode of operation of the ICE in which the gas exchange process needs to be initiated when there is still substantial energy content in the products of combustion. Consequently, a proportion of fuel enthalpy is transferred to the exhaust and cooling systems of the engine and is lost to the environment. This loss suggests a significant potential for improvements in the efficiency of the engine, if compact- and cost-effective methods can be conceived to recover even a proportion of this energy. Examples that have been presented in the literature include turbo-compounding devices,[2] bottoming cycles,[3] thermoelectric generators (TEGs),[4] thermo-acoustic systems[5] and chemical recuperation methods.[6] A recent study shows that the reduction of fuel consumption in passenger cars could be as high as 20% in Federal Test Procedure (FTP) tests simply by capturing the exhaust waste

RSC Energy and Environment Series No. 17
Thermoelectric Materials and Devices
Edited by Iris Nandhakumar, Neil M. White and Stephen Beeby

Published by the Royal Society of Chemistry, www.rsc.org

heat and converting 10% of it to electricity.[7] A number of research projects have demonstrated that the electric power produced by TEGs can be used as a practical vehicle power source and be able to contribute, significantly, to improving the efficiency of the internal combustion engine.[8–11] The understanding of WHR requires quantification of energy flows and in the following section, definitions and analysis will help create a framework for evaluating the potential for WHR.

The term *availability* refers to the availability of energy for work production. In order to quantify the energy that could actually be converted to useful work, we make use of the steady flow availability function, B defined for a flow of matter (in the gas or liquid state) as $B = H - T_0S$, where H is the enthalpy (typically expressed in kJ), S is the entropy (typically in kJ K^{-1}) and T_0 is a conceptual environment temperature at which heat transfer to the environment takes place. Whilst B is not a fluid property, since it is composed of both properties and the environmental temperature, it is of great use in understanding the potential for work production in a thermodynamic system. If a fluid in a steady-flow process is taken between two stable states, the difference in the availability function gives the amount of work that could be produced using reversible methods between those two states. In the case of the exhaust flow from an ICE, the available energy in the exhaust flow establishes the maximum work that could be produced by an energy recovery system with the exhaust flow as an energy source.[12,13]

Between two states 1 and 2, there is, in general, a change of steady flow available energy, $\Delta B = B_1 - B_2$

$$\Delta B = ((H_1 - H_0) - T_0(S_1 - S_0)) - ((H_2 - H_0) - T_0(S_2 - S_0)) = \Delta H - T_0\Delta S \tag{7.1}$$

By definition, the *steady flow exergy*, Ξ is the reversible shaft work that can be produced as the fluid passes from state 1 to the dead state (T_0, p_0).

$$\Xi_1 = B_1 - B_0. \tag{7.2}$$

An alternative form of eqn (7.1) is

$$\Delta B = \Xi_1 - \Xi_2 \tag{7.3}$$

Exergy is an important working concept being analogous to the property of a fluid in regard to its capacity to support the production of shaft work.

A comparison of the actual work produced, W with the ideal work, W_{rev} leads to the definition of the rational efficiency for an energy recovery process (see eqn (7.4))

$$\eta_r = \frac{W}{W_{\text{rev}}} \tag{7.4}$$

An example of the operating parameters of a typical heavy duty diesel engine illustrates the importance of considering the availability function in the context of energy recovery. The example is based on the operating parameters of a 9 litre turbocharged diesel engine operating at a mid-speed and mid-load condition.

At high engine load conditions, the exhaust temperature is 600 °C, while the coolant temperature is 110 °C. The coolant flow rate is 4 kg s^{-1} and the exhaust gas flow rate is 0.2 kg s^{-1}. For the purposes of the calculation we assumed that the coolant circuit is operating close to the environmental pressure, p_0 and that the properties of the exhaust gas are represented by $\left(c_p = 1.150 \text{ kJ kg}^{-1}\text{ K}^{-1},\ \gamma = \frac{c_p}{c_v} = 1.333\right)$. For the coolant, $c_p = 4.22$ kJ kg^{-1} K^{-1}. For a part-load condition, with an engine brake power of 200 kW, the exhaust enthalpy flux is 132 kW and the coolant enthalpy flux is 150 kW. Ignoring radiation losses, engine thermal efficiency at this condition is $\frac{200}{200+150+132} = 42\%$.

The specific exergy, ξ_f of the exhaust gas flow is given by eqn (7.5).[12]

$$\frac{\xi_f}{c_p T_0} = (\theta - 1) - \log\theta + \left(1 - \frac{1}{\gamma}\right)\log\beta \tag{7.5}$$

Where: $\theta = \frac{T}{T_0}$, $\beta = \frac{p}{p_0}$ and T_0 is the temperature of the dead state (usually taken to be ambient conditions).

The specific exergy of the coolant flow (which is considered incompressible and showing only very small changes in density) is given by:

$$\frac{\xi_f}{c_p T_0} = (\theta - 1) - \log\theta + \frac{3}{2}\nu T_0\left(\frac{p}{T} - \frac{p_0}{T_0}\right) \tag{7.6}$$

Where: ν is the specific volume of the coolant at temperature T.

Substituting for the typical operating conditions, gives, $\xi_{f,gas} = 348$ kJ kg^{-1} and $\xi_{f,coolant} = 1.3$ kJ kg^{-1}, and the available energy flux is $\Xi_{f,gas} = 70$ kW and $\Xi_{f,coolant} = 4$ kW, respectively.

The ratio between these two recovery sources reflects the thermodynamic significance of the higher temperature of the exhaust gas. In an engine, most thermal energy originates with the combustion process where temperatures are high. Heat transfer to coolant results in a substantial loss of available energy. This suggests that the exhaust gas flow is a primary target for energy recovery and that an increasing exhaust temperature is advantageous for energy recovery if not for the materials of the cylinder and exhaust system.

Figures 7.1 and 7.2 present data taken from laboratory engines ran under realistic load conditions. The figures show the exhaust manifold temperatures observed as the engine load is varied. Alongside the temperature data, the available energy is plotted.

As shown in Figures 7.1 and 7.2, the exhaust manifold temperatures of gasoline engines tend to be higher than those of diesel engines.

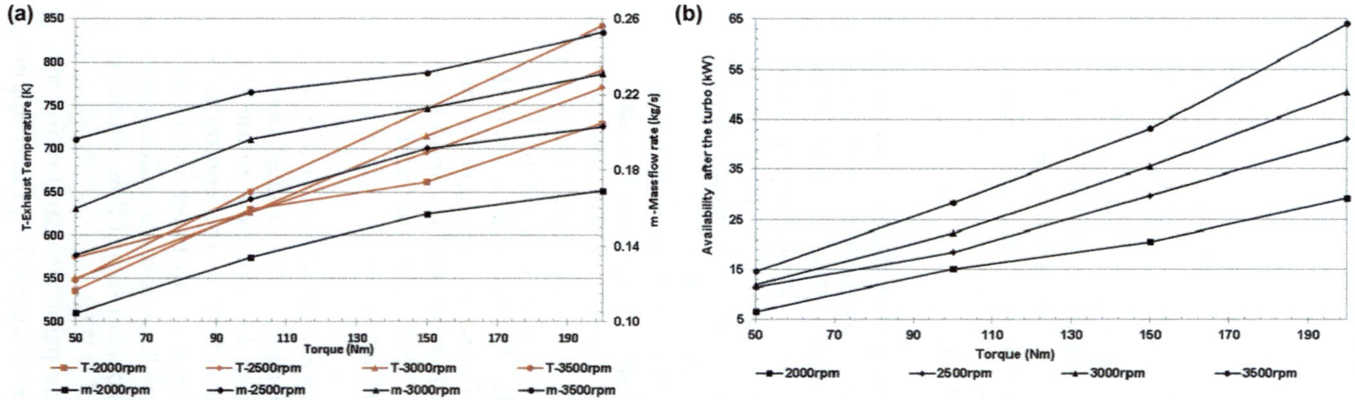

Figure 7.1 (a) Exhaust temperature and mass flow rate data and (b) the available energy of the exhaust gas flow of a VW 1.9TDi engine (1.9 litre turbocharged diesel engine).

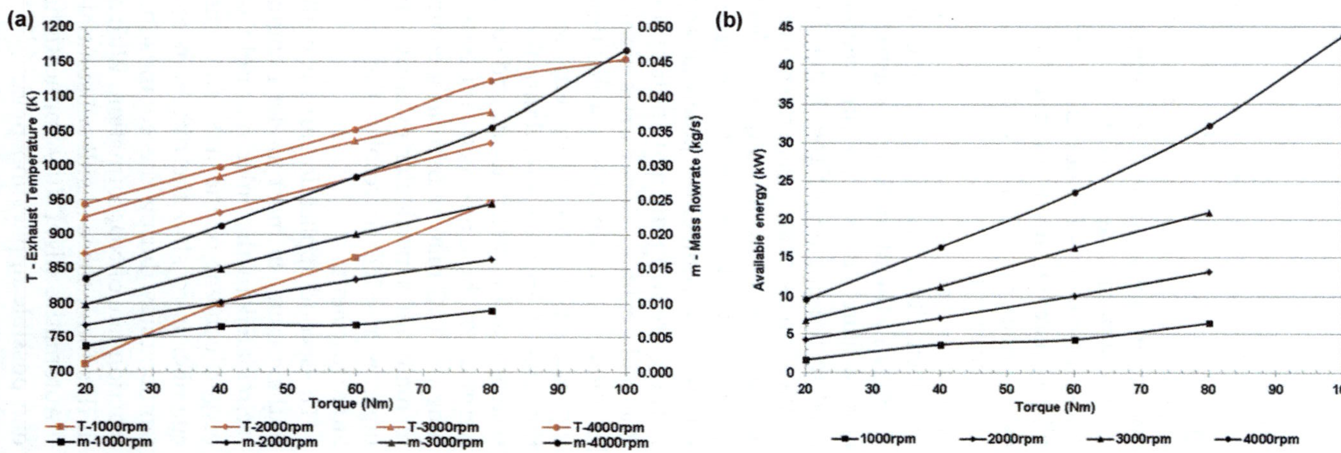

Figure 7.2 (a) Exhaust temperature and mass flow rate data and (b) the available energy of the exhaust flow upstream of the three way catalyst for a Ford Sigma (1.4 litre naturally aspirated gasoline engine).

However, the exhaust mass flow rates of naturally aspirated spark ignition (SI) engines tend to be lower. Diesel engines generally use a significant flow of excess air that tends to depress exhaust temperatures and utilize a large expansion ratio. Both contribute to the higher flow at lower temperatures.

The figures show that the available energy flux in automobile exhaust systems lies anywhere in the range of 5–70 kW for light duty vehicles and is dependent on the temperature and mass flow rate of the exhaust gas or the operating conditions of the engines.

TEG systems have a considerable advantage over other WHR methods. Those advantages include the solid-state nature of the device, which simplifies installation and maintenance and has no significant impact on the overall noise and vibration performance of the vehicle. Exhaust gas flow can be directly utilized to generate electricity to support the vehicle electrical system; as a replacement of an engine-driven alternator or, to contribute to the electricity supply needed for propulsion.[14]

Arguably the most desirable attribute of a WHR system is that its output is closely matched to a specific requirement—such as the electrical needs local to the engine, or the 'hotel' load of the vehicle. The most desirable application is that regenerated energy is consumed almost immediately without the need for any significant energy storage. The powering of electrical ancillaries on the engine (particularly oil and water pumps) represents a compelling potential application area of WHR.

There are two principal application areas for WHR in engine propulsion. The main exhaust flow carries significant available energy and is a strong candidate. However, the best positioning of the device, just downstream of the exhaust manifold of the engine could result in temperature changes in the exhaust flow that would interfere with the after-treatment system. A TEG system may be switched off, and can be used in a Peltier mode to assist the warm-up of the after-treatment system. Thereafter while the engine is running the TEG may be used in a control mode where its load is varied so as to maintain a particular exhaust temperature. Studies on integration are urgently needed to resolve this apparent conflict of WHR and exhaust after-treatment.

The second application area is in the exhaust gas recirculation (EGR) path of the engine. In most modern engines a proportion of the exhaust flow is recirculated to the inlet manifold of the engine to effectively dilute the oxygen content of the incoming gas flow. The effect in the combustion chamber is to slow the diffusion of oxygen, resulting in lower maximum temperatures during the combustion process.[15] This in turn reduces the reaction rate of NO_x formation. With about 20% of the exhaust flow available in this form, there is still substantial potential for WHR.

Automotive (light duty applications, including passenger cars) represent one possible application field.[1,4,14,16,17] Engines employed in other applications where engines run at high load for long periods of time—including machinery and power generation—also offer rich application possibilities.

The TEG is made up of a heat exchange device into which thermoelectric modules (TEM) are assembled connected through an interface electrical circuit to an external circuit. A TEM is a device that includes a number of thermocouples that converts thermal energy directly to electrical energy making use of the Seebeck effect. For a practical implementation, it is necessary to realize both a practical thermal efficiency and a cost effective configuration of module and heat exchanger.[18] In Section 7.2, the analysis that leads to a closed-form solution for the power output of a TEM is presented, but first the efficiency of the thermoelectric process is considered alongside its ramifications.[19] Figure 7.3 shows a simple module made up of a single thermocouple.[19]

The thermal efficiency (η_{max}) of the module is given in eqn (7.7) and is expressed in terms of the non-dimensional figure-of-merit (zT) of the TE material eqn (7.8), which in turn is a function of the Seebeck coefficient (α), thermal conductivity (κ) and electrical resistivity (ρ) of the material all of which are a function of temperature T. The subscripts refer to avg-average, h-hot and c-cold.

$$\eta_{\max} = \left[\frac{\sqrt{1+zT}-1}{\sqrt{1+zT}+\dfrac{T_c}{T_h}}\right]\left[\frac{T_h - T_c}{T_h}\right] \tag{7.7}$$

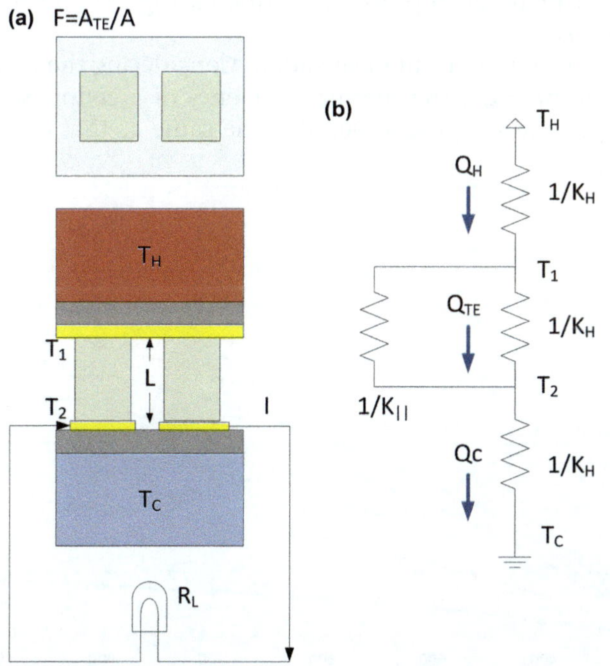

Figure 7.3 (a) Arrangement of a single module and (b) the equivalent thermal circuit. Adapted from ref. 19 with permission from The Royal Society of Chemistry.

Where

$$zT = \frac{\alpha^2}{k\rho} T_{\text{avg}} \tag{7.8}$$

Eqn (7.7) shows that the efficiency is made up of two terms: the Carnot efficiency $\left(1 - \frac{T_c}{T_h}\right)$, and the term, $\frac{\sqrt{1+zT}-1}{\sqrt{1+zT}+\frac{T_c}{T_h}}$, which can be interpreted as the rational efficiency. The figure-of-merit, zT, contains the physical parameters that are related to a direct loss (electrical resistance) or heat transfer across a finite temperature difference (thermal conductivity) both of which represent the irreversibility in the operation of the couple. The hot and cold junction temperatures, T_h and T_c both play a role in the rational efficiency because it is these temperatures that create the thermoelectric effect.

The efficiency of a couple can be increased by increasing the hot side temperature, T_h. The voltage generated will be higher and currents for a given power output are lower. TE materials with higher zT values will also show higher efficiencies for a given working temperature. A higher zT implies reduced irreversibility. The thermal efficiency of a single couple is shown in Figure 7.4 as the hot side temperature, T_h and zT are varied.

Figure 7.4 shows that the path to high efficiencies will require a combination of material improvements (increasing zT) and heat exchange design (higher T_h).

The data allow an important observation. Considering the respective plots for $zT = 1.0$ and $zT = 1.4$, the thermal efficiency of a couple with materials rated at $zT = 1.0$ and working at 900 K is the same as that of a couple with

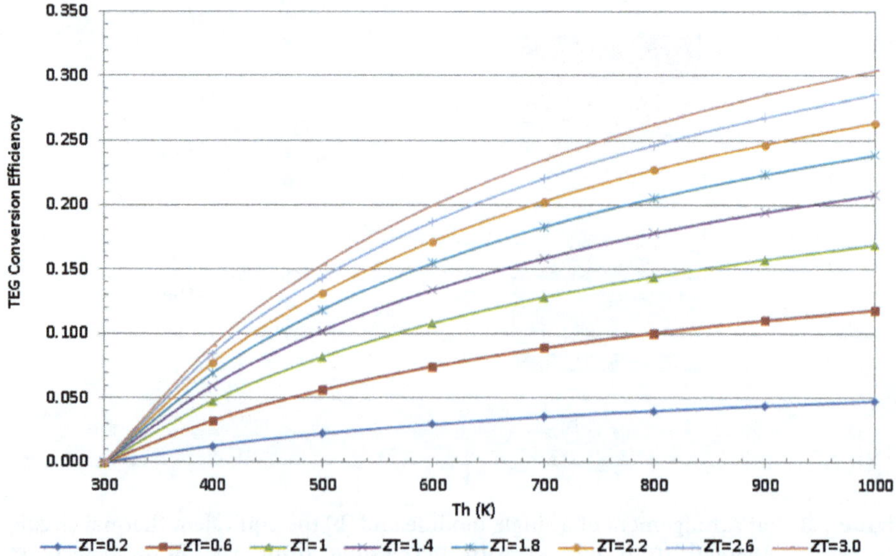

Figure 7.4 The thermal efficiency of a single couple.

$zT=1.4$ working at 700 K. In our context of ICEs, it is feasible to work at either temperature, but 900 K is commonly reached even in diesel engines. The current status of TE materials suggest a zT value close to 1.0. The implication is that achieving high temperature tolerance and creating a high hot side temperature is needed to extract the most benefit from current materials. The benefit of progress in heat exchange processes is that the needs of emerging material can be more readily met. Taking the broader view, the heat exchange performance for a TEG is both important and complementary with increases in the zT value of the materials.

7.2 Modelling TEG Performance

TEG performance is a function of materials, modules design and heat exchange design parameters. In vehicle applications, the design objective of the TEG system is to maximize the power output while regarding a number of constraints that will be application specific. Constraints include cost, back-pressure created in exhaust gas flows, volume and mass.

An understanding of the effect of design parameters (including material choice) is essential to calculate and optimize the performance of both the module and the integrated TEG. The estimation of optimal parameters has been reported for simple cases that do not include for example, the varying conditions that are commonly associated with energy recovery in power generation and vehicle propulsion. However, such analysis gives important design guidance and a starting condition for more detailed optimization studies that will be conducted using numerical methods.

To illustrate the scope and limitations of a closed-form analytical approach the derivation is presented.[19] Note that the analysis[19] has been extended to include the costs of manufacture.[20]

Eqn (7.9),[19] the electrical power generated by a module, can be expressed as

$$P=I^2R_L=\frac{S_{pn}^2(T_1-T_2)^2}{R}\left(\frac{m}{(m+1)^2}\right) \tag{7.9}$$

Where: $S_{pn}=S_p-S_n$, $I=\dfrac{S_{pn}(T_1-T_2)}{R+R_L}$, R_L is the load resistance, R is the module resistance and $m=\dfrac{R_L}{R}$.

If the heat flux through each element of the module can be considered equal, and this will be a reasonable assumption at the low efficiency levels of current devices, then the temperature drop across the module can be approximated as:

$$T_1-T_2=\frac{K_H K_C}{K_C K_T+K_H K_T+K_H K_C}(T_H-T_C) \tag{7.10}$$

Where K_T is the thermal conductance of the module.

If, in addition, the hot and cold sides of the heat exchanger have the same conductivity, that is

$$K_H = K_C = K = UA,$$

then the module output power is

$$P = \frac{S_{np}^2 \sigma (T_1 - T_2)^2 AF}{4} \left(\frac{m}{(m+1)^2} \right) \left(\frac{L}{(2L_T F + L)^2} \right) \tag{7.11}$$

The power is maximized when the leg length is chosen

$$L = 2FL_T,$$

giving a temperature distribution

$$T_1 - T_2 \approx \frac{1}{2}(T_H - T_C) \tag{7.12}$$

This is an important result.[19,21] Although an approximation, it represents a rule of thumb that may be applied during initial design work. The authors[21] refer to the result being counterintuitive. It would appear that maximizing the temperature drop across the module would be best—but this would imply a low thermal conductivity in turn diminishing the power output to a greater degree than the higher efficiency would support. This carries an important conclusion that realistic materials can be deployed to good effect. Given the relatively poor thermal conductivity of the modules, we may be able to continue in TEG design with lower conductivity metals typified by stainless steel, which can increase robustness to the TEG design while still achieving the maximum power criterion for the device.

This approximate solution faces a particular challenge in the context of engine WHR:

- Typically, the engine load varies resulting in varying temperature and flow rates in the exhaust system. This variation in flow rates in turn leads to changing heat transfer conditions between the hot exhaust gas and the hot side of the modules. Certain engine applications (power generation, marine propulsion) do offer consistent conditions to which a design can be adapted, but in general all energy recovery systems face the challenge of varying load. We may summarize this as the parameter U varying with operating conditions and, in general, being different between hot and cold sides of the heat exchange system.
- Modules are typically placed on a heat exchange surface. (See Figure 7.5 as an example.) Placing in this way—which is typical of TEG construction methods—leads to a constriction resistance. Because temperatures are varying, the pitch of mounting modules both across the flow and along the flow must change to meet the need for a maximum power output. In general, knowledge of constriction resistance will be required to select the appropriate module positions.[21,22]

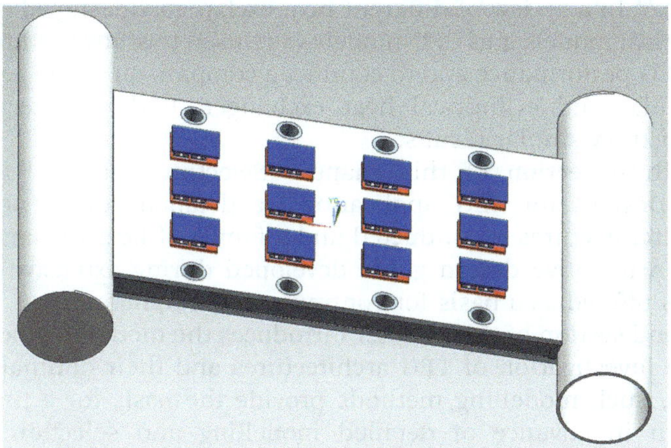

Figure 7.5 An illustration of the disposition of modules on a heat exchange surface.

- The choice of architecture will be dictated by the modal engine conditions and the corresponding gas conditions on the hot side of the heat exchanger. Whilst the design must take account of the changing conditions, the majority of applications will be oriented towards the engine's maximum torque condition and the corresponding temperatures and flow rates.

7.3 The Role of the Heat Exchanger in TEG Design

Since the state-of-the-art commercial TEMs are planar, heat exchangers for TEG applications have tended to be single plate or square channel in configuration. Some work has been reported with multi-split channels (rectangular topology) and hexagonal or octagonal cross-section heat exchangers.[10,11,23] Within the rectangular topology with longitudinal configuration in which the TEMs are placed with respect to the exhaust gas flow direction, there are two options for heat exchanger design: the heat exchanger can be made with the long side aligned to the exhaust flow direction (long plate heat exchanger) or the short side aligned to the exhaust flow direction (wide plate heat exchanger). Most rectangular TEGs follow the pattern of the long type heat exchanger.[10,11,23] However, there is no comparison evident in the literature between these two types of plate heat exchangers. Intuitively, the wide plate heat exchanger may have the advantage of a more even temperature distribution and reduced pressure drop compared with a long plate heat exchanger.

Fin types and fin density also have significant impact on heat exchanger performance.[24] Considering the constraints of back pressure in main exhaust pipe and the constraints on the pressure drop in the EGR path, open pipe with a short plate fin and turbulent bar or a vortex generator are good choices for TEG.[11,24] Plain fin can also be used in plate type heat exchangers.[11] In a review of different heat exchangers designed for WHR,[25] both empirical models and CFD models were used to study the impact of fin types on TEG performance and to conduct a comparison of the performance of rectangular and cylindrical heat exchanger. Both architectures were reported to show similar results.

In the third section of this chapter (Selecting and Evaluating Heat Exchange Designs for TEG Applications) a detailed comparison of heat exchange architectures is conducted and a family of heat exchange designs proposed. A tentative design guide developed during extensive modelling exercises is offered as a basis for continuing development.

The second section of this chapter introduces the modelling methods that permit the investigation of TEG architectures and their optimal operating conditions. Such modelling methods provide the basis for a first stage of TEG design in advance of detailed modelling and selection of design parameters.

7.4 Modelling a Thermoelectric Generator

In this section a model and a selection of results is used to demonstrate the value of modelling in the initial selection of the design parameters for a TEG.

In the TEG design process, the following factors are of prime importance:

- the hot side temperature and the transport properties of the hot side fluid
- the variation in fluid properties during the operation of the system
- Seebeck effect, Peltier effect, Joule heating, Fourier heat conduction and other heat transfer processes
- the influence of the geometric structure of both module and heat exchanger on the power output.

Using the model, the designer can begin to understand the trend of the optimal design parameters under changing operating conditions. Validation is required to provide credence to the model. Sometimes experimental equipment is available to check a particular aspect, but a validation may not be possible simply because hardware cannot be implemented. In this case, a form of validation through a different calculation route, utilizing a different set of assumptions, can be used as a check on the predicted power output.

The model description begins with the characterization of the structure of both heat exchanger and modules and discretization of the TEG (Section 7.4.1). Methods of calculating the temperature distribution are presented in Section 7.4.2. In Section 7.4.3, the thermal resistance network that represents the thermal function of the TEG is presented. Section 7.4.4 presents the electrical network calculations, while in Section 7.4.5 the model structure is reviewed. Section 7.4.6 covers the validation of the model. In Section 7.4.7, a series of applications of the models is presented including the optimization of key TEG design parameters.

7.4.1 Defining the Structure of TEG

A TEG with a finned heat exchanger on the hot side is shown in Figure 7.6.

It can be seen that a complete TEG consists of TEMs, a heat source, a heat sink and connecting wires. The heat source is connected to the exhaust flow from an internal combustion engine. The temperature of the heat sink is maintained by the engine coolant. A TEM consists of thermoelectric elements (legs), conductive tabs, and ceramic wafers. Positive-type (p-type) and negative-type (n-type) thermoelectric elements make up a thermocouple. All the thermoelectric legs are connected by the conductive tabs electrically in series and thermally in parallel between two ceramic wafers. The ceramic wafers are thermally conductive and electrically insulating. Both p-type and n-type thermoelectric elements are made with the same geometry dimensions but their material properties are different. α_P and α_N are the Seebeck coefficient, κ_P and κ_N are the conductivities, ρ_P and ρ_N are the electrical resistances.

Geometric parameters for the hot side exchanger are shown in Figure 7.7. b is the height of the exchanger, a is the width of the exchanger, s is the space between the fins, δ_f is the fin thickness and l_f is the length of the heat exchanger. The number of the fins is N_f. The fluid property variation along the flow direction in the heat exchanger is considered by discretizing the TEG into small control volumes (CVs) along the flow direction as shown in Figure 7.7. The number of CVs is n_{cv}.

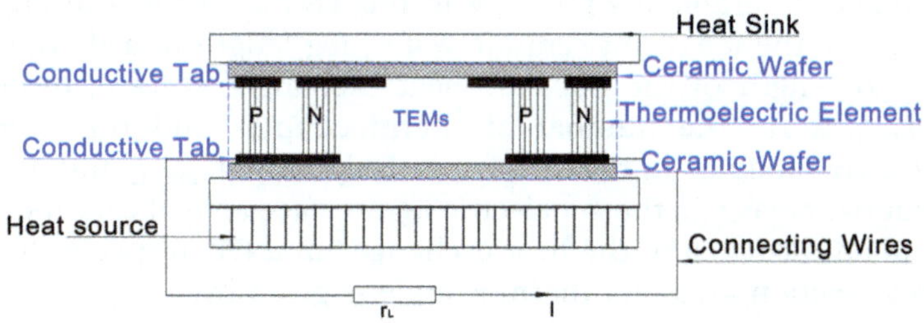

Figure 7.6 Structure of a TEG.

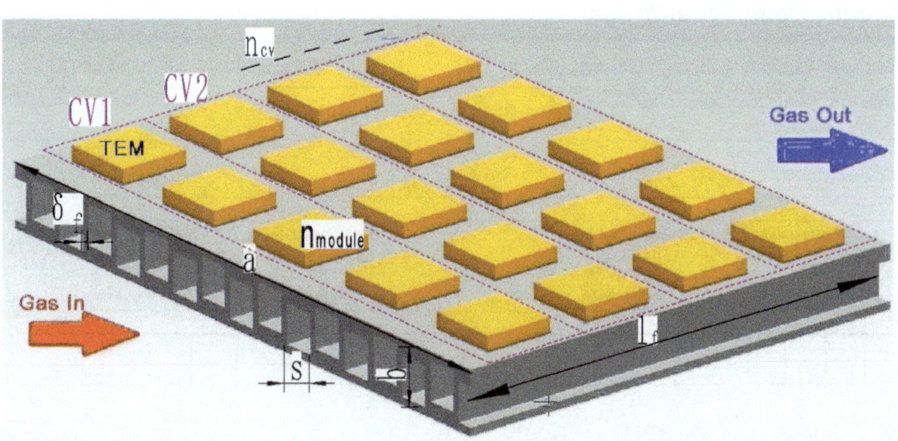

Figure 7.7 Geometric parameters of the hot side heat exchanger.

The surface area of the heat exchanger in a CV:

$$A_{CV} = \frac{a \cdot l_f}{n_{cv}} \tag{7.13}$$

The cross-sectional flow area A_x is given by:

$$A_x = b \cdot (a - N_f \delta_f) \tag{7.14}$$

The exposed area of one fin A_f in a CV:

$$A_f = \frac{b \cdot l_f}{n_{cv}} \tag{7.15}$$

Exposed area of the base plate A_b in a CV:

$$A_b = A_{CV} - \frac{N_f \delta_f l_f}{n_{cv}} \tag{7.16}$$

The total area of finned surface A_t in a CV:

$$A_t = 2N_f A_f + 2A_b \tag{7.17}$$

The geometric parameters for a CV in the TEG are shown in Figure 7.8. l_{cw}, l_{leg}, l_{hxr} are the length of ceramic wafer, leg length of a thermoelectric element and length of the hot side heat exchanger surface, respectively. n_{couple}, n_{module} are the number of thermocouples and the number of thermoelectric modules in the CV, respectively. A_{leg}, A_{module} and A_{cv} are the cross-sectional areas of a thermoelectric leg, surface area of ceramic wafer in a TEM and surface area of the heat exchanger in a CV, respectively.

The cross-section area of a thermocouple is given by:

$$A_{couple} = 2A_{leg} \tag{7.18}$$

The fill factor for a TEM, γ

$$\gamma = \frac{n_{couple} A_{couple}}{A_{module}} = \frac{n_{couple} \cdot 2A_{leg}}{A_{module}} \tag{7.19}$$

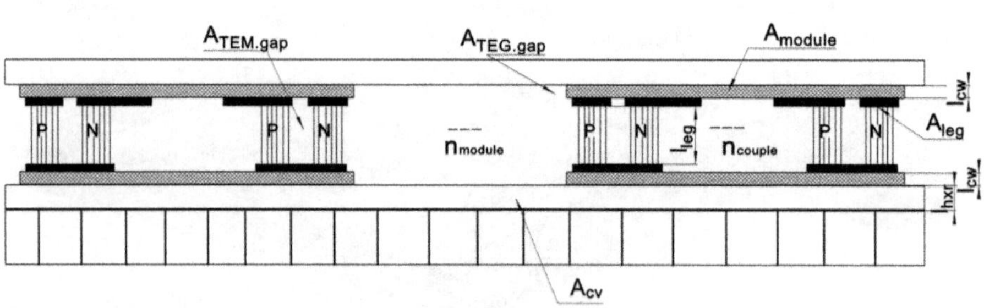

Figure 7.8 Geometric parameters of a TEG (front view).

The surface area of the air gap between the thermoelectric elements and ceramic wafers in a TEM, $A_{\text{TEM.gap}}$ is given by:

$$A_{\text{TEM.gap}} = A_{\text{module}} - n_{\text{couple}} A_{\text{couple}} = A_{\text{module}}(1-\gamma) \tag{7.20}$$

The surface area of air gap between the thermoelectric modules and heat exchangers $A_{\text{TEG.gap}}$ in a CV is given by:

$$A_{\text{TEG.gap}} = A_{\text{CV}} - n_{\text{module}} A_{\text{module}} \tag{7.21}$$

7.4.2 Calculating Temperature Distributions

The temperature distribution within the TEG is presented in Figure 7.9. The heat exchanger for the heat sink is not included, so the cold side boundary conduction is modelled as a constant temperature T_{ct} and constant coolant mass flow rate $\dot{m}_{\text{c}}$.

For the heat source, the exhaust gas mass flow rate is also assumed to be constant along the heat exchanger $\dot{m}_{\text{h}}$, while the exhaust gas properties vary as heat transfer takes place. The exhaust gas-in temperature of the heat exchanger is T_{in} and the exhaust gas-out temperature is T_{out}. Because the TEG is discretized into n_{cv} connected CVs, each CV has its own inlet and outlet exhaust temperature, which can be seen in Figure 7.9. The outlet gas temperature of a CV is also the inlet gas temperature for the adjacent CV. In the ith CV, the exhaust gas-in temperature is T_{gi} and the exhaust gas-out temperature is T_{gi+1}, which can be calculated as:[26]

$$T_{g\,i+1} = T_{gi} - (T_{gi} - T_{\text{ct}}) \frac{1 - e^{-\text{NTU}(1+C_{\text{R}})}}{1 + C_{\text{R}}} \tag{7.22}$$

NTU and C_{R} are the number of transfer units and the capacity ratio, respectively. NTU and C_{R} can be calculated using eqn (7.23) and (7.24):

$$\text{NTU} = \frac{1}{R_{\text{cv}i} \dot{m}_{\text{h}} C_{\text{ph}}} \tag{7.23}$$

$$C_{\text{R}} = \frac{\dot{m}_{\text{h}} C_{\text{ph}}}{\dot{m}_{\text{c}} C_{\text{pc}}} \tag{7.24}$$

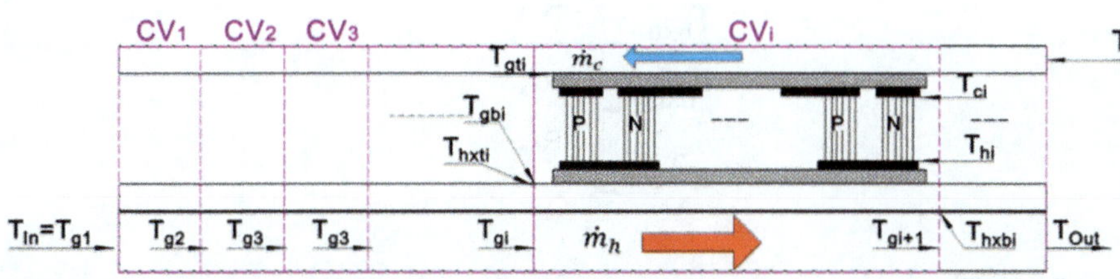

Figure 7.9 Temperature distribution of TEG (side view).

R_{cvi}, C_{ph}, C_{pc} are the total thermal resistances of the *i*th control volume, exhaust gas specific heat capacity at constant pressure and coolant specific heat capacity at constant pressure, respectively. C_{pc} is computed based on the coolant temperature T_{ct}. C_{ph} and R_{cvi} are calculated according to the average gas temperature of the CV T_{cvi}.

$$T_{cvi} = \frac{T_{gi} + T_{gi+1}}{2} \tag{7.25}$$

As can be seen in Figure 7.9, the temperature distribution of each CV is as follows. The temperature at the joint of the fins and the bottom surface of the heat exchanger is T_{hxbi}. It is assumed that the contact effect exists between the heat exchanger and the TEM, so T_{hxti} is the temperature at the top surface of the heat exchanger and T_{bti} is the temperature of the interface between TEM ceramic wafer and hot side exchanger surface. Hot side and cold side temperatures of TEM are T_{hi} and T_{ci}, respectively.

7.4.3 Thermal Resistance Network in a CV

The thermal resistance network and temperature distribution of the *i*th CV in a TEG is presented in Figure 7.10. Thermal resistances of the conductors connecting the thermo-elements and the joining materials are neglected as being very much smaller than the resistance of the materials themselves. Based on their different properties, thermal resistances are classified into four different groups: convection thermal resistance (R_{fin}), conduction thermal resistances (R_{hxr}, R_{cw}, $R_{TEG.gap}$), contact thermal resistance (R_{ce}) and equivalent resistance ($R_{TEM.eq}$ and $R_{load.eq}$).

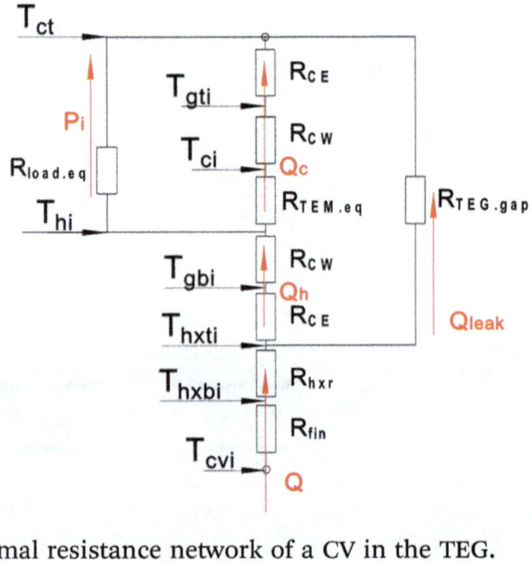

Figure 7.10 Thermal resistance network of a CV in the TEG.

7.4.3.1 Convection Thermal Resistance (R_{fin})

The heat convection thermal resistance between the finned heat exchanger and the heat reservoir is R_{fin},

$$R_{\text{fin}} = \frac{1}{A_t h_{\text{hxr}} \eta_o} \tag{7.26}$$

h_{hxr} and η_o are the coefficient of heat convection and the total heat transfer efficiency, respectively. h_{hxr} can be calculated by

$$h_{\text{hxr}} = \frac{\text{Nu} k_{\text{exh}}}{D_h} \tag{7.27}$$

where Nu_i is the Nusselt number in the ith control volume, k_{exh} is the thermal conductivity of exhaust gas and D_h is the hydraulic diameter of plate fin. Based on the structure of the plate fin exchanger, D_h and Nu_i can be calculated as follows:[27]

$$D_h = \frac{4(b - \delta_f)(s - \delta_f)}{2(b - \delta_f) + 2(s - \delta_f)} \tag{7.28}$$

$$\begin{cases} \text{Nu}_i = 1.233(\bar{x})^{-1/3} + 0.4 & \bar{x} \le 0.001 \\ \text{Nu}_i = 7.541 + 6.874 \times (1000\bar{x})^{-0.488} e^{-245\bar{x}} & \bar{x} > 0.001 \end{cases} \tag{7.29}$$

$\bar{x}$ is defined based on the discretization of the TEG

$$\bar{x} = \frac{i}{\text{RePr} D_h} \cdot \frac{l_f}{n_{cv}} \tag{7.30}$$

where Re and Pr are the Reynolds number and Prandtl number of the flow, respectively. They can be calculated as follows:

$$\text{Re} = \frac{v_{\text{exh}} D_h}{\mu_{\text{exh}}} \tag{7.31}$$

$$\text{Pr} = \frac{\mu_{\text{exh}} C_{\text{p.exh}}}{k_{\text{exh}}} \tag{7.32}$$

where μ_{exh}, $C_{\text{p.exh}}$, v_{exh} are exhaust gas kinematic viscosity, exhaust gas specific heat capacity at constant pressure and exhaust gas mass velocity, respectively. v_{exh} is defined by the exhaust gas mass flow rate m_h

$$v_{\text{exh}} = \frac{\dot{m}_h}{A_x} \tag{7.33}$$

All exhaust gas properties k_{exh}, μ_{exh}, $C_{\text{p.exh}}$, are temperature-dependent, which are defined by the average temperature of the control volume T_{cvi}. The temperature variation of these parameters can be seen in Appendix A.

The total heat transfer efficiency η_o of a CV can be calculated by:

$$\eta_o = 1 - \frac{A_f}{A_t}(1 - \eta_f) \tag{7.34}$$

η_f is the fin efficiency, which is modelled as:[27]

$$\eta_f = \frac{\tan h(m_f l)}{m_f l} \tag{7.35}$$

k_{fin} is the thermal conductivity of the fins. l and m_f can be calculated using,

$$l = \frac{b}{2} - \delta_f \tag{7.36}$$

$$m_f = \sqrt{\frac{2h_{hxr}}{k_{fin}\delta_f}\left(1 + \frac{\delta_f}{l_f}\right)} \tag{7.37}$$

7.4.3.2 Conduction Thermal Resistances (R_{hxr}, R_{cw}, $R_{TEG.gap}$)

The thermal resistance in a CV for the surface of heat exchanger R_{hxr} and ceramic wafer R_{cw}, is given by eqn (7.38) and (7.39). k_{hxr}, k_{cw} are the thermal conductivity of the heat exchanger surface and the thermal conductivity of the ceramic wafer, respectively.

$$R_{hxr} = \frac{l_{hxr}}{k_{hxr}A_{CV}} \tag{7.38}$$

$$R_{cw} = \frac{l_{cw}}{k_{cw}n_{module}A_{module}} \tag{7.39}$$

The heat transfer leakage is taken into consideration by the heat conduction thermal resistance of the air gap between the thermoelectric modules $R_{TEG.gap}$, which can be expressed by eqn (7.40)

$$R_{TEG.gap} = \frac{l_{leg} + 2l_{cw}}{k_{air}A_{TEG.gap}} \tag{7.40}$$

where k_{air} is the thermal conductivity of air.

7.4.3.3 Contact Thermal Resistance (R_{ce})

The thermal contact effect is reflected in the thermal contact resistance R_{ce} between the heat exchanger and the ceramic wafer. As shown in Figure 7.11, due to machining limitations, the solid surfaces of heat exchanger and ceramic wafer will never form a perfect contact when they are brought together.

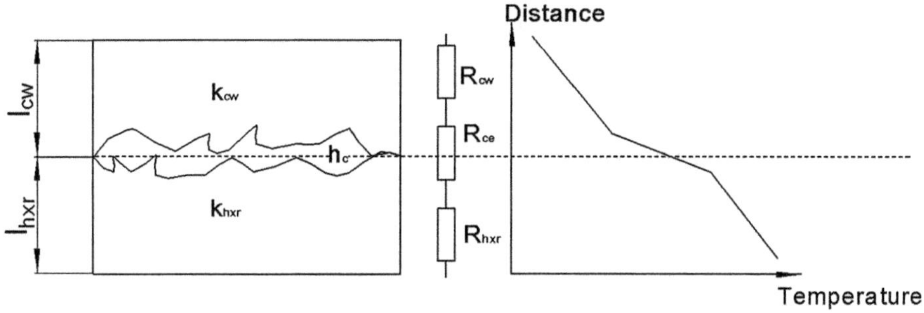

Figure 7.11 Thermal contact resistance.

Air gaps will always exist between the two contacting surfaces due to roughness. To calculate the contact thermal resistance, an interfacial conductance h_c is placed in series with the conducting media on both sides. R_{ce} can be written as

$$R_{ce} = \frac{1}{h_c n_{module} A_{module}} \tag{7.41}$$

The internal conductance h_c can be calculated by[28]

$$h_c = 1.25 k_m \frac{m}{\sigma}\left(\frac{P}{H}\right)^{0.95} \tag{7.42}$$

k_m, m, σ, P, H are the mean thermal conductivity of the interface, mean absolute asperity slope, contact surface roughness, pressure of contact surface, and the hardness of the softer of the two contacting solids, respectively. k_m can be calculated using

$$k_m = \frac{2k_{hxr}k_{cw}}{k_{hxr} + k_{cw}} \tag{7.43}$$

7.4.3.4 Equivalent Thermal Resistance ($R_{TEM.eq}$ and $R_{load.eq}$)

Figure 7.12 depicts the direction of heat flows through TEMs in a CV due to the Peltier effect (Q_P), the Joule heating effect (Q_J) and the Fourier effect (Q_f). I, r_{Ii}, P_i are the current, electrical resistance of the TEMs in a CV and power output of a CV, respectively (more details are included in Section 7.5).

The heat flows Q_p caused by the Peltier effect are absorbed at the hot junction $\alpha_{TE}IT_{hi}$ and released from the cold junction $\alpha_{TE}IT_{ci}$, respectively. Q_j is generated by the flow of electrical current through the TEM (I^2r_{Ii}) and power output (P_i). For the energy I^2r_I, half of it goes to the hot junction and half goes to the cold junction. Instead of going through TEMs, the energy P_i goes to the external load resistance. Q_f is the heat conduction transfer between the hot side and cold side is because of the Fourier effect. It is made

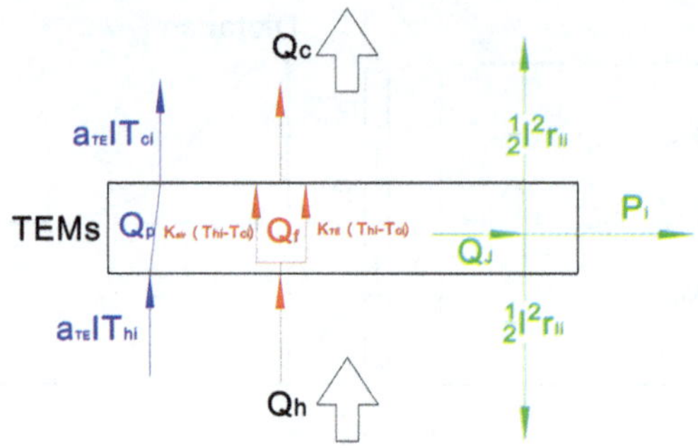

Figure 7.12 Energy transfer through TEMs in a CV.

up by heat conduction through the thermoelectric legs $K_{\mathrm{TE}}(T_{\mathrm{h}i} - T_{\mathrm{c}i})$ and the heat conduction through the air gap between the thermoelectric legs $K_{\mathrm{air}}(T_{\mathrm{h}i} - T_{\mathrm{c}i})$.

$$Q_{\mathrm{f}} = K_{\mathrm{TE}}(T_{\mathrm{h}i} - T_{\mathrm{c}i}) + K_{\mathrm{air}}(T_{\mathrm{h}i} - T_{\mathrm{c}i}) \tag{7.44}$$

K_{air} and K_{TE} are the thermal conductance of the air gap and the thermal conductance of the thermoelectric legs, respectively, which can be calculated as follows:

$$K_{\mathrm{air}} = \frac{k_{\mathrm{air}}\, n_{\mathrm{module}}\, A_{\mathrm{TEM.gap}}}{l_{\mathrm{leg}}} \tag{7.45}$$

$$K_{\mathrm{TE}} = \frac{k_{\mathrm{TE}}\, n_{\mathrm{couple}}\, n_{\mathrm{module}}\, A_{\mathrm{couple}}}{l_{\mathrm{leg}}} \tag{7.46}$$

where k_{air} and k_{TE} are thermal conductivities of air gap and TEM, which are decided by the average temperature $\frac{T_{\mathrm{hi}} + T_{\mathrm{ci}}}{2}$. Based on Figure 7.12, the heat balance equations at hot side and cold side and total energy balance are

$$Q_{\mathrm{h}} = n_{\mathrm{module}} n_{\mathrm{couple}}(\alpha_{\mathrm{P}} - \alpha_{\mathrm{N}}) T_{\mathrm{h}i} I - \frac{1}{2} I^2 r_{\mathrm{I}i} + Q_{\mathrm{f}} \tag{7.47}$$

$$Q_{\mathrm{c}} = n_{\mathrm{module}} n_{\mathrm{couple}}(\alpha_{\mathrm{P}} - \alpha_{\mathrm{N}}) T_{\mathrm{c}i} I + \frac{1}{2} I^2 r_{\mathrm{I}i} + Q_{\mathrm{f}} \tag{7.48}$$

$$Q_{\mathrm{h}} - Q_{\mathrm{c}} = \mathrm{P}_i \tag{7.49}$$

Since the thermal energy transfer through these TEMs is in parallel, their contribution in a CV can be summed up in an equivalent module resistance given as

$$R_{\mathrm{TEM.eq}} = \frac{T_{\mathrm{h}i} - T_{\mathrm{c}i}}{Q_{\mathrm{c}}} = \frac{T_{\mathrm{h}i} - T_{\mathrm{c}i}}{n_{\mathrm{module}} n_{\mathrm{couple}}(\alpha_{\mathrm{P}} - \alpha_{\mathrm{N}}) T_{\mathrm{h}i} I + \frac{1}{2} I^2 r_{\mathrm{I}i} + Q_{\mathrm{f}}} \tag{7.50}$$

In the same way, the equivalent thermal resistance for the thermal energy transfer through the external resistance can be given as

$$R_{\text{load.eq}} = \frac{T_{hi} - T_{ct}}{Q_h - Q_c} = \frac{T_{hi} - T_{ct}}{n_{\text{module}} n_{\text{couple}} (\alpha_P - \alpha_N)(T_{hi} - T_{ci}) I - I^2 r_{Ii}} \tag{7.51}$$

Based on energy conservation principle and the thermal resistance network in Figure 7.10, the temperature distribution of the CV and all thermal resistances can be solved simultaneously. Then the total thermal resistance of a control volume R_{CVi} can be calculated by

$$R_{CVi} = R_{\text{fin}} + R_{\text{hxr}} + \cfrac{1}{\cfrac{1}{R_{\text{TEG.gap}}} + \cfrac{1}{R_{cw} + R_{ce} + \cfrac{1}{\cfrac{1}{R_{\text{load.eq}}} + \cfrac{1}{R_{\text{TEM.eq}} + R_{cw} + R_{ce}}}}} \tag{7.52}$$

7.5 Electrical-resistance Network of a TEG

The electrical-resistance network of a TEG is shown in Figure 7.13.

The internal electrical resistance of TEMs in a CV is r_{Ii}, which is made up by the electrical resistance of the p-type semiconductor leg r_P and n-type semiconductor leg r_N.

$$r_{Ii} = r_P + r_N = \frac{n_{\text{module}} n_{\text{couple}} l_{\text{leg}}}{A_{\text{leg}}} (\rho_P + \rho_N) \tag{7.53}$$

where the electric resistance ρ_{TE} are decided by the average temperature $\frac{T_{hi} + T_{ci}}{2}$. The total internal electrical resistance of a TEG is r_{TEG}

$$r_{TEG} = \sum_{i=1}^{n_{cv}} r_{Ii} \tag{7.54}$$

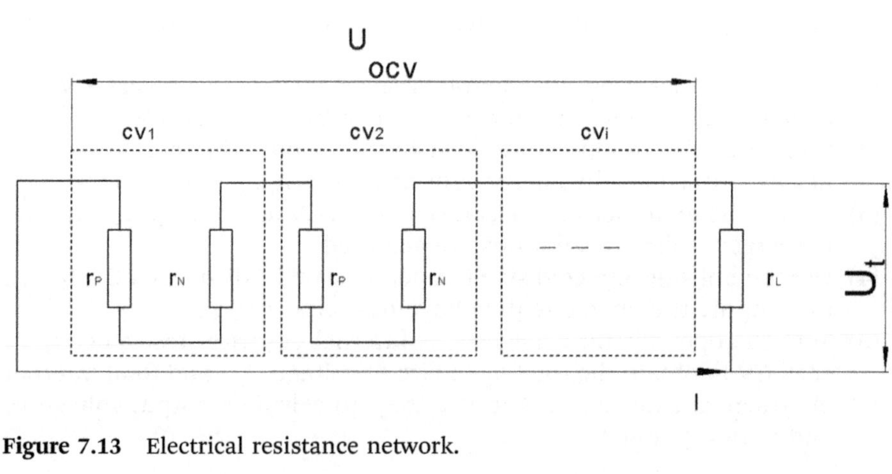

Figure 7.13 Electrical resistance network.

The open circuit voltage $U_{\mathrm{ocv}i}$ generated by the Seebeck effect in a CV is calculated by

$$U_{\mathrm{ocv}i} = n_{\mathrm{module}} n_{\mathrm{couple}} (\alpha_{\mathrm{P}} - \alpha_{\mathrm{N}})(T_{\mathrm{h}i} - T_{\mathrm{c}i}) \tag{7.55}$$

where the Seebeck coefficients α_{TE} are defined by the average temperature $\frac{T_{\mathrm{h}i} + T_{\mathrm{c}i}}{2}$. The total open circuit voltage $U_{\mathrm{ocv.TEG}}$ generated by the Seebeck effect can be calculated as

$$U_{\mathrm{ocv.TEG}} = \sum_{i=1}^{n_{\mathrm{cv}}} U_{\mathrm{ocv}i} \tag{7.56}$$

The external load resistance is r_{L}, so the current I is derived using Ohm's law

$$I = \frac{U_{\mathrm{ocv.TEG}}}{r_{\mathrm{L}} + r_{\mathrm{TEG}}} \tag{7.57}$$

The output voltage U_{t}

$$U_{\mathrm{t}} = U_{\mathrm{ocv.TEG}} - I r_{\mathrm{TEG}} = I r_{\mathrm{L}} \tag{7.58}$$

The total output power P is given by

$$P = I^2 r_{\mathrm{L}} = \sum_{i=1}^{n_{\mathrm{cv}}} P_i \tag{7.59}$$

7.6 Model Structure

The diagram of the TEG model structure is shown in Figure 7.14.

It can be seen that the exhaust gas-in temperature T_{In}, coolant temperature T_{ct}, the mass flow rate of the exhaust gas $\dot{m}_{\mathrm{h}}$, the mass flow rate of the coolant $\dot{m}_{\mathrm{c}}$ and the current I are regarded as inputs in this model. The output voltage U_{t} and output power P are considered outputs. Here I is used as the input instead of r_{L} because different r_{L} variables usually behave according to different currents I, which are usually easily evaluated. The solution procedure for the present coupled model is described briefly in the following:

(1) Make an initial guess for the total thermal resistance of each CV and compute the average exhaust gas temperature for each CV $T_{\mathrm{CV}i}$;
(2) Compute the temperature distribution and thermal resistance network based on the energy conservation principle in each CV;
(3) Return the total thermal resistance of each CV $R_{\mathrm{CV}i}$ to step (1) until the convergent iterative solutions are achieved;
(4) Use the hot side and cold side temperature of TEMs in each CV $T_{\mathrm{h}i}$ $T_{\mathrm{c}i}$, to compute the open circuit voltage in each CV $U_{\mathrm{ocv}i}$;
(5) Sum the open circuit voltage $U_{\mathrm{ocv}i}$, internal electrical resistance $r_{\mathrm{I}i}$ in each CV to obtain the total open circuit voltage U_{ocv} and total internal electrical resistance r_{TEG} and use the I to calculate output voltage U_{t} and output power P.

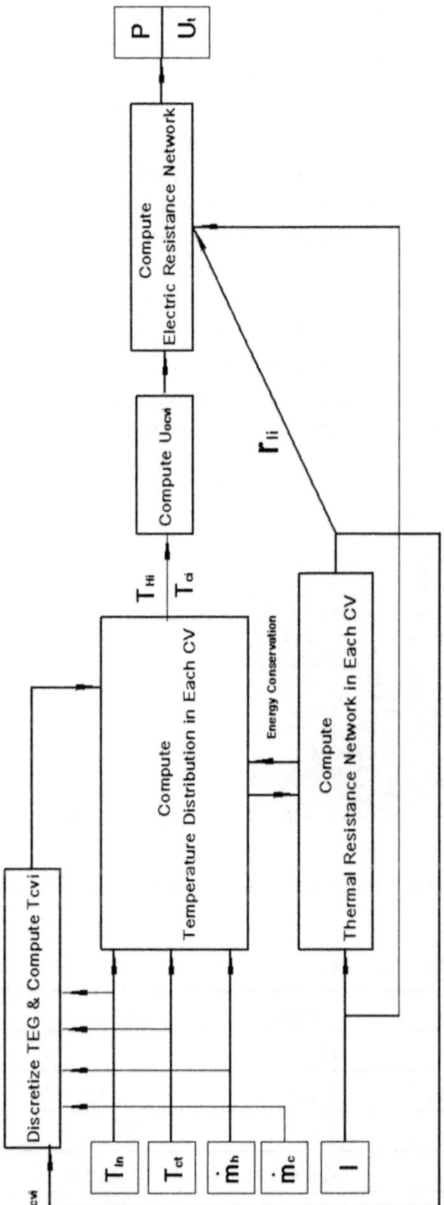

Figure 7.14 Diagram of the TEG model structure.

7.7 Validation Strategy for Models

In a validation exercise, a thermal generator with plate-fin heat exchanger that contains two HZ-14® TEMs is mounted on a CAT C6.6 engine, see Figure 7.15(a). The mounting location of the TEG is in the EGR path of the engine can be seen in Figure 7.15(b). The parameters of the TEG are presented in Table 7.1. The thermocouples making up the modules use p-type and n-type forms of bismuth telluride, respectively. The material properties can be seen in Appendix B as a function of temperature. In the experiment, the coolant flow is set at 4 °C and a volume mass flow rate of 5.9 litres min^{-1}. The engine was operated at different loads with different exhaust gas-in temperature (T_{In}) and volume mass flow rate (f), and the output voltage U_{t} and output power P were both recorded as the external electric resistance r_{L} changed. For validation, the TEG model was built based on the TEG parameters (Table 7.1) and material properties (Appendix B). The TEG model is discretized into two control volumes.

Figure 7.16 presents the experimental data compared with the model simulation results. It can be seen that, both the simulation results of voltage

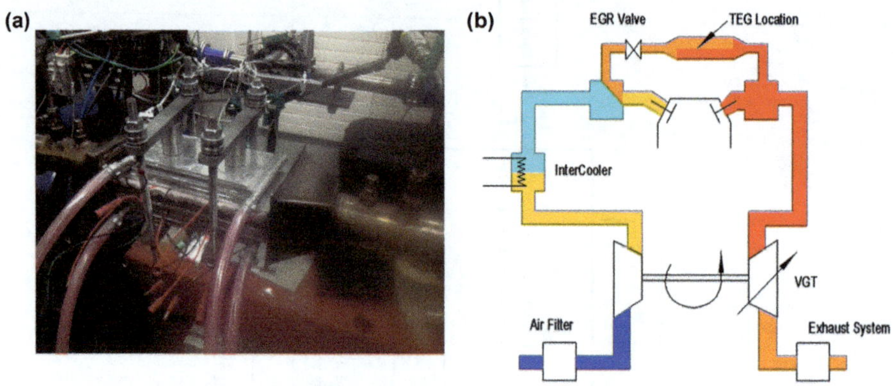

Figure 7.15 (a) The thermoelectric generator used in the lab and (b) its mounting location.

Table 7.1 Parameters for the TEG.

a/mm	l_{f}/mm	b/mm	N_{f}	δ_{f}/mm
110	220	12	14	1
s/mm	$k_{\text{fin}}/\text{m}^{-1}\ \text{K}^{-1}$	$k_{\text{hxr}}/\text{m}^{-1}\ \text{K}^{-1}$	$k_{\text{cw}}/\text{m}^{-1}\ \text{K}^{-1}$	n_{couple}
7.3	16	16	35	49
l_{leg}/mm	l_{hxr}/mm	l_{cw}/mm	$A_{\text{leg}}/\text{mm}^2$	$A_{\text{module}}/\text{mm}^2$
4.5	1	0.254	33.18	3931
σ/mm	H/MPa	m	P/MPa	n_{cv}
0.00036	300	0.08196	0.165	2
n_{module}				
1				

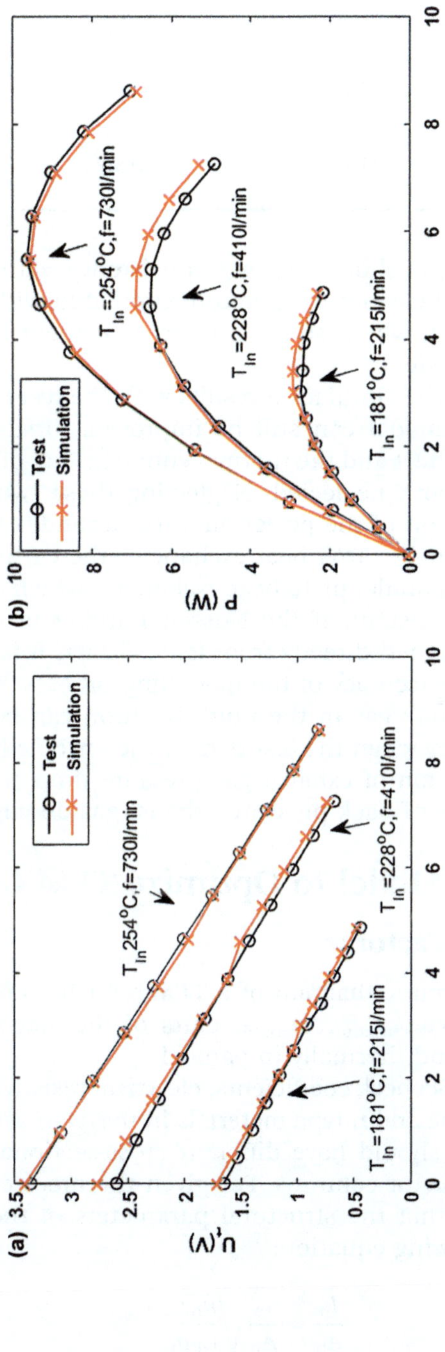

Figure 7.16 Comparison of simulation results with experiment data for voltage (a) and electrical power outputs (b).

Table 7.2 Simulation results for the gas-out temperature.

Operation environment f/litres min^{-1} and T_{In} (°C)	Measured gas-out temperature (°C)	Simulated gas-out temperature (°C)	Percentage error (%)
$f=210$ $T_{\mathrm{In}}=178$	124	119.7	3.5
$f=440$ $T_{\mathrm{In}}=220$	180	172.7	4.1
$f=733$ $T_{\mathrm{In}}=251.5$	216	213.7	1.1

U_t and power output P show a strong consistency with the experiment results. The simulated gas-out temperature compared with the test data is presented in Table 7.2, which shows a close agreement and supports the validity of the TEG model.

The comparison of the simulation results with the test data validates the TEG model, but the model can still be improved. Firstly, in the current model the Thomson effect and the thermal contact effect of the legs with the conductive tabs are both neglected. Neglecting these two influences may cause an over-estimation of the power outputs. Secondly, when calculating the Nusselt number of the TEG heat exchanger, the calculation equations used here are for the parallel plate heat exchanger, which assumes that the wall temperature is constant. If the Nusselt number of the specific heat exchanger can be obtained directly from its test data, it is possible to substantially improve the accuracy of the modelling process. Thirdly, the pressure drop of the exhaust gas in the fluid direction can be simulated with relatively high accuracy when the test data of the specific heat exchanger is available. The simulation of exhaust gas pressure drop can be used to investigate the influence of back pressure exhaust gas on engine efficiency.

7.8 Using the Model to Optimize TEM Geometry

7.8.1 Geometric Factors

Figure 7.17 is a schematic diagram of a TEM, which comprises two bases with module base area A_{module}, n_{couple} pairs of thermocouples connected electrically in series and thermally in parallel.

It is noted that the Seebeck coefficients, electrical resistivities and thermal conductivities of p-type and n-type materials in the TEM are different. Thus, a better TEM design should have different cross-sectional areas of p-type and n-type semiconductor columns. For given thermoelectric materials, to maximize z requires that the structural parameters of the semiconductor arms satisfy the following equation:[29]

$$\frac{l_{\mathrm{n}}}{A_{\mathrm{n}}} = \frac{l_{\mathrm{p}}}{A_{\mathrm{p}}}\sqrt{\frac{\kappa_{\mathrm{n}}\rho_{\mathrm{P}}}{\kappa_{\mathrm{P}}\rho_{\mathrm{N}}}} \tag{7.60}$$

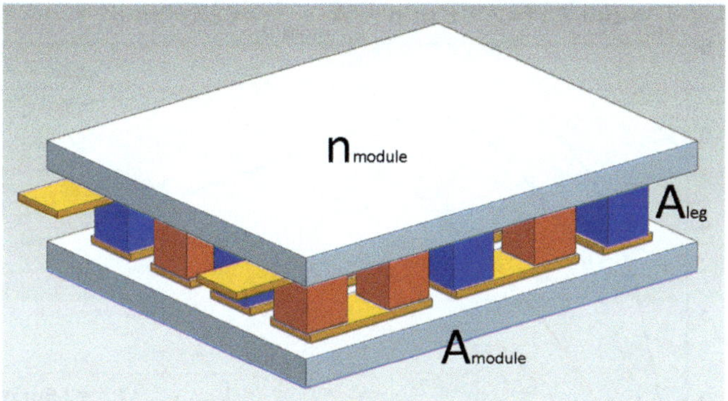

Figure 7.17 TEM schematic.

where l_n and l_p are leg length of p-type and n-type legs. A_n and A_p are the cross section area of the p-type and n-type legs. Due to manufacturing constraints it is likely that the length and cross sectional area of legs will be uniform in a module and very likely between modules. Hence, it is assumed that the p-type and n-type semiconductor columns have the same leg length and cross-sectional area.

$$l_n = l_p = l_{leg}, \quad A_n = A_p = A_{leg}$$

The fill factor γ for a TEM is defined as

$$\gamma = \frac{2n_{couple}A_{leg}}{A_{module}} \tag{7.61}$$

For the TEM with a fixed module base area A_{module}, once the fill factor of TEM (γ) and the leg length of thermoelectric legs (l_{leg}) are given, the TEM geometric structure is determined. Thus, these two parameters are chosen as the search variables and are optimized to reach maximum power output.

7.8.2 Influence of Geometric Parameters on Maximum Power

The influence of individual geometric parameters was analyzed for the operating environment $T_{In} = 200$ °C, $f = 200$ litres min^{-1}. And the module base area $A_{module} = 62 \times 62$ mm^2. When one geometric parameter is varied to evaluate its effect, the other parameter remains the same. The influence of γ and l_{leg} are presented in Figures 7.18 and 7.19, respectively.

It can be seen in Figure 7.18 that curves with different l_{leg} present two different trends of P_{max}. The blue curves are TEGs with relatively shorter l_{leg}. They increase initially, and then decrease with increasing γ. This means there is an optimal value of γ ($\gamma < 1$), which gives a maximum value of P_{max} but when the leg length is extended, the curves become flatter and peak value almost vanishes. For the red lines, which have comparatively longer

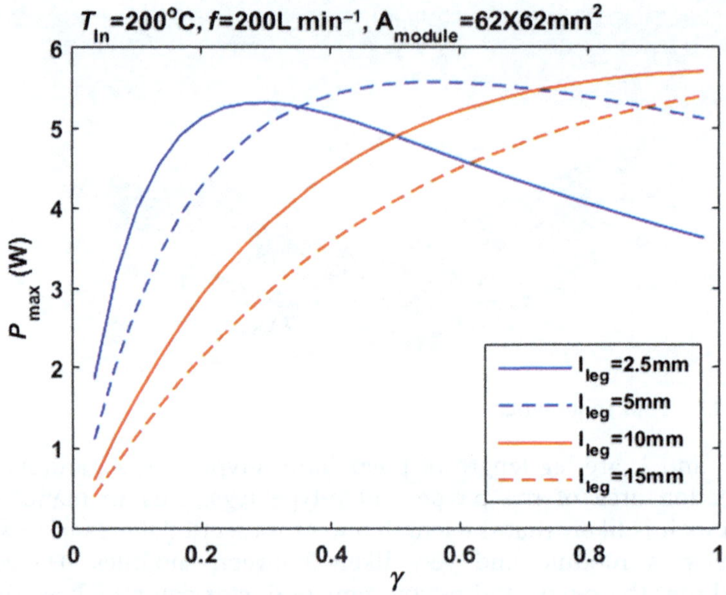

Figure 7.18 Maximum outputs of different leg lengths change with various fill factors.

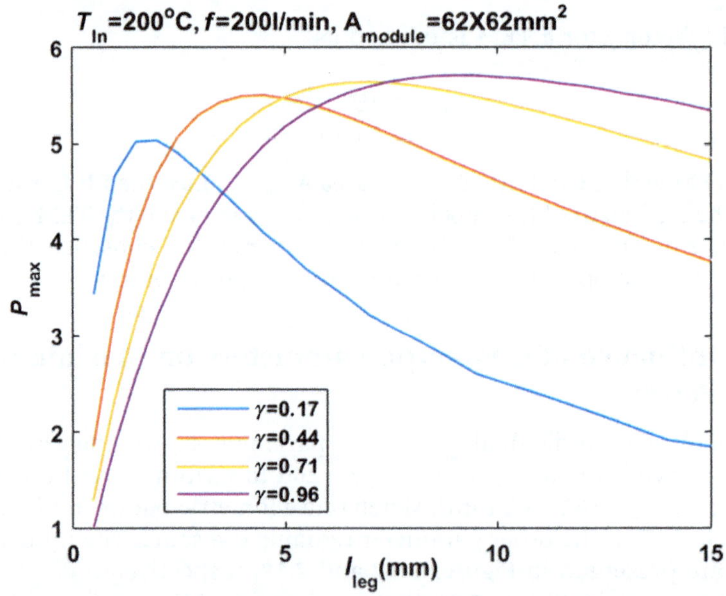

Figure 7.19 Maximum outputs of different fill factors change with various leg lengths.

l_{leg}, the values of P_{max} increase with the increasing value of γ. So the optimal γ is 1. The longer l_{leg} is, the smaller the value of P_{max}. In Figure 7.19, TEGs with different values of γ all share the same changing trend with the increasing value of l_{leg}. The peak value of P_{max} and the corresponding optimal value of l_{leg} both increase with increasing γ.

7.8.3 Influence of Operating Environments on Optimal Geometric Parameters

Figures 7.20 and 7.21 present the influence of operating environment on the optimal fill factor and leg length. All the optimal geometry values are tagged in the figures. The influence of gas-in temperature and exhaust flow rate on the optimal fill factor is analyzed at fixed leg length ($l_{leg} = 2.5$ mm) and fixed module base area ($A_{module} = 62 \times 62$ mm^2). The influence of gas-in temperature and exhaust flow rate on the optimal leg length is analyzed at fixed fill factor ($\gamma = 0.44$) and fixed module base area ($A_{module} = 62 \times 62$ mm^2). When one operating environment is varied to evaluate its effect, the other remains the same. As seen in Figure 7.20, when either exhaust gas-in temperature or flow rate increases, the optimal fill factors increase. While in Figure 7.21, when exhaust gas-in temperature or flow rate increases, the optimal leg length decreases. It is concluded that when the input energy of the exhaust gas increases, the optimal leg length value decreases and the optimal fill factor increases.

7.8.4 Three-dimensional Figures to Identify the Optimal Geometry Parameters

Figure 7.22(a) presents a 3D figure showing how maximum power changes with fill factor and leg length in the operating environment of $T_{In} = 200$ °C, $f = 500$ litres min^{-1}. Figure 7.22(b) shows the maximum power output in the dark red triangle region.

Therefore, the optimal geometry parameters of the TEM under this operation environment are $\gamma = 0.7 - 1$, $l_{leg} = 7 - 10$ mm. The 3D relationship, Figure 7.22, forms a useful visual tool to help engineers to choose the optimal geometry parameters of TEM under different operating conditions.

7.9 Selecting and Evaluating Heat Exchange Designs for TEG Applications

7.9.1 Identifying the Design Parameters of a Plate Fin Heat Exchanger

The plate fin heat exchanger is a form often used in vehicle applications and particularly where high heat transfer rates are required. The plate fin design therefore provides a useful starting point for the exploration of heat exchange design for thermoelectric applications. When the plate-fin type heat exchanger is considered for a TEG application, a detailed selection of heat exchanger parameters such as channel width, channel height, fin type,

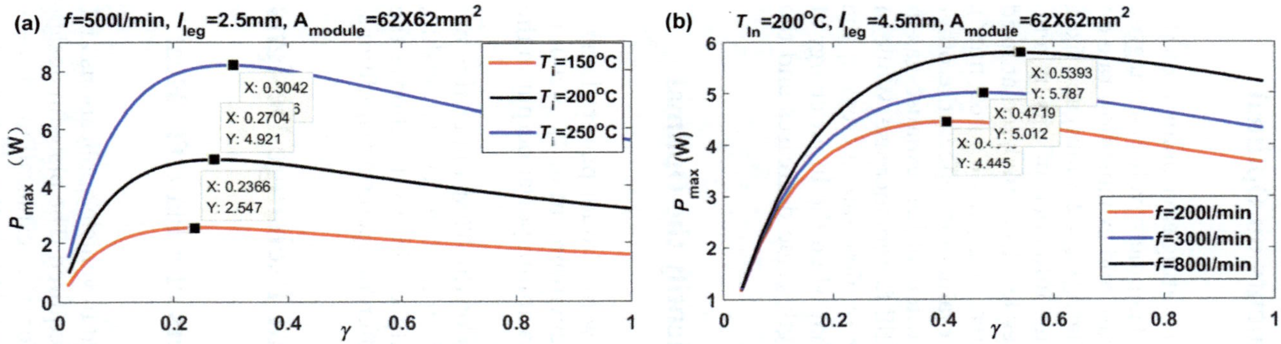

Figure 7.20 (a) Influence of gas-in temperature and (b) influence of flow rate on optimal fill factor.

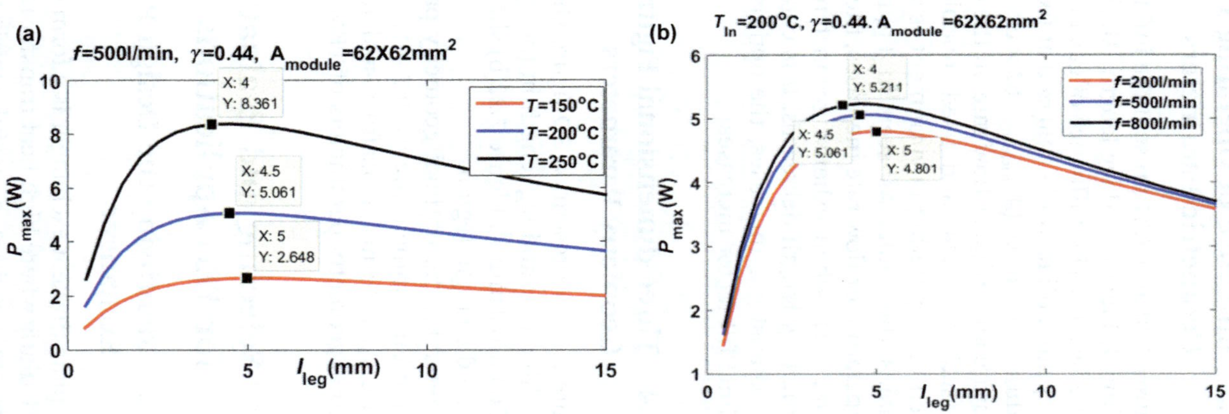

Figure 7.21 (a) Influence of gas-in temperature and (b) influence of flow rate on optimal leg length.

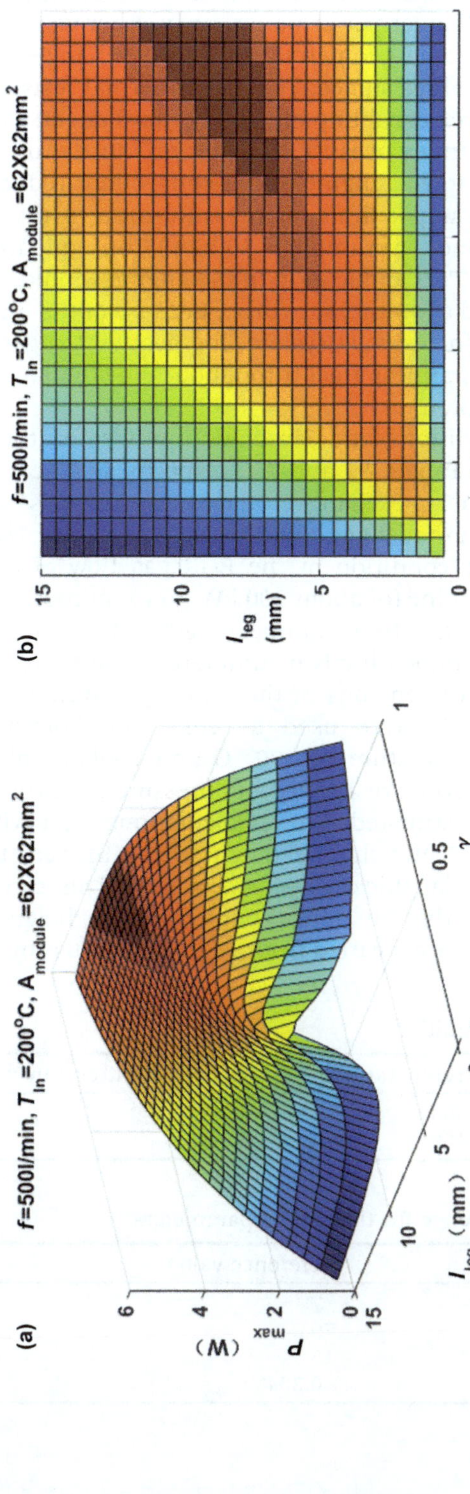

Figure 7.22 Maximum electrical powers changing with fill factor (a) and leg length (b).

fin density and fin thickness needs to be made. An example demonstrates the effect of channel width, channel height, fin type, fin density and fin thickness on heat exchanger performance, or in other words, the sensitivity of the performance to the geometry. The heat exchanger performance is evaluated using thermal resistance. The lower thermal resistance represents the higher heat transfer capability of the heat exchanger. Pressure drop is another performance index because a higher pressure drop will result in a higher back pressure to the engine and a consequent reduction in engine efficiency. (A sensible upper limit for pressure drop in the EGR path of a heavy duty diesel engine for example is 10 kPa.) Consequently a 'good' heat exchange design for a TEG application in the engine should have both a low thermal resistance and a low pressure drop. Because of the high temperature conditions of the engine exhaust system, the heat exchanger is assumed to be constructed using stainless steel throughout the simulation and analysis presented in this chapter.

The boundary conditions and simulation range for the working parameters are shown in Tables 7.3 and 7.4, respectively. The outline of a two-channel plate-type heat exchanger and two fin-types are shown in Figure 7.23. The inlet exhaust gas condition of 500 °C and 1800 litres min^{-1} is the maximum flow condition of the EGR gas flow of a representative medium duty diesel engine (of about 200 kW power output). A configuration of two split channels with a fixed channel length of 130 mm was chosen with the intention of accommodating two commercially available TEMs on each channel surface. The dimensions of the TEM is 60 mm × 60 mm × 4 mm. Three coolant channels were used as cold side boundary conditions with coolant inlet temperature of 100 °C and coolant inlet flow rate of 15 litres min^{-1}. The thermal resistance and pressure drops of each candidate heat exchanger were estimated. These calculations start with a plain fin, 70 mm channel width, 7 mm channel height, with a fin density of 16 fins per inch (fpi) and a material thickness of 0.254 mm. This can be taken as a reference case. Then the fin density, the channel height, the channel width and fin thickness were varied from 14–20 fpi, 1–50 mm, 65–80 mm and

Table 7.3 Boundary conditions.

Items	Temperature (°C)	Flow rate/litres min^{-1}	Pressure/bar
Inlet gas conditions	500	1800	1.6
Coolant conditions	100	15	N/A

Table 7.4 Simulation range for the tuning parameters.

Items	Reference value	Variation range
Channel height/mm	7	1–50
Channel width/mm	70	65–80
Fin density/fpi	16	14–20
Fin thickness/mm	0.254	0.1–0.3

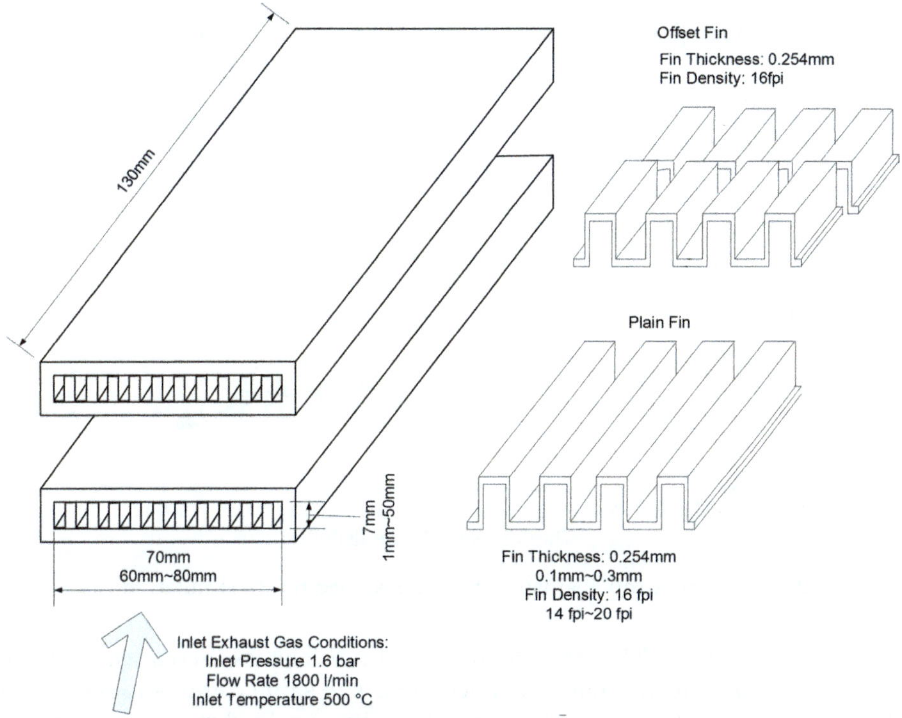

Figure 7.23 Dimensions of two gas channel plate type heat exchanger.

0.1–0.3 mm, respectively, with other parameters kept at the values of the reference case. In the last simulation case, the offset strip fin is used to replace the plain fin but all the other geometry parameters are kept at their reference values. There are in total 39 simulation cases. The simulation results are plotted in a single figure showing pressure drop against the thermal resistance (Figure 7.24). Since when channel height is less than 3 mm, the pressure drops are much higher than 7 kPa, the two simulation cases with channel heights of 1 mm and 2 mm are not shown.

It can be seen from Figure 7.24 that when channel height is varied, thermal resistance reaches a peak value at 7 mm. When channel height is less than 3 mm, the pressure drop will be higher than 7 kPa, which implies that narrow channels are not suitable for the TEG application. When channel height is between 7 mm and 50 mm, the wider the channel is in a vertical sense, the less thermal resistance and the smaller the pressure drop. So the other constraints such as heat exchanger volume and weight should be considered in order to choose the channel height. It is also very obvious from Figure 7.24 that using offset fins substantially reduces the thermal resistance but also increases the pressure drop from approximately 1.2 kPa to 6.4 kPa, which implies that offset fins are not a good choice. As channel width increases, both pressure drop and thermal resistance fall. However, because the dimensions of TEMs are fixed, simply increasing channel width

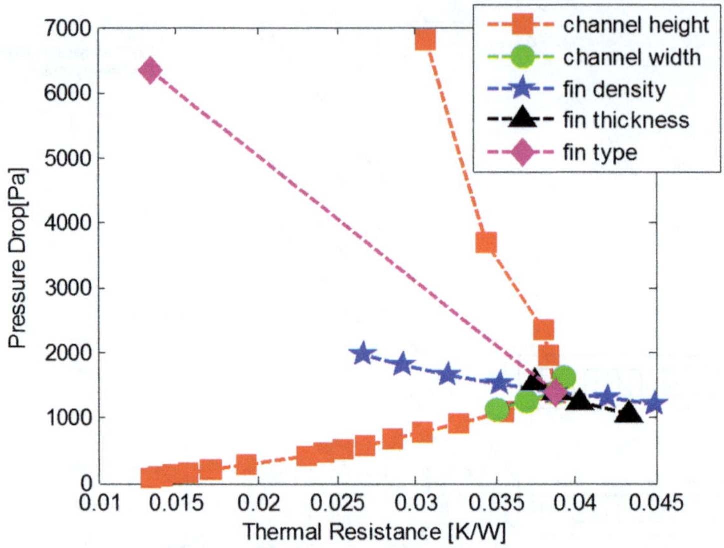

Figure 7.24 Pressure drop against thermal resistance for 37 simulation cases.

may not increase the TEG performance. As fin thickness increases, pressure drop increases, but thermal resistance falls. The same trend is found as fin density increases. However, the slope of the line as fin density varies is less steep than that of the fin thickness. This demonstrates that using a higher fin density is better than using greater fin thickness because it leads to a lower pressure drop. A guide to the design for a longitudinal plate type heat exchanger in TEG applications in either the EGR or the main exhaust path can be drawn from these simulation results:

1. Use plain fins.
2. The selection of channel width should consider the TEM dimensions.
3. Channel height should not be less than 10 mm.
4. Fin thickness can be set at about 0.2 mm.
5. Use a high fin density under the constraints of pressure drop and heat exchanger mass. (The choice of fin density must also take into account the manufacturing processes available.)

The thermal resistance computed here is a parameter to indicate the total heat convection performance of the heat exchanger. It is not, however, a suitable index to represent TEG performance. If the number of TEMs is fixed, the decrease in thermal resistance created by increasing the channel width may not increase the TEG output. In the following simulation work, conducted using a CFD code, the average temperature difference between hot and cold sides of all TEMs within a TEG is used as an indicator of the TEG performance. This average temperature difference is closely correlated to the TEG electrical power output because the measured output is the cumulative result of each element. Even though the Seebeck coefficient, electrical conductivity and thermal conductivity of each p and n element will change with temperature,[24] the tendency is to moderately increase TEG output at higher temperatures. To simplify the analysis, the effect of the variation of these three physical parameters with operating temperature was considered negligible.

7.9.2 Comparing Heat Exchange Architectures

Since the performance of a TEG relies on the cumulative power output of all connected TEMs, the balance of thermal conditions of each TEM provides the opportunity to improve the total TEG power output capability. The question posed is: will the TEG have a higher power output if the TEMs are mounted thermally in parallel (using wide configuration heat exchanger) or thermally in series (using a long configuration heat exchanger)? By using CFD simulation methods, the question can be answered before carrying out any experimental comparisons.

A commercial CFD code (STAR-CCM+ version 9.06) was used in this simulation work. For comparison purposes, a three single-channel rectangular plate-type heat exchanger model with no fins inside the channels was developed and simulated (see Section 7.9.2.1). Then three complete heat exchangers with two split-channels whose dimensions are the same as those of the three single-channel rectangular plate-type heat exchangers, but having inside fins, are developed and simulated (see Section 7.9.2.2). The average delta temperature and the heat transfer rate have been compared among these six simulation models. The pressure drop will be compared with the octagonal form of heat exchangers introduced in Section 7.10.

7.9.2.1 Single Channel Without Fins

The benchmark heat exchanger for the single channel test is a long plate type heat exchanger of which the channel length is 130 mm, the channel width is 70 mm, the channel height is 12 mm. Two wide heat exchangers were set up for comparison. Both of them have the same channel length and channel width. Their channel length and channel width equals the channel width and the channel length of the benchmark long heat exchanger, respectively. The difference between these heat exchangers is that one has the same channel volume, the other has the same cross-section flow rate as the benchmark long heat exchanger. The boundary conditions for these three simulation models are the same. Inlet exhaust gas temperature is 500 K and the flow rate is 500 litres min^{-1}. Four identical TEMs with dimensions: 60 mm × 60 mm × 4 mm were placed on both heat exchanger surfaces of each single channel model. The thermoelectric material is bismuth telluride (Bi_2Te_3). The temperature of the cold side of each TEM was set to a constant 300 K. The boundary conditions and the dimensions of these three simulation models are listed in Tables 7.5 and 7.6, respectively. The exhaust gas flow is assumed to be evenly distributed across the channel.

There are four TEMs mounted on a single channel of the heat exchanger. Perfect contact between TEMs and the surface of heat exchanger is assumed. Only the middle temperature distributions of two TEMs on the same top surface of each heat exchanger are investigated (see Figure 7.25). The middle hot side temperature distributions for the two TEMs in the benchmark long heat exchanger are very different. The first TEM on the exhaust gas flow

Table 7.5 Boundary conditions.

Items	Temperature (K)	Flow rate/litres min^{-1}	Other conditions
Inlet gas conditions	500	500	Laminar flow
Cold side of TEMs	300	N/A	Constant temperature
Environment	300	N/A	Natural convection

Table 7.6 Dimensions of simulation models (without fins).

Name of the model	Length/ mm	Width/ mm	Height/ mm	Material thickness/ mm
Benchmark long heat exchanger (HXR)	130	70	12	1
Wide HXR with same channel volume	70	130	12	1
Wide HXR with same cross-section area	70	130	6.46	1

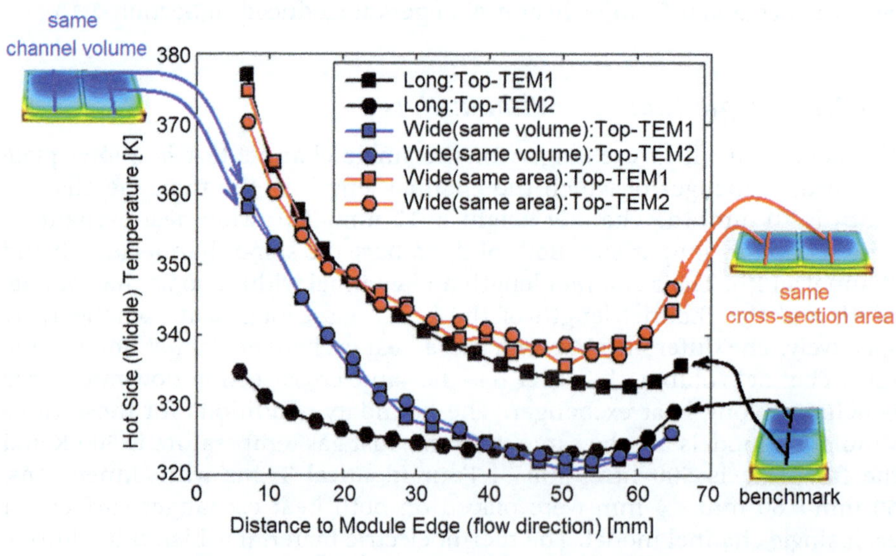

Figure 7.25 Temperature distributions for long and wide plate-type heat exchangers (without fins).

direction has a much higher hot side temperature than the second TEM. As expected, the middle hot side temperature distributions for the two TEMs are the same in both of the two wide heat exchangers. However, the wide heat exchanger with the same cross-section flow rate has higher hot side temperatures than that of the heat exchanger, which has the same channel volume as the benchmark long heat exchanger. This can be explained as the wide heat exchanger with the same channel volume has a bigger cross-section flow area and, hence, a smaller flow velocity under the same volume flow rate conditions, which results in a lower heat transfer rate. The average delta temperature against total heat transferred is then scaled up to eight TEMs in each simulation model and plotted in Figure 7.27 together with the results from all the simulation cases. It can be seen that among the single channel simulation models with no fins, the wide heat exchanger with the same cross-section area as the benchmark has the highest average delta temperature and heat transfer. The wide heat exchanger with the same channel volume as the benchmark, has the lowest average delta temperature and heat transfer which implies that if there are no fins, the long type TEG is essentially turned by 90°, and the resulting wide TEG has poor output performance. Otherwise, if the height of this wide TEG is reduced to make the cross flow rate equal to the benchmark, the resulting wide TEG output will be much higher than that of the benchmark long TEG because of the higher flow velocities.

7.9.2.2 Two Split-channel Complete Heat Exchanger With Fins

Another three simulation models were created to investigate the same problem. This time, a complete heat exchanger CAD model is developed for each case. This complete heat exchanger includes plain fins, inlet and outlet chambers. The fin density was chosen to be 2.9 fpi and is the same in all three cases. These three simulation models also include three cases: benchmark long heat exchanger; wide heat exchanger with the same channel volume as the benchmark long heat exchanger; wide heat exchanger with the same cross-section flow area to the benchmark long heat exchanger. The boundary conditions are the same as in Table 7.5 except the flow rate is 1000 litres min^{-1} as two channels are now modelled. Eight TEMs are mounted on each heat exchanger because two channels are simulated. The channel length, width and height are the same for each case listed in Table 7.4. The thickness of fins is 1 mm for each case. A perfect contact condition between TEMs and surface of heat exchanger is also assumed here.

The temperature distributions of two TEMs mounted on the top surface of the top channel of these three heat exchangers are extracted and depicted in Figure 7.26. The comparison of the total TEG output performance can be viewed in Figure 7.27 in terms of average delta temperature and heat transferred from the heat exchanger to the TEMs. It can be seen that the benchmark long heat exchanger with fins has the best performance; the wide heat exchanger with the same channel volume has a lower output while

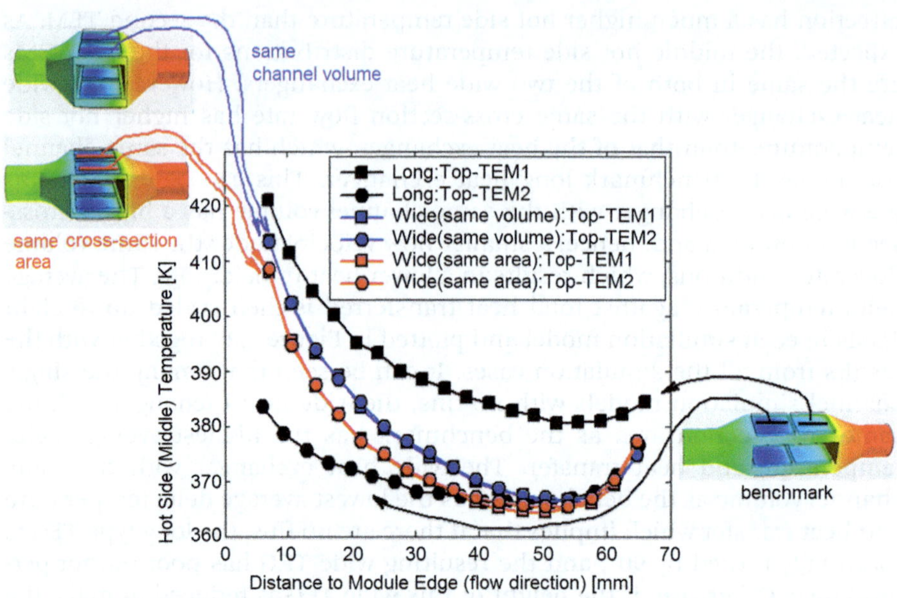

Figure 7.26 Temperature distributions for long and wide plate type heat exchanger (with fins).

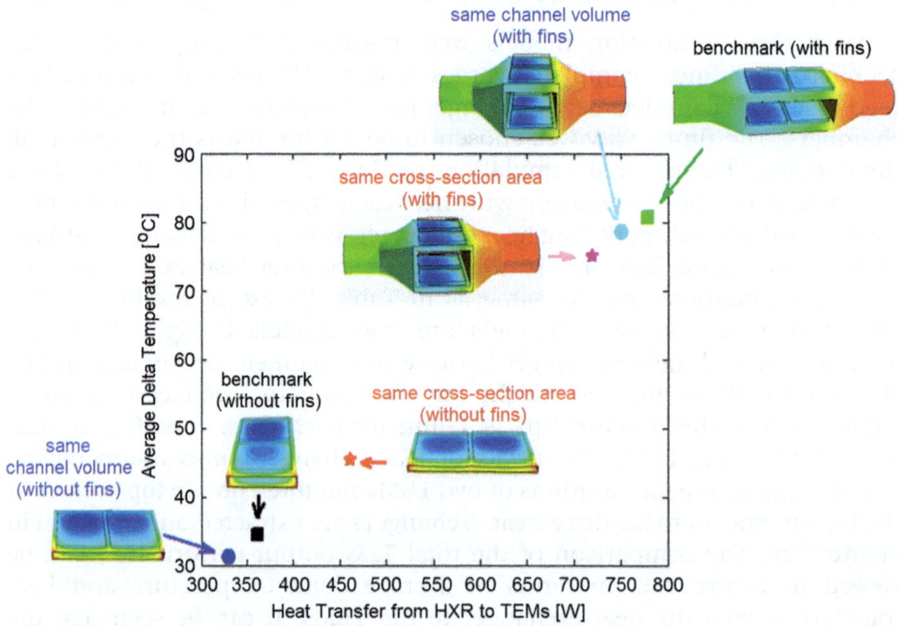

Figure 7.27 Average delta temperatures against transferred heat from the heat exchanger to four TEMs.

the wide heat exchanger with the same cross-section area has the worst performance among the three cases. This performance order is different to the results from the single channel models. The fins play a vital role in heat exchanger performance. The performance of heat exchanger can be seen to have been vastly improved using inside fins.

Considering just the power output, the long format heat exchanger is clearly preferable. It is easy to predict that the pressure drop for a wide heat exchanger with the same channel volume will be lower than that of the benchmark long heat exchanger. Hence the design of rectangular channel heat exchanger should maximize the TEG output using a long heat exchanger with fins but keeping the pressure drop within the constraints by limiting the fin density and making the channel wider or using multiple split channels.

7.10 An Example of a Family of Heat Exchangers

A multi-sided (polygonal format) heat exchanger offers substantial advantages as the basis of a TEG. TE modules may be mounted on the outside surfaces and cooling jackets clamped in place. The ease of access makes this kind of heat exchange architecture convenient for research purposes. For production, the inner construction could take different forms including that of an exhaust after-treatment catalyst. We have chosen to use an octagonal format to illustrate the principles of a multi-sided design. In order to understand the TEG performance when using this architecture, eight different octagonal heat exchanger models with different inner fin structures were developed and evaluated using the STAR-CCM + 3D code. These eight octagonal heat exchangers are named after their inner fin structures: thin channel, no fins, cross fins, square cells, two channel spider net fins, four channel spider net fins, six channels spider net fins and thin channels with fins (see Figure 7.28). These inner structures were selected considering both the ease of manufacturing and following the principles established earlier in the chapter for limiting the pressure drop through the device. The central parts (except for the device without any inner fin structure) are assumed to be manufactured using the same grade of stainless steel. This central part of a future practical octagonal heat exchanger could be a bypass pipe resulting in a lighter heat exchanger and a controllable TEG power output.

The first octagonal heat exchanger only has eight narrow channels close to the outside surface. The sum of the eight channel cross-section flow area equals the sum of the cross-section area of the two channels of the long type benchmark heat exchanger without fins. The last octagonal heat exchanger has eight thin channels but with fins inside the channels. The height of each channel and the fin density are the same as in the benchmark long heat exchanger. From the outside, all eight octagonal heat exchangers have the same dimensions and would have the same external appearance. The gas flow is assumed to be in a steady state laminar condition.

The dimensions of each side of octagonal heat exchanger are 70 mm $\times$ 70 mm. Eight bismuth telluride TEMs with dimensions of

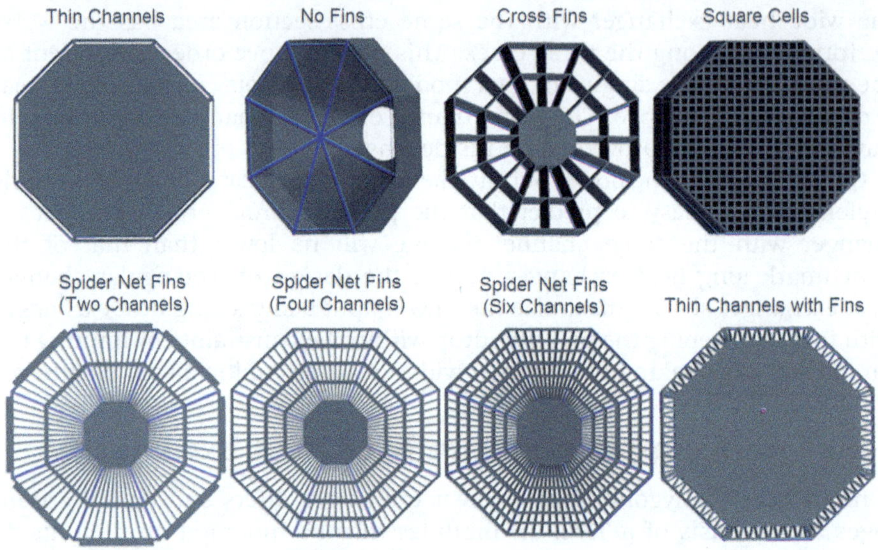

Figure 7.28 Eight inner fin structures for octagonal cross-section heat exchangers.

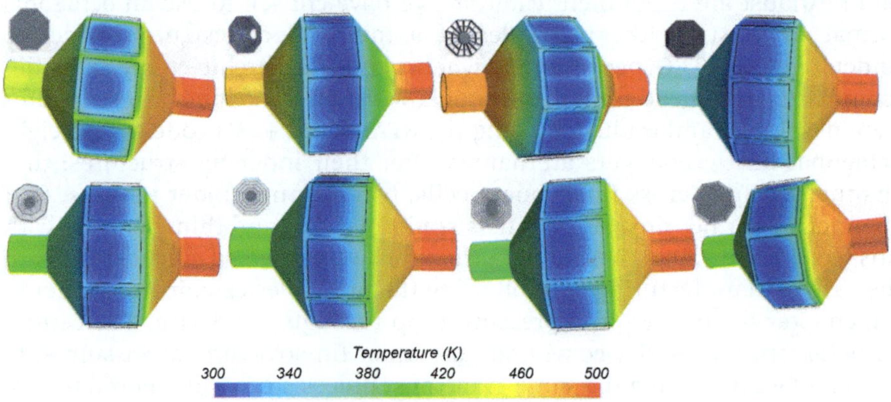

Figure 7.29 Surface temperature distributions for eight octagonal heat exchangers.

62 mm × 62 mm × 4 mm are placed on each side of the octagonal heat exchanger. The two channel spider net fins in Figure 7.28 show the cross sectional part of eight TEMs. Surface temperature distributions for the eight heat exchangers are shown in Figure 7.29. There is an obvious surface temperature distribution pattern caused by cross fins which can be seen in Figure 7.29. Two temperature distribution lines along the hot side of the TEM are shown in Figure 7.31. They are along the middle line of the TEM along the flow direction and a line just to the left.

The temperature distribution patterns on the cross-sectional surface at the middle location of the eight octagonal heat exchangers are shown in

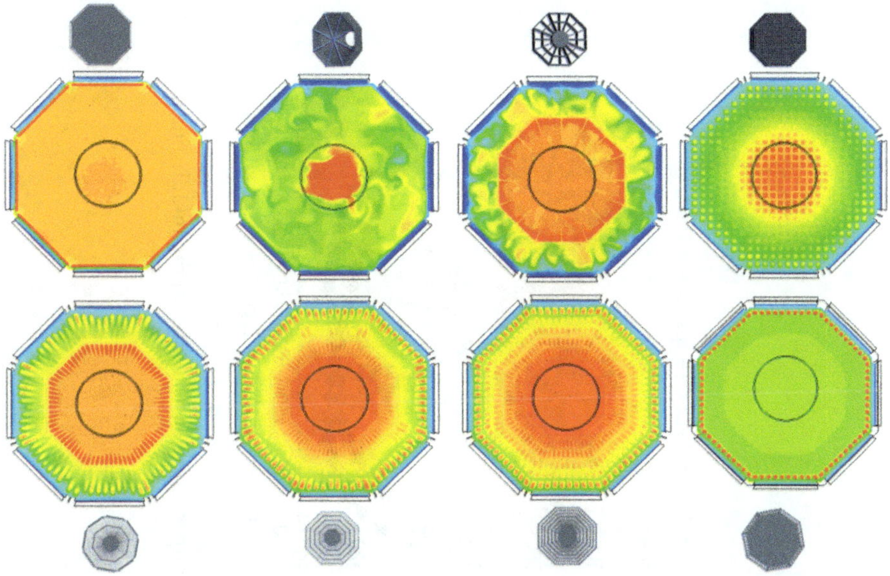

Figure 7.30 Temperature distribution patterns on the cross-section surface at the middle location of the eight octagonal heat exchangers.

Figure 7.30. It can be seen that without the inner fins, the gas flow will be concentrated in the middle of the pipe. It can be expected that this type of heat exchanger will give poor output performance. In Figure 7.31 it can be seen that two octagonal heat exchangers without fins show very low temperatures. The octagonal heat exchanger with thin channels and plain fins demonstrates the highest temperatures along both lines.

The average hot to cold side temperature differences plotted against heat transfer from the heat exchanger to eight TEMs are shown in Figure 7.32 together with the benchmark long-type TEG performance. It can be seen that the long-type benchmark TEG has better performance over all eight octagonal TEGs and also has the best performance among all simulated TEGs. The performance of the octagonal TEG with thin channels and fins is very close to that of the benchmark long heat exchanger. This comparison result is very encouraging as it shows that it is possible to make an octagonal type TEG that can keep the traditional pipe-like outline and that has similar performance to the benchmark long rectangular TEG. The performance of the octagonal TEG with square cells as the inner fin structure ranks fourth among the eight octagonal TEGs. It has even better TEG output performance than the thin channel type octagonal TEG, which suggests that the integration of the TEG with an after-treatment system, such as diesel particulate filter (DPF) or three-way catalyst (TWC), could offer an integration path for thermal energy recovery and after-treatment.

The pressure drops across different TEG designs are plotted in Figure 7.33. It is shown that both the benchmark long-type heat exchanger and the thin

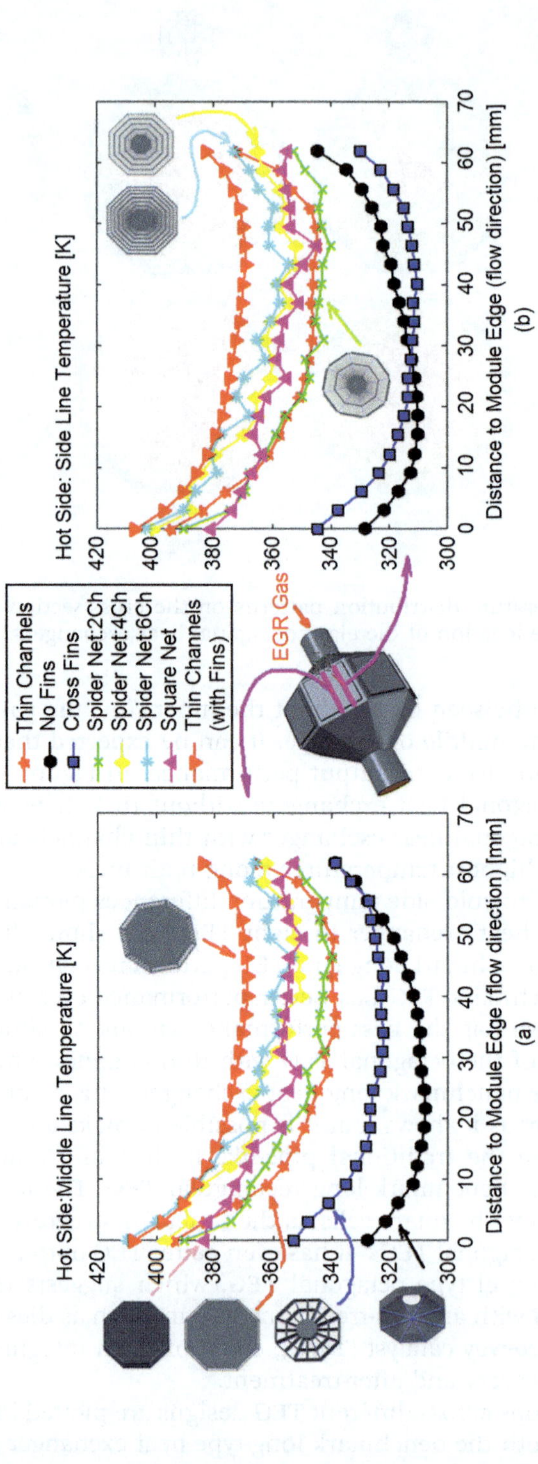

Figure 7.31 (a) Temperature distributions on the middle line of the top TEM; (b) temperature distributions on the middle line of the left half part of the top TEM.

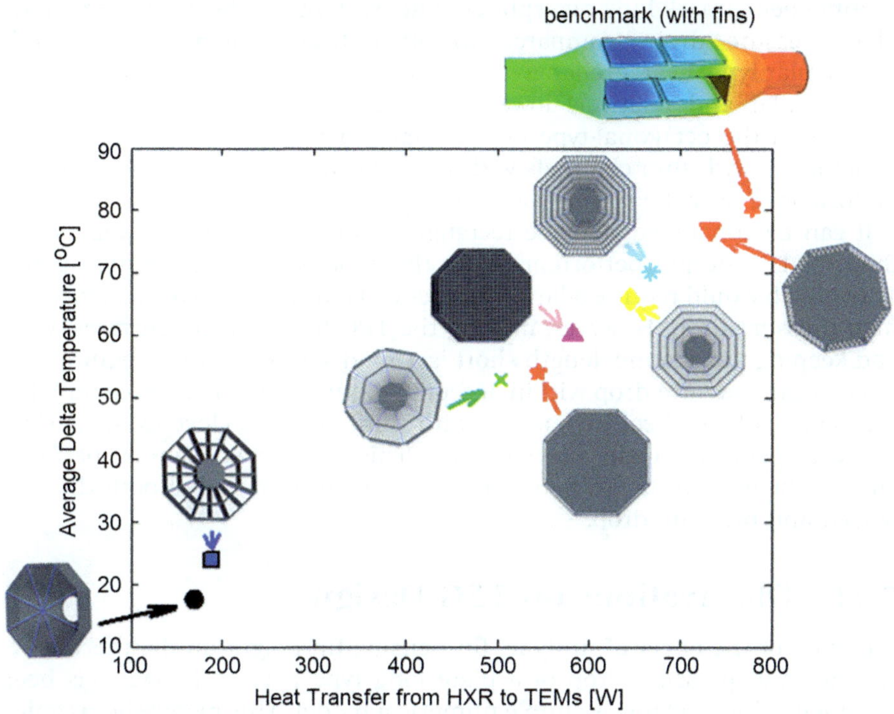

Figure 7.32 Average delta temperature against heat transfer from heat exchanger to TEMs.

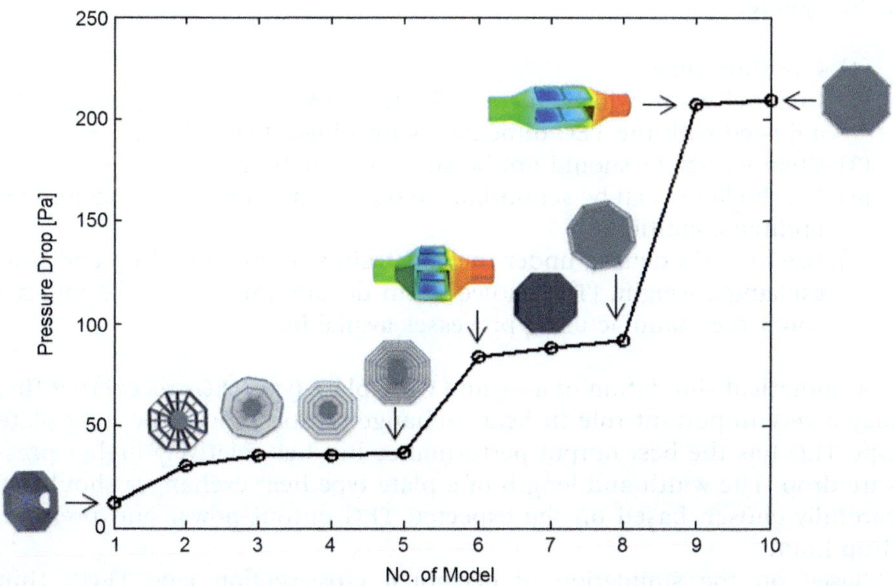

Figure 7.33 Comparison of pressure drops.

channel benchmark long two split channel rectangular heat exchanger have high pressure drops compared to other heat exchangers. The wide rectangular-type heat exchanger and the octagonal-type heat exchangers have similar and relatively lower pressure drops. This comparison also shows that the octagonal-type heat exchangers with the square cells inner structure or with thin channels with fins have advantages in pressure drop performance over the simple long rectangular heat exchanger.

It can be concluded that the rectangular long-type heat exchangers give the best TEG output performance but this type is not fit for a high power TEG, which would require a large number of TEMs and the consequence of a high pressure drop. However, making the TEG heat transfer surfaces wider and keeping the channel length short is a good solution for the requirement to keep the pressure drop within acceptable limits. This is demonstrated in the octagonal-type heat exchangers with thin channels close to the surface and with plain fins inside the channels. This device has a similar performance to the wide rectangular-type heat exchanger in terms of both the TEG output and pressure drop.

7.11 Observations on TEG Design

A picture of the effect of fin type, fin density, fin height on the thermal resistance and pressure drop of a plate long type heat exchanger has been obtained using a relatively simple numerical model. This picture helps guide the design of a long plate heat exchanger in TEG applications in either the engine EGR path or the main exhaust path. For a medium or heavy duty diesel engine application, a set of design rules emerge, which form a simple design guide:

(1) Use plain fins;
(2) Channel width will be dictated by the TEM dimensions, but can be adjusted with the TEG dimensions forming a lower limit;
(3) Channel height should not be smaller than 10 mm;
(4) Fin thickness can be set initially to 0.254 mm and adjusted to find an optimal condition.
(5) Use high fin density under the constraints of pressure drop and heat exchanger weight. (The choice of fin density must also take into account the manufacturing processes available.)

A numerical simulation of long and wide plate-type TEGs reveal that fins play a very important role in heat exchange performance. The long plate type TEG has the best output performance but has relatively higher pressure drop. The width and length of a plate type heat exchanger should be carefully chosen based on the expected TEG output power and pressure drop limit.

Based on the simulation of octagonal cross-section type TEGs, thin channels with fins represent a good choice for the inner structure of the octagonal TEG. This octagonal TEG has a comparable performance to the wide plate-type TEG both in power output and pressure drop. It was also found that the octagonal TEG with square cells as the inner fin structure has acceptable output performance and relatively low pressure drop. This implies that the integration of a TEG with after-treatment devices such as a diesel particulate filter (DPF), and a three-way catalyst (TWC) has a feasible starting point.

7.12 Concluding Remarks

Waste heat recovery has a substantial part to play in the improvement of energy efficiency of all types of equipment. All fuel-burning engines reject heat to the environment and represent potential for recovery processes. The value of energy flows may be assessed and their potential judged against the implementation requirements of a recovery process. Thermoelectricity shows considerable potential as an energy recovery process because of its relative simplicity. However, a successful design must bring together TE materials in the form of thermocouples, which in turn are integrated with a heat exchanger to form a complete TEG. Successful implementation will match the design of the TEG with the application and ensure that the TEG continues to deliver an acceptable output power even as conditions change. The design process requires an accurate, yet flexible, modelling process. The process may start with models based on empirical correlations and a simple analysis structure. Such an approach acts as a filter to sift a set of designs that may be analyzed in greater depth before a design configuration is established. The initial model will take account of the source of energy, the behaviour of the heat exchanger as well as the characteristics of the external circuits. As the design proceeds, finer resolution is needed in the design parameters, and CFD methods alongside, for example, mechanical and thermal finite element models will allow the final selection of design parameters. The application of modelling methods has provided initial guidance in the formulation of design guidelines. More experience will be needed in TEG design to build a body of expertise sufficient to be regarded as a design guide. The plate-fin heat exchanger outperforms other designs to the extent that it can be regarded as a benchmark. This form of design, however, has practical limitations while other designs, such as an octagonal format, reach a level of performance close to this standard and offer functional advantages, such as being able to provide a by-pass function or integrate exhaust gas after-treatment.

Appendix A: Exhaust Gas Properties Employed in the Simulation Model

The exhaust gas properties are presented in Figure 7.34.

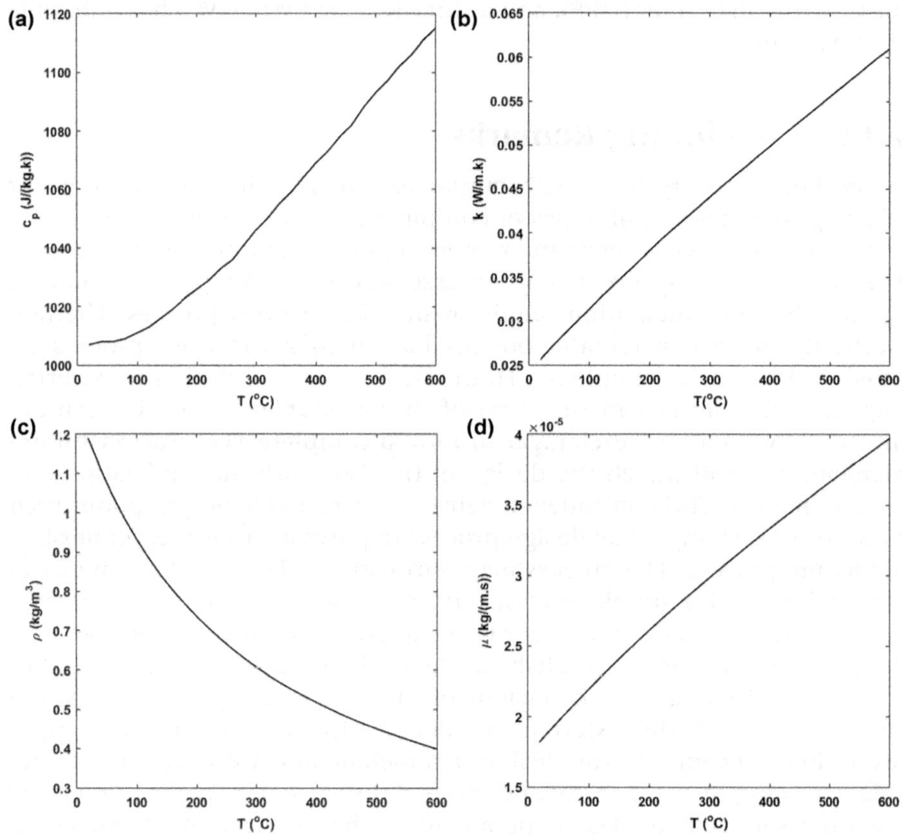

Figure 7.34 Exhaust gas properties: (a) Specific heat of exhaust gas at constant pressure; (b) Thermal conductivity of exhaust gas; (c) Exhaust gas density and (d) Exhaust gas viscosity.

Appendix B: Properties of the Simulated Thermoelectric Module

The properties of the thermoelectric modules HZ-14® are shown in Figure 7.35.

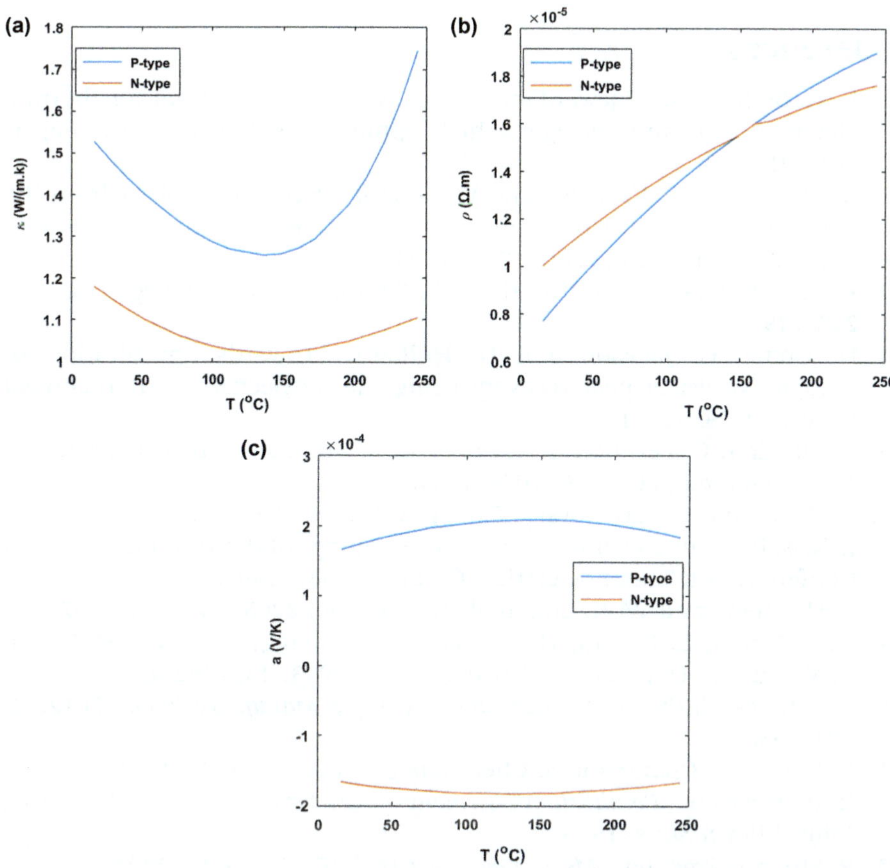

Figure 7.35 (a) Thermal conductivity; (b) Electric resistance and (c) Seebeck coefficient.

Acknowledgements

The authors would like to gratefully acknowledge the UK Engineering and Physical Sciences Research Council (EPSRC) for funding this research work under Grant Number: EP/K026658/1. They would like to thank Dom McKean, Graham Smith, Iain Harber, Steve Horner and Drew Mason for their excellent technical support for experimental setup and engine testing in Powertrain lab in Aeronautical and Automotive Engineering Department, Loughborough University. The authors also want to thank the project partners: Hervé Palanchon from Dana GmbH, John Burgers from Dana Canada, Matthew Phillips, Min Gao for University of Cardiff, Jesús Pradogonja, Anthony Powell, Paz Vaqueiro-Rodriquez from Reading University for very helpful and constructive discussions during every project meeting. They want to thank Dr Anusha Wijewardane now at the University of Muratowa, Sri Lanka for her significant previous work.

References

1. R. K. Stobart, A. Wijewardane and C. Allen, The potential for thermoelectric devices in passenger vehicle applications, SAE Technical Paper, 2010-01-0833.
2. D. Hountalas, C. Katsanos and V. Lamaris, Recovering energy from the diesel engine exhaust using mechanical and electrical turbocompounding, SAE Technical Paper, 2007-01-1563.
3. P. Bombarda, C. M. Invernizzi and C. Pietra, *Appl. Therm. Eng.*, 2010, **30**, 212–219.
4. D. Crane, G. Jackson and D. Holloway, Towards optimization of automotive waste heat recovery using thermoelectrics, SAE Technical Paper, 2001-01-1021.
5. C. Haddad, C. Périlhon, A. Danlos, M.-X. François and G. Descombes, *Energy Procedia*, 2014, **50**, 1056–1069.
6. A. Poran and L. Tartakovsky, *Energy*, 2015, **88**, 506–514.
7. J. Bass, R. Campana and N. Elsner, Proceedings of the 10th International Conference on Thermoelectrics, Cardiff, Wales, 1991.
8. Y.-H. Cheng and W.-K. Lin, *Appl. Therm. Eng.*, 2005, **25**, 2983–2997.
9. H.-C. Chien, E.-T. Chu, H.-L. Hsieh, J.-Y. Huang, S.-T. Wu, M.-J. Dai, C.-K. Liu and D.-J. Yao, *J. Electron. Mater.*, 2013, **42**, 2362–2370.
10. M. Hodes, *IEEE Trans. Compon., Packag., Manuf. Technol.*, 2012, **2**, 483–495.
11. Y. Hsiao, W. Chang and S. Chen, *Energy*, 2010, **35**, 1447–1454.
12. R. W. Haywood, *Equilibrium Thermodynamics for Engineers and Scientists*, John Wiley & Sons, 1980.
13. R. Stobart, *Proc. Inst. Mech. Eng., Part D*, 2007, **221**, 1107–1124.
14. E. Thacher, B. Helenbrook, M. Karri and C. J. Richter, *Proc. Inst. Mech. Eng., Part D*, 2007, **221**, 95–107.
15. N. Winkler and H.-E. Ångström, Simulations and measurements of a two-stage turbocharged heavy-duty diesel engine including egr in transient operation, SAE Technical Paper, 2008-01-0539.
16. J. Yang, FY2008 DOE Vehicle Technologies Annual Merit Review, 2008.
17. A. Eder and M. Linde, Second Thermoelectric Applications Workshop, San Diego, 2011.
18. G. P. Meisner, Proceedings of the 2d Thermoelectrics Applications Workshop, 2011.
19. S. K. Yee, S. LeBlanc, K. E. Goodson and C. Dames, *Energy Environ. Sci.*, 2013, **6**, 2561–2571.

20. S. LeBlanc, S. K. Yee, M. L. Scullin, C. Dames and K. E. Goodson, *Renewable Sustainable Energy Rev.*, 2014, **32**, 313–327.
21. P. Mayer and R. Ram, *Nanoscale Microscale Thermophys. Eng.*, 2006, **10**, 143–155.
22. S. Song, V. Au and K. P. Moran, Proceedings of the 4th ASME/JSME thermal engineering joint conference, 1995.
23. J.-Y. Jang and Y.-C. Tsai, *Appl. Therm. Eng.*, 2013, **51**, 677–689.
24. M. Yovanovich and E. Marotta, *Heat Transfer Handbook*, 2003, vol. 1, pp. 261–394.
25. S. Kumar, S. D. Heister, X. Xu, J. R. Salvador and G. P. Meisner, *J. Electron. Mater.*, 2013, **42**, 944–955.
26. J. Esarte, G. Min and D. Rowe, *J. Power Sources*, 2001, **93**, 72–76.
27. W. M. Rohsenow, J. P. Hartnett and Y. I. Cho, *Handbook of Heat Transfer*, McGraw-Hill, New York, 1998.
28. E. E. Marotta and M. M. Yovanovich, Thermal Spreading and Contact Resistances, *Heat Transfer Handbook*, John Wiley & Sons, Inc., USA, 2003.
29. J. Holman, New York, 1974.

CHAPTER 8

Electrodeposition of Thermoelectric Materials

A. J. NAYLOR,*[a] N. M. WHITE[b] AND I. NANDHAKUMAR[c]

[a] University of Oxford, Department of Materials, Parks Road, Oxford OX1 3RQ, UK; [b] University of Southampton, Department of Electronics and Computer Science, Southampton SO17 1BJ, UK; [c] University of Southampton, School of Chemistry, Southampton SO17 1BJ, UK
*Email: andy.naylor@materials.ox.ac.uk

8.1 Introduction

8.1.1 Electrodeposition of Nanostructured Materials

8.1.1.1 Principles of Electrodeposition

The general principles of electrodeposition are briefly described here in the context of fabrication of thermoelectric materials. This introduction is not exhaustive and for more detail, readers are referred to dedicated texts.[1–5]

Electrodeposition, or electrochemical deposition, is a technique used to form a solid phase at an electrode from an aqueous or non-aqueous electrolyte. Such solid phases can include metals, semiconductors, and polymers in bulk, mesoporous or nanostructured form, for example. Electrodeposition can be induced by applying a constant direct current across the external circuit of the electrochemical cell (galvanostatic electrodeposition) or by applying a constant voltage, or potential, to the working electrode (substrate) with respect to a reference electrode (potentiostatic deposition).[1] Pulse electrodeposition is another technique where the current

RSC Energy and Environment Series No. 17
Thermoelectric Materials and Devices
Edited by Iris Nandhakumar, Neil M. White and Stephen Beeby

Published by the Royal Society of Chemistry, www.rsc.org

is stepped at intervals between two values in order to achieve a higher quality deposit.[6]

In potentiostatic deposition, the potential at the working electrode is stepped from a value where no current flows to one where a reaction occurs and held for a fixed period of time. The measurement of the current passing through the external circuit as a function of time (chronoamperometry) allows for detailed quantitative and qualitative analysis of the deposition process.[1,2] This analysis may include determining the mechanism for nucleation and growth of the deposit (instantaneous or progressive; two- or three-dimensional),[7] calculating the expected thickness of a deposit, or calculating the diffusion coefficient from the Cottrell equation.[2] Cyclic voltammetry can be used to determine the potential range over which deposition will occur. In cyclic voltammetry the potential is swept between two limits and a current density response is recorded; a peak indicates where an electrochemical reaction is occurring.[2]

The rate of a reaction at a working electrode is dependent on the applied potential. The overpotential, η, is defined as the difference between the applied potential, E, and the reaction equilibrium potential, E_e: ($\eta = E - E_e$), as governed by the Nernst equation (eqn (8.1)).[2] At low overpotentials, the rate of electron transfer is very slow, the current density varies strongly with potential and is independent of the mass transport regime. While at higher overpotentials, the rate of electron transfer at the working electrode becomes very fast and mass transport becomes the rate determining step in the reaction.

$$E_e = E_e^o + \frac{2.3RT}{nF} \log c_{M^{n+}} \tag{8.1}$$

Nernst equation: E_e^o is the equilibrium potential when the concentrations of the oxidised species and the reduced species are equal under standard conditions, R is the gas constant (8.314 J K^{-1} mol^{-1}), T is the absolute temperature (K), n is the number of electrons involved in the reaction, F is the Faraday constant (96 485 C mol^{-1}) and $c_{M^{n+}}$ is the concentration of the metal cation in the electrolyte.

In electrodeposition, however, a new solid phase is being formed at the surface of the working electrode and this requires nucleation of the new species to occur before any further growth. The initial nucleation process tends to require a greater overpotential, which is usually dependent on the substrate, than would be expected for the phase to continue growing under mass transport limited conditions. To illustrate this, a cyclic voltammogram expected for the typical deposition and stripping processes of a new phase is shown in Figure 8.1. At point B, an increase in the current density is observed indicating the initiation of nucleation at the electrode. Peak A is associated with the deposition of the new phase, while on the reverse scan, mass transport limited deposition continues to higher potentials than the initial potential required for nucleation. Point D can be considered as the

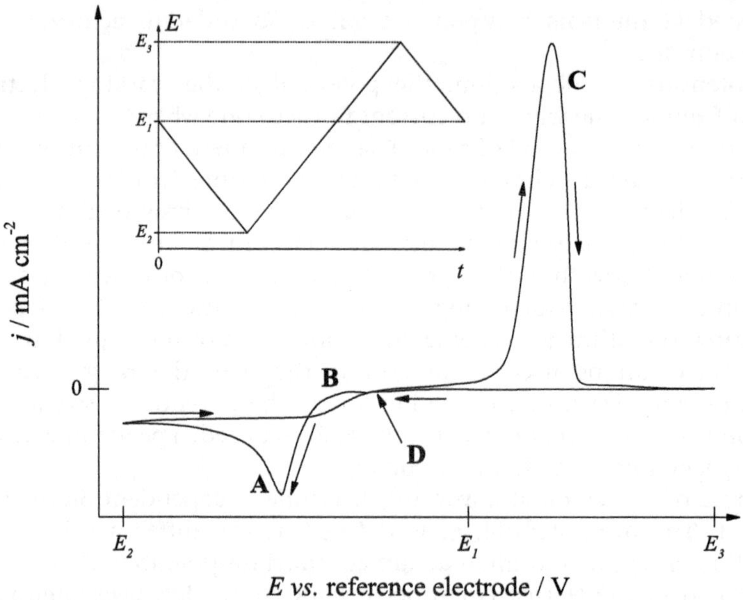

Figure 8.1 A typical cyclic voltammogram expected for the formation of a solid phase from an aqueous electrolyte on an electrode and the subsequent stripping. Refer to the text for explanation of the labelled features. Inset: typical potential-time profile employed.

$$M^{n+}_{\text{electrode}} + ne^- \xrightarrow{\text{electron transfer}} M_{\text{electrode}}$$

Scheme 8.1

equilibrium potential for the M^{n+}/M couple and the anodic peak (C) is associated with the stripping of the phase from the electrode.

Deposition can also be carried out galvanostatically where the current remains constant and the potential of the working electrode, with respect to a reference electrode, is allowed to vary.[1] This technique is generally more often used in industry for large-scale electroplating purposes, since only two electrodes are required.

In a simple system, solvated metal cations undergo mass transport from the bulk solution to the electrode, usually by diffusion or forced convection. The metal is then electrodeposited as in Scheme 8.1 where M^{n+} is reduced by electron transfer to form the elemental oxidation state of the metal at the substrate surface.

This reaction will usually occur if a negative current or a sufficient overpotential to induce the electron transfer reaction is applied to the working electrode. The reverse reaction (stripping) can occur by oxidation of the deposit by a positive current or a potential greater than the equilibrium potential of the M^{n+}/M couple.

Most systems, however, are much more complicated; the electrodeposition process and the resulting deposit are inherently affected by the slightest variation in any parameter. For example, the applied potential, substrate material, duration, cell geometry, temperature, concentration of ions or additives in solution, pH, and impurities could all influence the quality, morphology, crystal structure, or composition of the resulting deposit. This affords a highly complex and fragile, yet powerful, technique for the fabrication of nanostructured materials.

Fabricating new and low-dimensional thermoelectric materials, with higher zT, the thermoelectric figure-of-merit, values, at low-cost and with high throughput, crucial for commercialization, are the main challenges faced by this field of research.[8] A range of fabrication techniques are currently employed such as single crystal growth, bulk powder syntheses and thin film technologies (metal–organic vapour phase diffusion (MOVPD) deposition, molecular beam epitaxy (MBE) including liquid phase epitaxy).[8–10] These methods have proven to be costly and challenging, requiring elevated temperatures or high vacuum conditions.

On the contrary, electrodeposition offers high deposition rates and scalability, as well as the ability to tune multiple parameters for the desired output. In addition, it is widely recognised as one of the most cost-effective processes, operating at room temperature and pressure.[1,8,9,11]

Electrodeposition of thin films and nanostructures of complex thermoelectric materials and the ability to show that they demonstrate equal or greater efficiency than their analogues fabricated by more expensive and challenging techniques could eventually make thermoelectric materials commercially viable for a wide range of emerging applications.

8.1.1.2 *Application of Electrodeposition to the Fabrication of Thermoelectric Materials*

Most instances of electrodeposition of thermoelectric materials are much more complex than the relatively simple deposition of a pure metal phase. Often, at least two or three elements will be co-deposited simultaneously to form semiconductor compounds or alloys. Therefore, it is important to consider at what potentials the individual elements are deposited. Underpotential deposition (UPD), first proposed by Kröger,[12] is the mechanism used to describe the situation where the reduction potential of two species deposited simultaneously is more positive than their individual reduction potentials.[13] This phenomenon occurs due to the free energy gain for the formation of the compound semiconductor.[14]

There are a few methods by which the deposition potentials of the two elements can be brought closer together, in the case that the UPD mechanism cannot be induced. By lowering the concentration of the more noble metal in solution, small shifts in the deposition potential can be achieved; the potential is dependent on the concentration of the solvated ion according to the Nernst equation.[2,15] Complexing the electroactive ions is a further technique for shifting the deposition potential. In the electrodeposition of the thermoelectric semiconductor antimony telluride, the addition of 0.5 M of tartaric acid aids in the dissolution of antimony and brings the reduction potential closer to that of tellurium.[15]

A well-studied thermoelectric material, used in commercial devices, is bismuth telluride. It is often fabricated by techniques such as chemical vapour deposition (CVD),[16,17] sputtering,[18] evaporation,[19] pulse laser deposition (PLD),[20] and molecular beam epitaxy (MBE).[21] However, these techniques often require specialised equipment, high temperatures and vacuum conditions, causing them to be costly and sometimes difficult to achieve. However, as mentioned previously, electrodeposition can be used as a relatively cost-effective and simple method for the fabrication of such materials.

Bismuth telluride has been electrodeposited by many different mechanisms and methods,[22] including electrochemical atomic layer epitaxy (ECALE),[23] galvanostatic electrodeposition,[24,25] pulse electrodeposition,[26] and potentiostatic electrodeposition.[13,27–33] The potentiostatic reduction of Bi^{3+} and $HTeO_2^+$ ions from a nitric acid electrolyte is considered as the most popular method and has been extensively studied by Martín-González *et al.*[13] The overall cathodic reaction proceeds by Scheme 8.2.

Figure 8.2 presents a Pourbaix diagram calculated for a 7.5 mM Bi and 10 mM Te in 1 M HNO_3 electrolyte at standard temperature and pressure showing the thermodynamic stability of Bi_2Te_3 as a function of potential and pH.[13] The calculations were made using the Nernst equation for the relationship between concentrations and either the standard electrode potential E° or pH.

The Pourbaix diagram shows that the Bi_2Te_3 bulk material is stable over the pH range of between approximately −1 and 14 and can be produced by reduction of its cations (Bi^{3+}, $HTeO_2^+$) at a pH of −0.07.[13] It was pointed out that there are two possible mechanisms for the reduction of $HTeO_2^+$.[13] Between −0.443 V and −0.093 V *vs.* SCE (saturated calomel reference electrode), $HTeO_2^+$ ions are directly reduced to elemental Te. Between −0.643 V and −0.443 V *vs.* SCE, $HTeO_2^+$ are first reduced to H_2Te followed by immediate transformation to elemental Te.

In electrodeposition, a substrate should be selected that is stable in the electrolyte over the electrochemical potential window used during the deposition. Reports for the electrodeposition of bismuth telluride on different metallic substrates (stainless steel, Ni, Au, Pt, Mo, Ti, Cu) have been made.[11,31,34–36]

Stainless steel substrates have been investigated for the electrodeposition of bismuth telluride.[34,35] It was found that tellurium deposition is kinetically hindered, exhibiting a large nucleation overpotential (700 mV) relative to that for bismuth (80 mV). However, stoichiometric Bi_2Te_3 thin films were deposited

$$3HTeO_2^+ + 2Bi^{3+} + 9H^+ + 18e^- \rightarrow Bi_2Te_3 + 6H_2O$$

Scheme 8.2

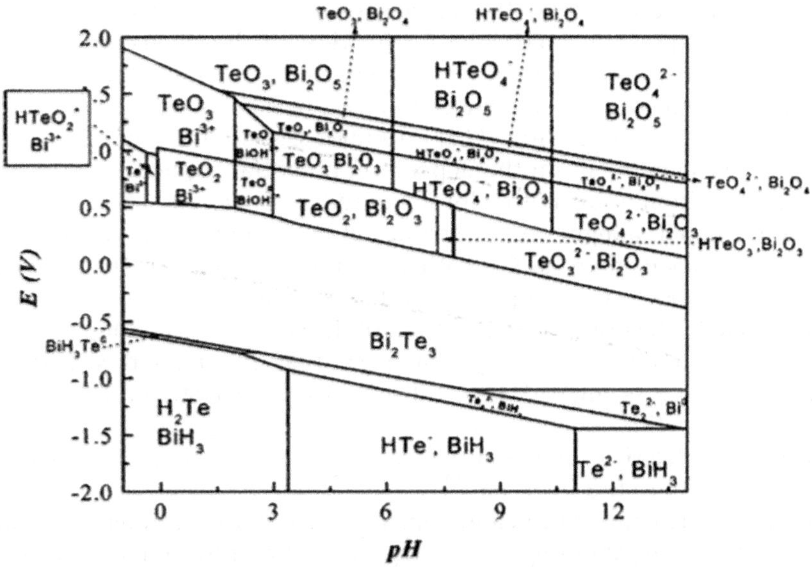

Figure 8.2 A Pourbaix diagram for the electrodeposition of Bi_2Te_3 from a 7.5 mM Bi and 10 mM Te in 1 M HNO_3 electrolyte (25 °C, 1 atm) showing the thermodynamic stability of the dominant species as a function of potential and pH.[13]
Reproduced by permission of The Electrochemical Society.

and a Seebeck coefficient of $-51.9\ \mu V K^{-1}$ was measured. Use of titanium substrates by Miyazaki and Kajitani[31] resulted in strong (1 1 0) orientation of the crystallites and a maximum Seebeck coefficient of $-63\ \mu V K^{-1}$.

Tittes *et al.*[36] found that copper substrates are unsuitable for the electrodeposition of bismuth telluride after finding traces of copper in the deposited thin films. However, they demonstrated the use of nickel substrates.

Typically, concentrations of bismuth and tellurium close to those used in the Pourbaix diagram are reported (7.5 mM Bi and 10 mM Te in 1 M HNO_3). However, a few interesting variations are described here.

In one instance,[27] ethylenediaminetetraacetic acid (EDTA) was used to complex bismuth in solution, while both ions were used in relatively low concentrations (0.85 mM Bi and 1 mM Te). Using titanium substrates, it was found that Bi_2Te_3 could be deposited within a 300 mV potential window. At a potential of -0.20 V *vs.* (Ag/AgCl/3.33 M KCl), a strong (1 1 0) orientation was observed and a maximum Seebeck coefficient of $-70 \pm 10\ \mu V K^{-1}$ was measured.

Concentrations as high as 93 mM and 80 mM for bismuth and tellurium, respectively, have been used.[37] However, bismuth is typically used in lower concentrations than tellurium due to the lower diffusion coefficient for bismuth.

Michel *et al.*[29] showed that the composition of bismuth telluride deposits depends not only on the electrodeposition conditions but also on the concentration of the electroactive species in solution. Where a Bi : Te ratio of 2 : 3 was used ([Te] = 20 mM), the Te fraction in the deposits was found to decrease with more cathodic (more negative) potentials but was always Te-rich compared with stoichiometric Bi_2Te_3, irrespective of the deposition potential. However, when [Bi] = [Te] = 10 mM, Te-rich or Bi-rich films could be deposited, dependent on the deposition potential applied.

The composition of bismuth telluride found to give the best thermoelectric properties, such as a high Seebeck coefficient and high electrical conductivity, Bi_2Te_3,[11] is known to crystallise with a rhombohedral symmetry.[38] For the optimum thermoelectric properties of Bi_2Te_3 when combined with other relevant factors, the preferred orientation of the compound is along the (1 1 0) plane.[39] This orientates the crystallites so that the layered structure of bismuth telluride is aligned perpendicular to the substrate, as shown in Figure 8.3. For this arrangement, in the direction perpendicular to the substrate the electrical and thermal conductivities are higher than in the direction parallel to the substrate.[40]

The electrolyte composition and deposition conditions can have profound effects upon the crystallinity of the deposits, as demonstrated by Erdoğan and Demir.[41] They showed that deposition of bismuth telluride from an acidic electrolyte (pH 1.5) resulted in a predominant orientation along the (0 1 5) plane, whilst deposition from a basic medium (pH 9.0) using EDTA as a Bi complexing agent, (1 1 0) is found to be the predominant growth orientation.

Boulanger reports that Seebeck coefficients for Bi_2Te_3 thin films typically are negative and range from −60 to −100 μV K^{-1}.[11] Annealing the samples can lead to higher absolute values of the Seebeck coefficient (greater

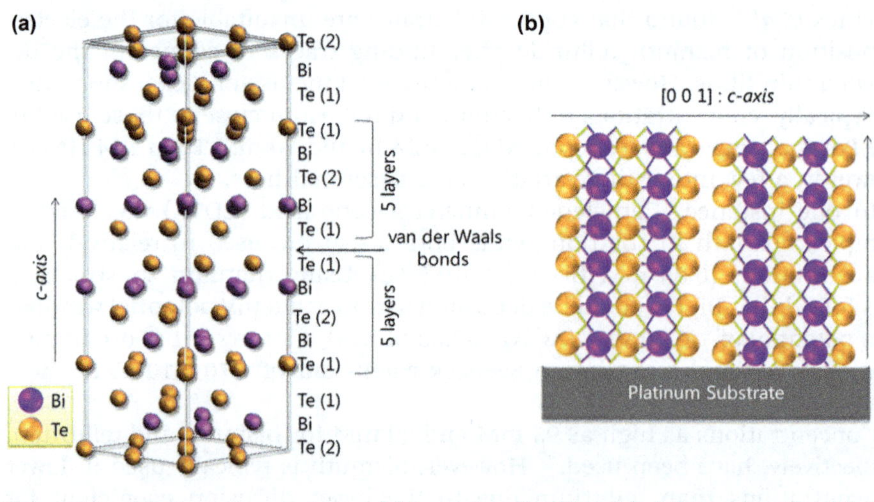

Figure 8.3 (a) Crystal structure of bismuth telluride. (b) Schematic of a highly oriented (1 1 0) Bi_2Te_3 film, deposited by electrodeposition.

than −100 μV K^{-1}). However, the resistivity of the thin films remain high resulting in low power factors (200 μW m^{-1} K^{-2}).

Electrodeposition of bismuth telluride is not limited to thin films or bulk fabrication but can also be applied to nanostructured materials. A growing number of studies are emerging of nanostructured materials including nanowires.[24,25,39,42–45] The use of nanostructured materials in thermoelectric devices is thought to induce quantum confinement effects due to the reduced dimensionality, and also enhance the phonon scattering effect due to the presence of heterointerfaces, leading to improved thermoelectric efficiency.

Nanostructuring techniques such as using ion-track etch polycarbonate membranes,[42] and anodic alumina membranes (AAM)[24,43] as templates for electrodeposition have been reported. These methods result in nanowire bundles exhibiting strong (1 1 0) orientation, and enhanced thermoelectric performance.

8.1.2 Recent Advances in the Electrodeposition of Thermoelectric Materials

Whilst there are various methods through electrodeposition for improving the microstructural properties in order to enhance the thermoelectric properties of a material, this chapter deals with two in particular. This section investigates recent use of additives and dopants in the electrodeposition of bismuth telluride-based thermoelectric materials.

8.1.2.1 Additives in the Electrodeposition of Thermoelectric Materials

Additives, such as surfactants, can be used in electrodeposition to tune the microstructural properties of the deposited material. They are added to the electrolyte and their concentration can significantly vary the properties of the deposit. The introduction of organic additives to the electrolyte solution has been shown to alter the physical properties, such as brightness, smoothness, hardness, and ductility, of electrodeposited films.[30] They affect the deposition and crystal-building process as adsorbates at the surface of the substrate. An adsorbate reduces the free energy surface tension, as energy is spent in bonding the adsorbate to the surface.[1] However, since they also form part of the bulk electrolyte, they may additionally affect other parameters, for example the diffusion of the electroactive species, which is dependent on the viscosity of the electrolyte. Adsorbed additives affect the kinetics of deposition and the growth mechanism by changing the concentration of adions (adsorbed ions) on the surface, the diffusion coefficient, D, and the activation energy for the surface diffusion of adions.[46] In many cases of electrodeposition, the additive is not used up by incorporation into the deposit. The amount of incorporation is determined by the rate of adsorption and desorption processes of the additive and the species being deposited from solution. If the rates are of the same order of magnitude, no incorporation or entrapment will occur. Incorporation can result in a poor quality of the resulting deposit,[1] but can often be difficult to detect by such methods as energy-dispersive X-ray spectroscopy (EDX) due to there often being organic contaminants already on the surface of the sample after exposure to the atmosphere.

There have been few reported cases on the use of additives in the electrodeposition of n- and p-type bismuth telluride alloys. In 2006, Li *et al.*[30] used the organic additive ethylene glycol (EG, 0–40% v/v) in the electrodeposition of bismuth telluride. It was found that the adhesion and average film thickness were increased (~300 μm per 24 h) with higher concentrations of EG and that the morphology of deposits is strongly affected by the deposition potential and electrolyte composition. Stoichiometric Bi_2Te_3 films were deposited at 40% v/v EG over all potentials ($-200 \leq E \leq -20$ mV *vs.* SCE).

In 2010, Qiu *et al.*[10] reported the use of polydiaminourea (PUB) and polyaminosulphone (PAS) as additives in a basic electrolyte for the electrodeposition of bismuth telluride. It was found that PUB and PAS bond ionically with TeO_3^{2-} and feature strong adhesion at the cathode, thereby increasing the concentration of TeO_3^{2-} at the cathode surface. An ideal quality film with near-amorphous crystal structure was produced.

In the same year, Kuleshova *et al.*[9] reported the use, in the electrodeposition of p-type Bi-Sb-Te alloys, of sodium lignosulfonate, Triton X100, EG and sodium dodecyl sulfate, of which sodium lignosulfonate led to a smooth, dense and pinhole-free deposit. At a concentration of 60 mg dm^{-3} of sodium lignosulfonate, the deposits were smooth and compact compared to those deposited in the absence of the surfactant. The homogeneity in the chemical composition and the Seebeck coefficients of the Bi-Sb-Te samples prepared in the presence of the surfactant showed significant improvement. Whilst films deposited without surfactant displayed modest Seebeck coefficient values of between 50 and 120 $\mu V K^{-1}$, deposition in the presence of the surfactant improved these measured values to between 220 and 250 $\mu V K^{-1}$.

In 2014, Caballero-Calero *et al.*[47] reported on the electrodeposition of bismuth telluride films with the addition of up to 120 mg dm^{-3} of sodium lignosulfonate. Seebeck coefficients of approximately -80 $\mu V K^{-1}$ were measured for the films deposited with sodium lignosulfonate, a 33% increase on those deposited without.

Sodium lignosulfonate (Figure 8.4), which comes from the group of anionic sulfonated lignin compounds, is a water-soluble polyelectrolyte

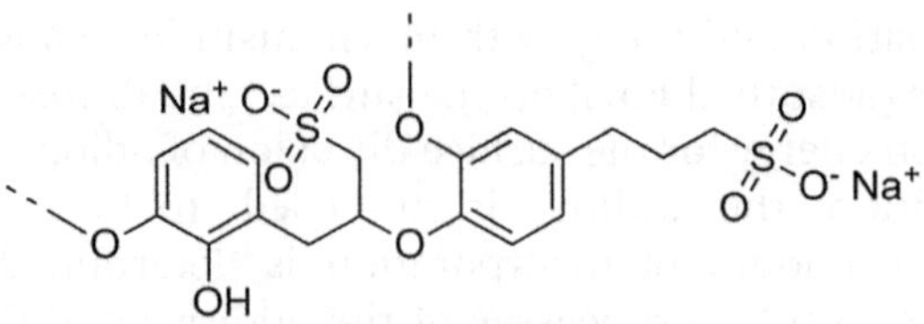

Figure 8.4 Molecular structure of sodium lignosulfonate.

polymer. Lignosulfonates usually have very broad ranges of molecular mass ranging from 5000–400 000 $g\,mol^{-1}$ and have complex sulfonated organic backbones.[48] It is known to act as a complexing agent for heavy metals.[48] This polymer has not been widely reported for its use as an additive in electrodeposition.

One of the first references to the use of sodium lignosulfonate in electrodeposition is in a paper by Hazza *et al.*[49] This involved the use of the surfactant in a lead acid battery to significantly improve the quality of lead deposits without decreasing charge and energy efficiency of the battery.[49,50] Although the additive was used up to a concentration of 5 $g\,dm^{-3}$, at a concentration of just 1 $g\,dm^{-3}$, sodium lignosulfonate is an effective levelling agent for lead, inhibiting dendritic growth even when using high current densities over long deposition durations. This, in turn, enhances the energy storage capacity of the battery. The surfactant is thought to prevent nucleation of the electroactive species at low energy sites on the electrode surface such as vacancies or kink sites. Sodium lignosulfonate is, therefore, considered a suitable candidate as an additive for further study into its effects on the electrodeposition of thermoelectric materials.

8.1.2.2 Doping of Thermoelectric Materials by Electrodeposition

As well as introducing additives to the electrolyte in order to modify the growth mechanism of thermoelectric materials, additional elements can be doped into the structure. This may have the effect of introducing more crystal defects and crystallite boundary as a result of an increased level of disorder in the crystal structure. This can cause a more severe scattering of phonons leading to a reduction of the lattice thermal conductivity, desirable for enhanced thermoelectric properties, and also affect electrical properties including conductivity and charge carrier concentration.[51]

New thermoelectric materials based on ternary and quaternary alloy systems such as Cu-doped Bi-Se-Te[52,53] and In-doped Bi-Sb-Te-Se[54] have shown excellent thermoelectric properties over binary and simple ternary alloys, reaching zT values of up to 1 and 15% increases in efficiency over un-doped analogues.

In these examples, there are three or four main elements being included in the semiconductor compound and a small amount of another dopant element. When so many elements are introduced, it can make the electrodeposition mechanism very complicated. However, these ternary or quaternary compound materials are fabricated as such to introduce either some n- or p-type character, both of which are required as materials in a full thermoelectric device. For instance, $Bi_2Te_{2.7}Se_{0.3}$ is an n-type thermoelectric material and $Bi_{0.5}Sb_{1.5}Te_3$ is a p-type material, both known as the best commercially available thermoelectric materials for use at room temperature.[9]

Table 8.1 shows some thermoelectric properties of electrodeposited bismuth tellurium selenide thin films from nitric acid electrolytes, reported in the literature.

Table 8.1 Summary of thermoelectric and electrical properties for bismuth tellurium selenide thin films, prepared by electrodeposition, reported in the literature. (SS = Stainless steel.)

Ref.	Seebeck coefficient/ $\mu V\ K^{-1}$	Resistivity/ $m\Omega$ cm	Power factor/ $mW\ m^{-1}\ K^{-2}$
55	−40	5.9	0.027
56	−70	—	—
57	−80	—	—
58 (Au)	−62	1.4	0.274
58 (SS)	−48	0.8	0.288

The Seebeck coefficients of these electrodeposited materials are not as good as $Bi_2Te_{2.7}Se_{0.3}$ fabricated by other methods. A maximum Seebeck coefficient of $-208\ \mu V\ K^{-1}$ was reported of materials fabricated by such methods as sputtering, flash evaporation and molecular beam epitaxy.[59–61]

Doping of electrodeposited thin films is carried out during the deposition process. Small amounts of a salt of the dopant is simply added to the electrolyte from which the thin films are deposited. However, little work has been performed in this area for the preparation of thermoelectric materials. The only example of electrochemical doping of bismuth telluride or bismuth tellurium selenide found in the literature is by Zhou *et al.*[62] They investigated the effect of doping Bi_2Te_3 with phosphorus by adding 1 mM phosphorus acid to a nitric acid electrolyte. The deposition potential used was −50 mV *vs.* SCE and the deposit was made onto a gold-sputtered aluminium substrate over 24 h at room temperature. The resultant deposit was made up of 2% phosphorus, had a crystallite size of 14–16 nm, and was orientated towards the (1 1 0) plane. Whilst a Seebeck coefficient of $-35\ \mu V\ K^{-1}$ was measured for the deposit, no comparison against undoped deposits was made.

Although there have been few reported instances of the electrodeposition of doped thermoelectric materials, there have been many promising studies of doping by other methods.[52,53,63–66]

Chromium intercalated and doped Bi_2Te_3 materials, synthesized by a high temperature solid state reaction under vacuum, were reported to exhibit improved zT values of up to 0.8 and 0.65, respectively.[63] Various compositions of the In-doped p-type thermoelectric material $(Sb_{0.75}Bi_{0.25})_{2-x}In_x Te_{2.8}Se_{0.2}$ were synthesized by a solid-state technique;[53] the zT was improved by over 15% where $x = 0.02$, compared with the undoped analogue.

Cui *et al.*[52] investigated the copper-doping of bismuth tellurium selenide by a spark plasma sintering technique. Compositions of $(Bi_2Te_3)_{0.9}(Bi_{2-x}Cu_xSe_3)_{0.1}$ where $x = 0$–0.2 were synthesized and characterized. The optimum value of x was found to be 0.1 where the electrical conductivity was measured as $8.1 \times 10^4\ S\ m^{-1}$ at 318 K and the Seebeck coefficient was $-170\ \mu V\ K^{-1}$ at 417 K. The thermal conductivity ($0.88\ W\ m^{-1}\ K^{-1}$ at 318 K) was decreased at the same composition and a maximum zT of 0.98 was calculated at 417 K. In addition, at every composition, the X-ray diffraction patterns showed the (0 1 5) peak as the most intense peak.

This chapter presents experimental results to demonstrate the use of the surfactant sodium lignosulfonate in the electrodeposition of bismuth telluride, and the doping with copper of the ternary n-type thermoelectric material bismuth tellurium selenide. It is shown how the use of additive or dopants can affect the microstructural properties of electrodeposited thermoelectric materials and enhance their performance.

8.2 Experimental

Aqueous electrolytes for the electrodeposition of bismuth telluride were prepared from 7.5 mM Bi (elemental, Alfa Aesar, 99.999%), 10 mM Te (elemental, AlfaAesar, 99.999%) in 1 M HNO_3 (Fisher, 70%).[13] For studies investigating the effect of sodium lignosulfonate on the electrodeposition of bismuth telluride, up to 1.5 $g\,dm^{-3}$ of sodium lignosulfonate (Aldrich, Product Number 370975, CAS Number 8061-51-6) was added to the electrolyte. For the electrodeposition of ternary bismuth tellurium selenide thin films, electrolytes comprising 8 mM $Bi(NO_3)_3 \cdot 5H_2O$ (Acros Organics, 99.999%), 9 mM TeO_2 (Acros Organics, 99.9995%), 1 mM SeO_2 (Acros Organics, 99.999%), and 1 M HNO_3 (Fisher, 70%) were employed.[51] For the electrodeposition of copper-doped bismuth tellurium selenide thin films, up to 0.75 mM $Cu(NO_3)_2$ (Aldrich, 99.999%) was added to the electrolyte.

Cyclic voltammetry and electrodeposition were performed at room temperature using a standard three-electrode (working, counter and reference electrodes) cell design, controlled by an Auto-Lab PGSTAT30 potentiostat/galvanostat and AutoLab General Purpose Electrochemical System (GPES) software. The cell volume was approximately 10 cm^3.

In cyclic voltammetry experiments, a gold working electrode was used (2 mm diameter disc [CHI101, CH Instruments, Inc.] or 1 cm^2 100 nm-thick gold-coated archival DVD-R disc [Delkin Devices]). A large area platinum gauze electrode (approximately 2 cm^2) and a saturated calomel electrode (SCE), both prepared in-house, were used as the counter and reference electrodes, respectively.

For the electrodeposition of thin films either 1 cm^2 area pieces of 100 nm-thick gold-coated archival DVD-R disc (Delkin Devices) or 200 nm-thick evaporated gold on silicon wafer substrates were used as working electrodes. A large area platinum gauze electrode and an SCE, both prepared in-house, were used as the counter and reference electrodes, respectively.

A JSM-6500F field emission gun-scanning electron microscope (FEG-SEM) and a Zeiss EVO LS25 ESEM were used to determine thin film thicknesses and observe the morphology of electrodeposited materials. Accelerating voltages of up to 30 kV were used with a secondary electron detector. Cross-sectioned samples were made using a diamond scribe, in order to determine the thickness of deposits.

Both the JSM-6500F FEG-SEM and the Zeiss EVO LS25 ESEM incorporated EDX detectors which were used to analyse the composition of the thin films. The measurements were referenced against Bi_2Te_3 and Bi_2Se_3 standards.

A Thermo Scientific Wavelength-Dispersive X-ray Spectroscopy (WDS) MagnaRay Spectrometer mounted on a Philips XL30 ESEM was used for the quantitative analysis of copper-doped bismuth tellurium selenide thin films. The measurements were referenced against Bi, Te, Se and Cu elemental standards.

Siemens D5000 and Bruker GADDS powder X-ray diffractometers with Cu K radiation ($\lambda = 1.5406$ Å) were used to determine the crystal structure and for the calculation of crystallite sizes of the thin films. The instruments were set up in a symmetrical theta–theta configuration.

Seebeck coefficients, S (μV K^{-1}), for the thin films were determined using a custom-made Seebeck measurement unit calibrated against a polycrystalline bismuth foil reference standard. A photograph of the equipment is shown in Figure 8.5. The Seebeck coefficient measurements were completed on the as-deposited thin films without removal from the gold substrate. There is negligible contribution from the gold substrate (~0.7 μV K^{-1}) relative to that of the thin film.[67]

The transport properties (electrical resistivity/conductivity, Hall coefficient, carrier concentration, and Hall mobility) of the films were measured using the standard van der Pauw technique with a DC current of 10–20 mA and a permanent magnetic field of 0.55 T at room temperature using a commercial Hall effect measurement system (HMS 3000, Ecopia), schematically represented in Figure 8.6. Contacts to the sample board were made to the four corners of the sample using 0.2 mm-thick copper wires and a conductive silver epoxy (CircuitWorks CW2400).

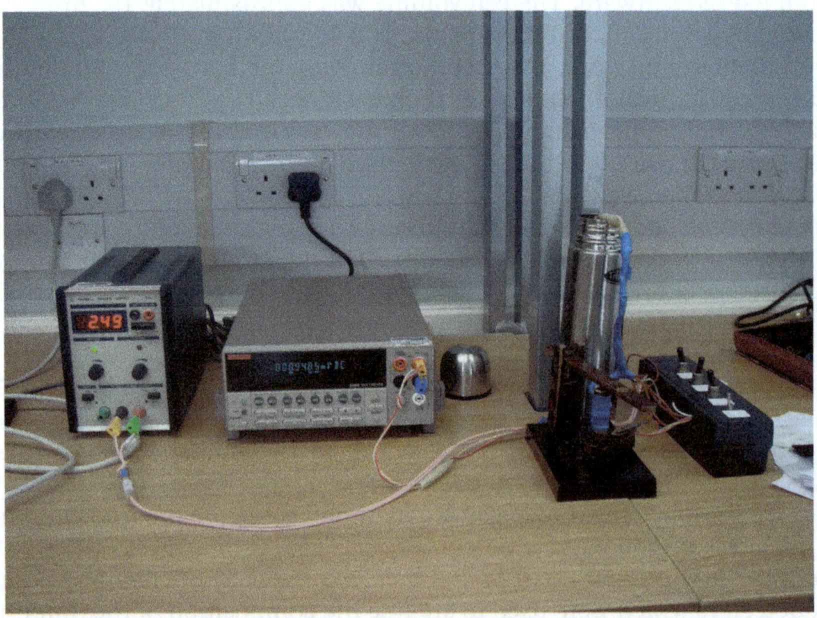

Figure 8.5 Photograph of the custom-made Seebeck coefficient measurement unit.

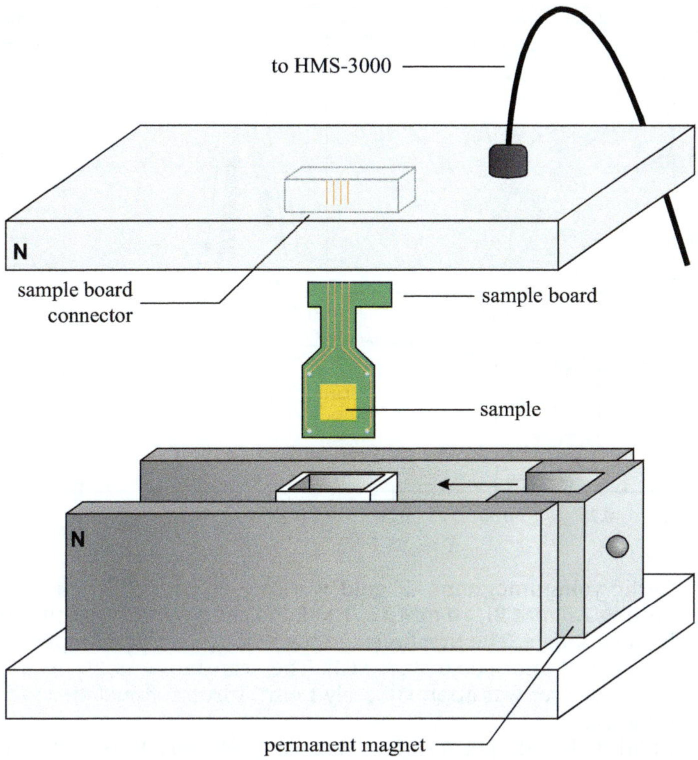

Figure 8.6 Schematic diagram of the Hall effect measurement magnet set.

8.3 Results and Discussion

8.3.1 Sodium Lignosulfonate as an Additive in the Electrodeposition of Bismuth Telluride

Sodium lignosulfonate was used as a surfactant additive in the electrodeposition of bismuth telluride. The effect of the surfactant on the electrochemistry of the deposition and stripping of bismuth telluride at a gold working electrode was investigated by cyclic voltammetry. Figure 8.7 shows a series of voltammograms for the electrodeposition of bismuth telluride from solutions containing 7.5 mM Bi and 10 mM Te in 1 M HNO_3 with up to 1.5 $g\,dm^{-3}$ of sodium lignosulfonate. The inset presents the data between 0 and −0.20 V *vs.* SCE for clarity.

The general shape of the CV remains similar with increasing concentrations of sodium lignosulfonate added to the electrolyte; however, there are some trends to be noted. An addition of 1.5 $g\,dm^{-3}$ sodium lignosulfonate to the electrolyte is shown to affect the onset potential for deposition, shifting it almost 90 mV negative. This suggests that the presence of the surfactant is causing an increase in the overpotential required for stable nuclei to form on the electrode, a requirement for the further growth

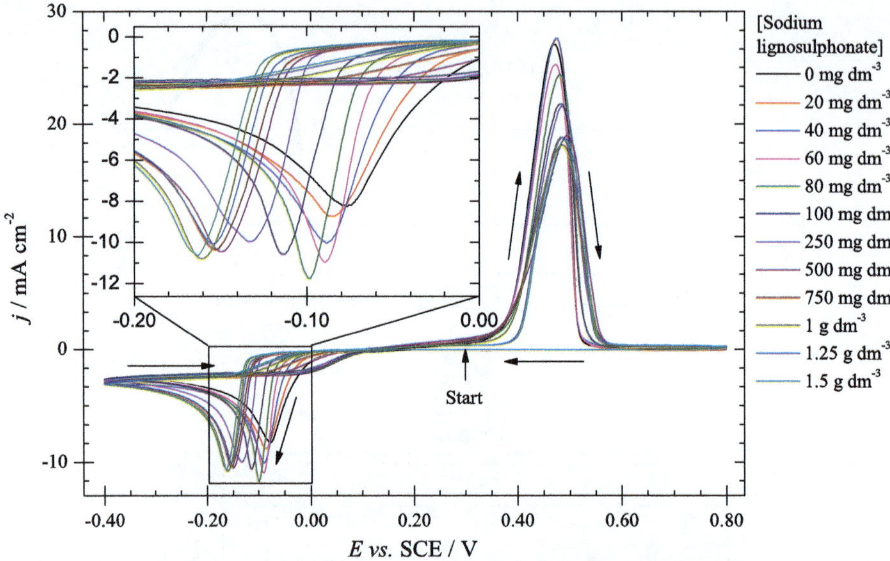

Figure 8.7 Cyclic voltammograms of gold working electrodes in electrolytes containing 7.5 mM Bi, 10 mM Te, 1 M HNO_3 and up to 1.5 g dm^{-3} of sodium lignosulfonate. The first cycle is shown in each case, with potential limits of −0.40 V and +0.80 V *vs.* SCE. The scan rate was 20 mV s^{-1} and the electrode area was approximately 1 cm^2. Inset: zoomed area (0 to −0.20 V *vs.* SCE).
Reprinted with permission from A. J. Naylor, E. Koukharenko, I. S. Nandhakumar, N. M. White, *Langmuir*, 2012, **28**, 8296.[68] Copyright 2016 American Chemical Society.

of bismuth telluride layers. Additionally, whereas at concentrations of 250 mg dm^{-3} and below the reverse scans follow a similar profile until the zero-current crossover, at higher concentrations the drop-off of current towards zero shifts negative but maintains a similar gradient. This indicates a generally similar kinetic profile for the reaction independent of the surfactant concentration. However, the negative shift suggests that the equilibrium potential becomes more cathodic with an increase in the surfactant concentration. This is likely due to the high additions of the surfactant in the solution increasing the thermodynamic favourability for the dissolution of the solid phase. A decrease in the current density associated with the anodic stripping peak is also observed as a function of increased surfactant concentration. This decrease is likely caused by the reduced time of deposition due to the negative shift of the deposition potential on the forward scan and, as a result, there not being as much material to strip from the electrode. Therefore, the surfactant offers a method for controlling the deposition potential and the nucleation and growth processes during the Electrodeposition of bismuth telluride.

Electrodeposition of bismuth telluride without a surfactant demonstrated that thin films could be fabricated over a wide potential range

(−0.60–0 V *vs.* SCE), with all films displaying the preferred stoichiometric Bi_2Te_3 composition. However, films deposited at more anodic (more positive) potentials showed improved adhesion to the substrate, crystallinity, and thermoelectric properties.

The addition of up to 100 mg dm^{-3} of sodium lignosulfonate to the electrolyte, at a deposition potential of −0.10 V *vs.* SCE, resulted in bismuth telluride thin films with the stoichiometric Bi_2Te_3 composition. The surfactant, therefore, is not thought to alter the composition of bismuth telluride films.

SEM images of cross-sections of bismuth telluride thin films deposited in the absence and with 100 mg dm^{-3} of sodium lignosulfonate are shown in Figure 8.8. Without the surfactant, dendritic growth of 18.35 μm thickness

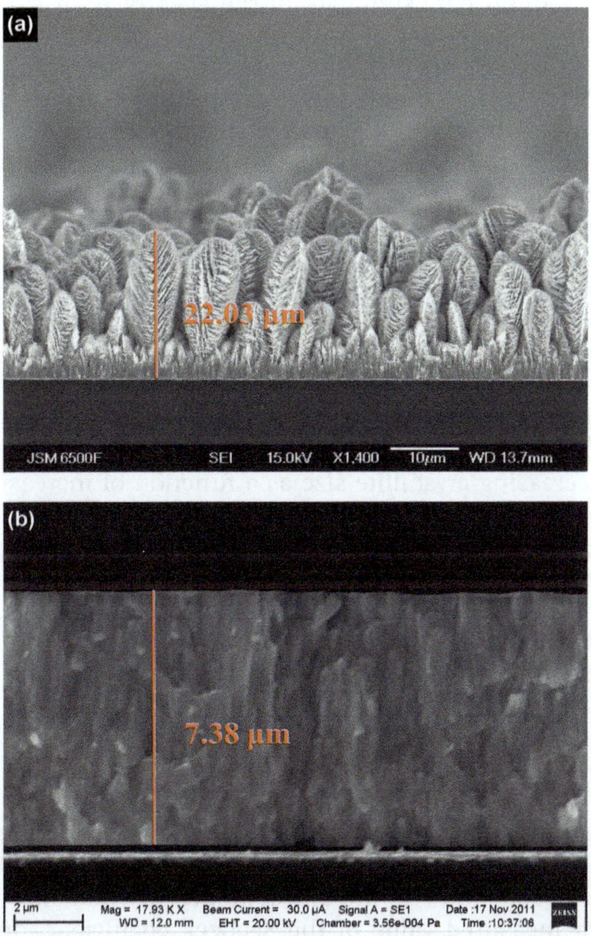

Figure 8.8 SEM (secondary electron) images of cross-sections of bismuth telluride thin films electrodeposited at a potential of −0.10 V *vs.* SCE (a) in the absence of and (b) with 100 mg dm^{-3} of sodium lignosulfonate.

occurs upon a more compact layer of 3.68 μm. This suggests that there is uniform nucleation and growth for the beginning of the deposition. However, with time, some growth centres dominate resulting in a rough and disjointed deposit comprising many different sized growths. The addition of 100 mg dm^{-3} to the electrolyte produces a compact deposit of 7.38 μm thickness. The presence of the surfactant is thought to reduce the surface tension at the electrolyte–electrode interface during growth. The surfactant, therefore, compacts the growth of the bismuth telluride, preventing dendritic growth and creating a more coalescent deposit. This method of levelling thermoelectric materials by the addition of a surfactant could be advantageous for when employing the materials in a device.

Figure 8.9 shows the XRD patterns for bismuth telluride deposits formed by electrodeposition at −0.10 V *vs.* SCE in the presence of between 0 and 120 mg dm^{-3} sodium lignosulfonate in the electrolyte. It is observed from the diffraction patterns that an increase in the concentration of sodium lignosulfonate up to 80 mg dm^{-3} increases the crystallographic orientation towards the (1 1 0) plane. Orientation towards other planes such as (1 0 10) is inhibited relative to the (1 1 0) plane which is not observed to such a great extent in previous studies such as that by Ma *et al.*[35] An optimum concentration of 80 mg dm^{-3} is required to give the greatest degree of texturing, but at higher concentrations of sodium lignosulfonate this tendency is reduced, thought to be due to an unfavourable arrangement of the surfactant at the electrode at higher concentrations. This XRD data shows that the orientation of bismuth telluride films deposited at −0.10 V *vs.* SCE may be optimised by employing 80 mg dm^{-3} in the electrolyte. The orientation towards the (1 1 0) plane is important since it is the preferred orientation for improved thermoelectric efficiency.[39]

The crystallite size, as calculated by the Scherrer equation (eqn (8.2)), indicates a decreasing crystallite size as a function of increasing surfactant concentration. The crystallite size drops by 16 nm between the surfactant concentrations of 0 (36 nm) and 60 mg dm^{-3} (20 nm) suggesting that, between these concentrations, the surfactant is having a large effect in inhibiting the growth of the crystallites in the deposit. This demonstrated that the crystallite size can be adjusted by varying the concentration of sodium lignosulfonate in the electrolyte.

$$D = \frac{K\lambda}{\beta \cos \theta} \tag{8.2}$$

Scherrer equation: D is the crystallite size (nm), λ is the wavelength of the incident X-rays (nm), β is the full-width at half-maximum (FWHM) intensity of the peak (radians) and θ is the diffraction angle (radians).

The highest measured value of the Seebeck coefficient for the bismuth telluride films deposited with sodium lignosulfonate was −15 μV K^{-1} at a concentration of 20 mg dm^{-3}. Hall coefficient measurements for the thin films shows that they are n-type semiconductors. Electrical resistivity for

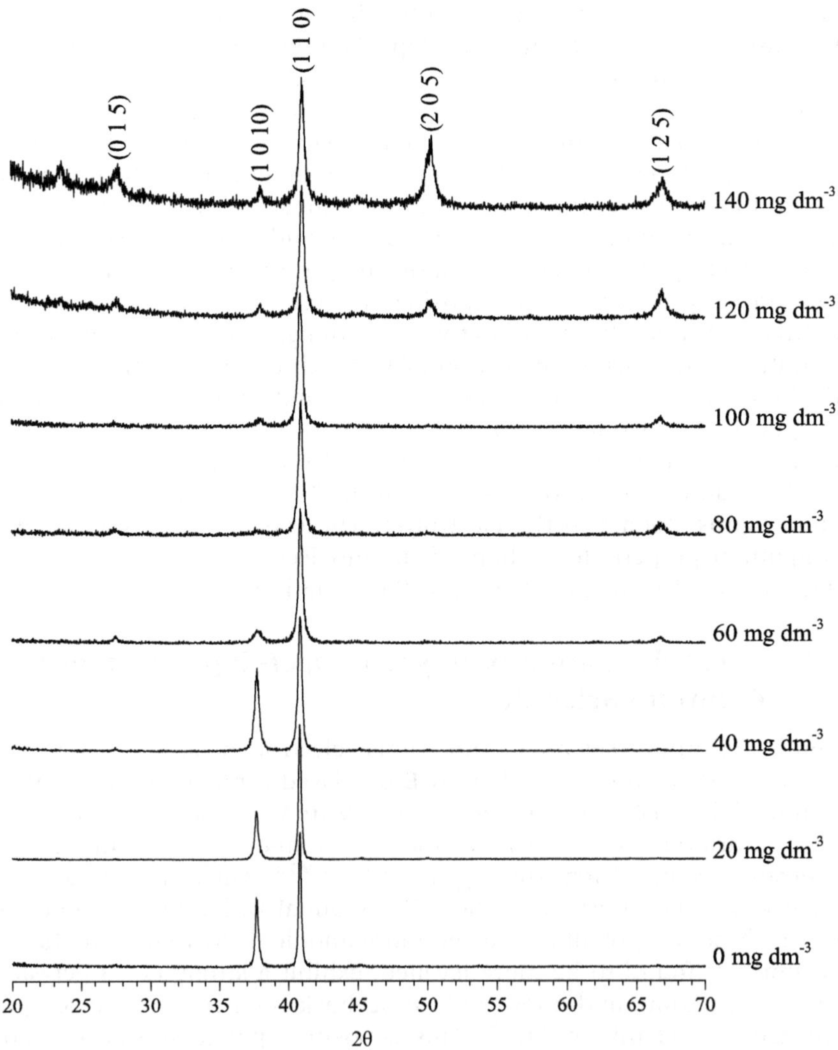

Figure 8.9 XRD patterns for as-deposited bismuth telluride thin films electrodeposited at −0.10 V *vs.* SCE with up to 140 mg dm^{-3} of sodium lignosulfonate in the electrolyte.

thin films deposited with 60 and 80 mg dm^{-3} of sodium lignosulfonate were measured as 3×10^{-4} Ω cm and 1.3×10^{-3} Ω cm, respectively. The Hall mobility and the carrier concentration were calculated as 4 cm^2 V^{-1} s^{-1} and 4.5×10^{21} cm^{-3} for 80 mg dm^{-3} and 35 cm^2 V^{-1} s^{-1} and 2.5×10^{20} cm^{-3} for 60 mg dm^{-3}. These transport property characteristics are either similar or improved in comparison with other electrodeposited bismuth telluride thin films reported.[11,32] The power factor ($S^2\sigma$) was calculated for the bismuth telluride thin films deposited at −0.10 V *vs.* SCE in the presence of sodium lignosulfonate. The thin film deposited with 60 mg dm^{-3} of sodium lignosulfonate in the electrolyte has a calculated power factor of 0.0563 mW m^{-1} K^{-2}. The thin film deposited with 80 mg dm^{-3} of sodium lignosulfonate in the electrolyte has a calculated power factor of 0.0111 mW m^{-1} K^{-2}. Therefore, the addition of 60 mg dm^{-3} of sodium lignosulfonate to the electrolyte is the optimum to obtain the best power factor.

Electrodeposition of bismuth telluride at a potential of −0.20 V *vs.* SCE required higher concentrations of sodium lignosulfonate to induce similar effects to those seen here for the crystallinity and crystallite size. However, the films were generally cracked, more rough and exhibited poorer thermoelectric properties to those deposited at −0.10 V *vs.* SCE.

In this study it has been shown that sodium lignosulfonate can be used as an additive in the electrodeposition of bismuth telluride thin films. It can shift the deposition potential more negative and influence the nucleation and growth behaviour of this commercially significant thermoelectric material. At a deposition potential of −0.10 V *vs.* SCE, levelled Bi_2Te_3 thin films can be grown to almost 10 μm in 2 h with just 100 mg dm^{-3} of sodium lignosulfonate in the electrolyte. The crystallographic orientation, an important property for enhanced thermoelectric performance, can also be tailored to offer the preferred (1 1 0) orientation.

8.3.2 Electrodeposition of n-type Copper-doped Bismuth Tellurium Selenide

Bismuth tellurium selenide thin films electrodeposited at potentials between −0.10 V and +0.05 V *vs.* SCE displayed a range of compositions, determined by EDX. At a potential of −0.10 V *vs.* SCE, the thin films were bismuth-rich but tellurium- and selenium-deficient compared with the desired composition ($Bi_2Te_{2.7}Se_{0.3}$),[51,56,69,70] which was successfully deposited at +0.05 V *vs.* SCE. There is a gradual variation in composition between these two potentials. At the more anodic deposition potential, the thin films are found to be more levelled, exhibit a strong (1 1 0) crystallographic orientation, and offer the best Seebeck coefficient (65 μV K^{-1}) and power factor (2.11 mW m^{-1} K^{-2}). This deposition potential was selected to be used to study the effect of copper doping on electrodeposited bismuth tellurium selenide thin films.

Cyclic voltammetry of a 2 mm diameter gold disc working electrode for the electrodeposition of copper-doped bismuth tellurium selenide is shown in Figure 8.10.

Whilst the cyclic voltammograms are generally similar in shape, with increasing quantities of $Cu(NO_3)_2$ added to the electrolyte, the deposition peak is shifted to more cathodic potentials. The addition of 0.75 mM $Cu(NO_3)_2$ results in a negative shift of approximately 20 mV. This is likely due to a thermodynamic effect of including copper in the co-deposition process.

With the addition of 0.50 mM $Cu(NO_3)_2$ to the electrolyte, a selenium-rich, bismuth-deficient phase ($Bi_{1.5}Te_{2.7}Se_{0.6}Cu_{0.2}$) is deposited compared with that deposited without $Cu(NO_3)_2$ ($Bi_2Te_{2.7}Se_{0.3}$). This may suggest that the

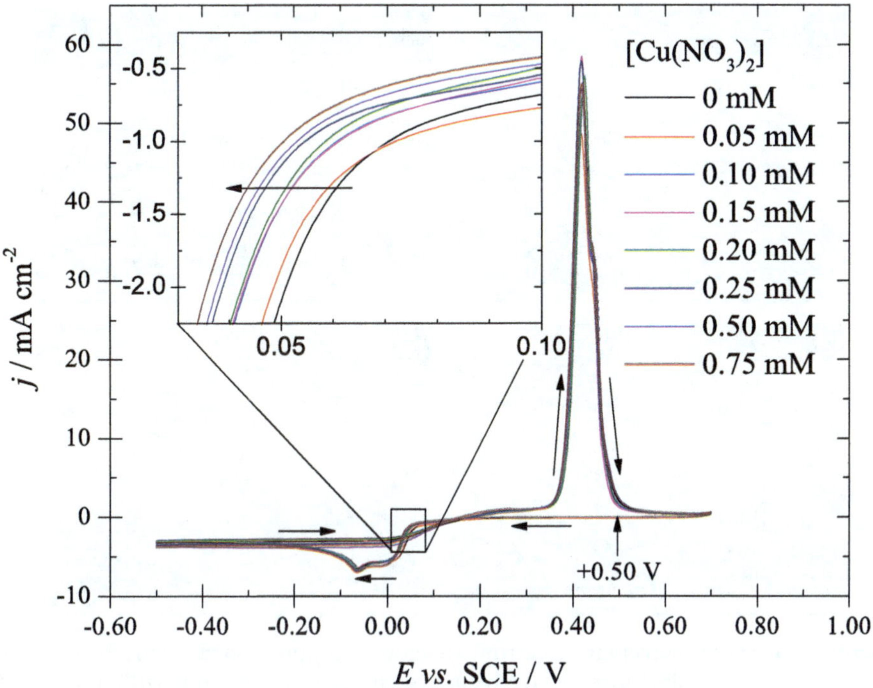

Figure 8.10 Cyclic voltammograms of a 2 mm diameter gold disc working electrode in 8 mM $Bi(NO_3)_3 \cdot 5H_2O$, 9 mM TeO_2, 1 mM SeO_2 and up to 0.75 mM $Cu(NO_3)_2$ in 1 M HNO_3. The first cycle is shown in each case, with limits at −0.50 V and +0.70 V *vs.* SCE. The scan rate was 20 mV s^{-1}.

doping of copper is causing selenium and copper to replace bismuth sites in the crystal structure.

Figure 8.11 shows the SEM images for as-deposited copper-doped bismuth tellurium selenide thin films deposited at +0.05 V *vs.* SCE with an electrolyte $Cu(NO_3)_2$ concentration of up to 0.75 mM.

The addition of copper nitrate to the electrolyte is seen to affect the morphology of the thin films in a significant way. Whilst the addition of 0.25 mM $Cu(NO_3)_2$ does not affect the shape of the grains, the addition of 0.75 mM of $Cu(NO_3)_2$ in the electrolyte results in a morphology which features smaller grains and a wire-like structure. The thin film deposited with 0.50 mM of $Cu(NO_3)_2$ is not uniform in morphology.

Figure 8.12 presents images of as-deposited copper-doped bismuth tellurium selenide thin film cross-sections, electrodeposited at +0.05 V *vs.* SCE with up to 0.75 mM of $Cu(NO_3)_2$ in the electrolyte.

The images of the thin film cross-sections show that a reasonable thickness can be deposited over the 2 h deposition period with up to 0.75 mM of $Cu(NO_3)_2$ added to the electrolyte. However, there is a noticeable variation in thickness with changing $Cu(NO_3)_2$ concentration. A minimum thickness of 3.25 μm is measured for the thin film deposited

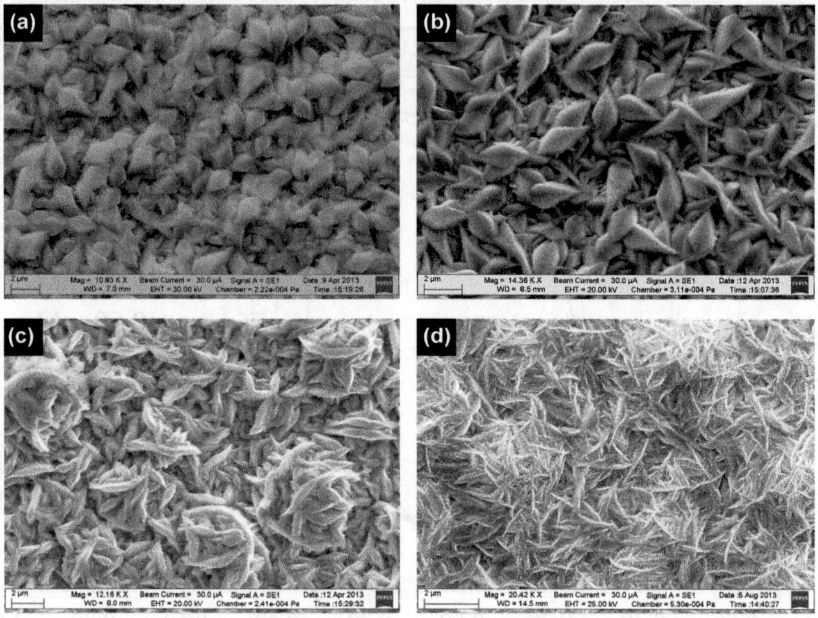

Figure 8.11 SEM (secondary electron) images of copper-doped bismuth tellurium selenide films electrodeposited at +0.05 V *vs.* SCE with (a) 0 mM, (b) 0.25 mM, (c) 0.50 mM, (d) 0.75 mM of $Cu(NO_3)_2$ in the electrolyte.

with 0.50 mM of $Cu(NO_3)_2$ in the electrolyte, more than half that measured for thin films deposited in the absence of $Cu(NO_3)_2$. This could be due to the film becoming more compacted, rather than as a result of a slower rate of deposition.

XRD experiments indicate a strong preferential orientation towards the (0 1 5) plane for the thin film deposited with 0.50 mM of $Cu(NO_3)_2$ in the electrolyte. While with 0.25 mM or 0.75 mM $Cu(NO_3)_2$, the (1 1 0) peak is dominant. Normally, orientation of crystallites towards the (1 1 0) plane would be considered preferential for enhancing thermoelectric efficiency. However, copper-doped bismuth tellurium selenide thin films fabricated by Cui *et al.*[52] all show the (0 1 5) peak as the most intense, suggesting that this may also result in high thermoelectric efficiency. Therefore, the doping of Cu into the crystal structure has a big effect on the crystallite orientation.

Whilst for most of the copper-doped thin films, the crystallite size was determined to be approximately 23 nm, for a $Cu(NO_3)_2$ concentration of 0.50 mM a crystallite size of 18 nm was calculated. It is speculated that the introduction of copper and additional selenium sites in the crystal structure promotes the number of crystallite boundaries, accounting for the decrease in crystallite size.

The Seebeck coefficient, measured as an average of three spots on each of five samples, is observed to increase in absolute value to a maximum

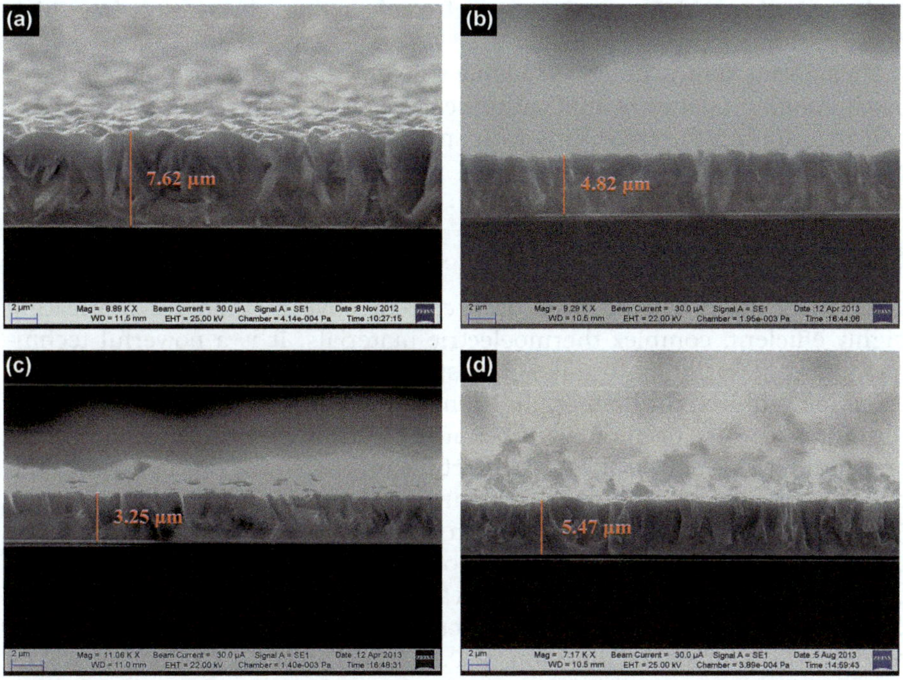

Figure 8.12 SEM (secondary electron) images of copper-doped bismuth tellurium selenide thin films cross-sections electrodeposited at +0.05 V *vs.* SCE with (a) 0 mM, (b) 0.25 mM, (c) 0.50 mM, (d) 0.75 mM of $Cu(NO_3)_2$ in the electrolyte.

of -227 ± 20.2 μV K^{-1} at a $Cu(NO_3)_2$ concentration of 0.50 mM. In comparison with the literature, this value is nearly 60 μV K^{-1} greater than that measured by Cui *et al.*[52] for copper-doped bismuth tellurium selenide thin films fabricated by spark plasma sintering and still statistically greater, by 20 μV K^{-1}, than that measured by Peranio *et al.*[61] for undoped bismuth tellurium selenide thin films grown by molecular beam epitaxy.

The measured electrical conductivity doubles in value with the addition of 0.50 mM of $Cu(NO_3)_2$ as compared with the undoped thin film. A maximum electrical conductivity of $115 \pm 2.43 \times 10^4$ S m^{-1} is achieved at a $Cu(NO_3)_2$ concentration of 0.50 mM. At the higher $Cu(NO_3)_2$ concentration of 0.75 mM, the electrical conductivity decreases. The maximum electrical conductivity achieved here is over 14 times greater than that reported in the literature for copper-doped bismuth tellurium selenide thin films.[52]

The power factor ($S^2\sigma$), calculated from the Seebeck coefficient and the electrical conductivity, shows a general increase with $Cu(NO_3)_2$ concentration in the electrolyte. A maximum power factor of 59.9 ± 12.5 mW m^{-1} K^{-2} is calculated for the thin films deposited in the presence of 0.50 mM $Cu(NO_3)_2$. This demonstrates that, whilst the thin film deposited with 0.50 mM $Cu(NO_3)_2$ did not display the preferential (1 1 0) peak in the X-ray diffraction pattern, the high Seebeck coefficient and electrical conductivity values result in a high power factor.

It has been shown here that highly-efficient thermoelectric materials for use in room temperature applications can be fabricated by electrodeposition and that doping the thin films with small fractions of copper dramatically enhance thermoelectric performance.

8.4 Conclusions

Electrodeposition can be used as a cost-effective method for fabricating highly efficient, complex thermoelectric materials. It is a powerful technique, offering flexibility to tailor nanostructured materials to the exact specification required for their application. In the fabrication of thermoelectric thin films, the use of different substrates, electrolytes, deposition potential, and additives have been shown to affect the microstructural properties of the deposit. In the case of bismuth telluride, the addition of approximately 80 mg dm^{-3} of the surfactant sodium lignosulfonate can produce levelled deposits with the stoichiometric composition Bi_2Te_3 and the preferred (1 1 0) crystallographic orientation, both desired to achieve enhanced thermoelectric properties. For the electrodeposition of bismuth tellurium selenide, doping with copper, by addition of 0.50 mM $Cu(NO_3)_2$ to the electrolyte, was found to reduce crystallite size and dramatically enhance the Seebeck coefficient and electrical conductivity.

Acknowledgements

Dr Elena Koukharenko is acknowledged for help with Seebeck coefficient measurements. This work was funded by the Engineering and Physical Sciences Research Council (EPSRC), the School of Chemistry, and the School of Electronics and Computer Science at the University of Southampton.

References

1. M. Paunovic and M. Schlesinger, *Fundamentals of Electrochemical Deposition*, John Wiley and Sons, New York, 2nd edn, 2006.
2. D. Pletcher, *A First Course in Electrode Processes*, Royal Society of Chemistry, Cambridge, 2nd edn, 2009.
3. A. J. Bard and L. R. Faulkner, *Electrochemical Methods: Fundamentals and Applications*, Wiley, 6th edn, 2000.
4. C. G. Zoski, *Handbook of Electrochemistry*, Elsevier, 2007.
5. Southampton Electrochemistry Group, Instrumental Methods in Electrochemistry, *Ellis Horwood series in physical chemistry*, Ellis Horwood Limited, Chichester, 1985.
6. R. Rostek, N. Stein and C. Boulanger, *J. Mater. Res.*, 2015, **30**, 2518.
7. B. Scharifker and G. Hills, *Electrochim. Acta*, 1983, **28**, 879.

8. F. Xiao, C. Handgarter, B. Yoo, Y. Rheem, K. Lee and N. Myung, *Electrochim. Acta*, 2008, **53**, 8103.
9. J. Kuleshova, E. Koukharenko, X. Li, N. Frety, I. S. Nandhakumar, J. Tudor, S. P. Beeby and N. M. White, *Langmuir*, 2010, **26**, 16980.
10. W.-J. Qiu, S.-N. Zhang, T.-J. Zhu and X.-B. Zhao, *Int. J. Miner., Metall. Mater.*, 2010, **17**, 489.
11. C. Boulanger, *J. Electron. Mater.*, 2010, **39**, 1818.
12. F. A. Kröger, *J. Electrochem. Soc.*, 1978, **125**, 2028.
13. M. S. Martín-González, A. L. Prieto, R. Gronsky, T. Sands and A. M. Stacy, *J. Electrochem. Soc.*, 2002, **149**, C546.
14. R. J. Hudson, Master's thesis, University of Southampton, 2008.
15. A. Brenner, *Electrodeposition of Alloys: Principles and Practice*, Academic Press, 1963, vol. 1.
16. A. Giani, A. Boulouz, F. Pascal-Delannoy, A. Foucaran and A. Boyer, *Thin Solid Films*, 1998, **315**, 99.
17. R. Venkatasubramanian, T. Colpitts, E. Watko, M. Lamvik and N. El-Masry, *J. Cryst. Growth*, 1997, **170**, 817.
18. X. Wang, H. He, N. Wang and L. Miao, *Appl. Surf. Sci.*, 2013, **276**, 539.
19. H. Zou, D. Rowe and G. Min, *J. Cryst. Growth*, 2001, **222**, 82.
20. A. Bailini, F. Donati, M. Zamboni, V. Russo, M. Passoni, C. Casari, A. L. Bassi and C. Bottani, *Appl. Surf. Sci.*, 2007, **254**, 1249.
21. S. Cho, Y. Kim, A. DiVenere, G. K. Wong, J. B. Ketterson and J. R. Meyer, *Appl. Phys. Lett.*, 1999, **75**, 1401.
22. F. Xiao, C. Handgarter, B. Yoo, Y. Rheem, K. Lee and N. Myung, *Electrochim. Acta*, 2008, **53**, 8103.
23. W. Zhu, J. Y. Yang, X. H. Gao, S. Q. Bao, X. A. Fan, T. J. Zhang and K. Cui, *Electrochim. Acta*, 2005, **50**, 4041.
24. C. Jin, X. Xiang, C. Jia, W. Liu, W. Cai, L. Yao and X. Li, *J. Phys. Chem. B*, 2004, **108**, 1844.
25. S. A. Sapp, B. B. Lakshmi and C. R. Martin, *Adv. Mater.*, 1999, **11**, 402.
26. L. Li, Y. Yang, X. Huang, G. Li and L. Zhang, *Nanotechnology*, 2006, **17**, 1706.
27. M. Takahashi, M. Kojima, S. Sato, N. Ohnisi, A. Nishiwaki, K. Wakita, T. Miyuki, S. Ikeda and Y. Muramatsu, *J. Appl. Phys.*, 2004, **96**, 5582.
28. P. Magri, C. Boulanger and J.-M. Lecuire, *J. Mater. Chem.*, 1996, **6**, 773.
29. S. Michel, S. Diliberto, C. Boulanger, N. Stein and J. Lecuire, *J. Cryst. Growth*, 2005, **277**, 274.
30. S. Li, M. S. Toprak, H. M. A. Soliman, J. Zhou, M. Muhammed, D. Platzek and E. Müller, *Chem. Mater.*, 2006, **18**, 3627.
31. Y. Miyazaki and T. Kajitani, *J. Cryst. Growth*, 2001, **229**, 542.
32. B. Y. Yoo, C. K. Huang, J. R. Lim, J. Herman, M. A. Ryan, J. P. Fleurial and N. V. Myung, *Electrochim. Acta*, 2005, **50**, 4371.
33. P. Heo, K. Hagiwara, R. Ichino and M. Okido, *J. Electrochem. Soc.*, 2006, **153**, C213.
34. Y. Ma, A. Johansson, E. Ahlberg and A. E. Palmqvist, *Electrochim. Acta*, 2010, **55**, 4610.

35. Y. Ma, E. Ahlberg, Y. Sun, B. B. Iversen and A. E. C. Palmqvist, *Electrochim. Acta*, 2011, **56**, 4216.
36. K. Tittes, A. Bund, W. Plieth, A. Bentien, S. Paschen, M. Plotner, H. Grafe and W. J. Fischer, *J. Solid State Electrochem.*, 2003, **7**, 714.
37. W. Glatz, L. Durrer, E. Schwyter and C. Hierold, *Electrochim. Acta*, 2008, **54**, 755.
38. H. Chaouni, J. Bessières, A. Modaressi and J. Heizmann, *J. Appl. Electrochem.*, 2000, **30**, 419.
39. E. Koukharenko, X. Li, I. Nandhakumar, N. Frety, S. P. Beeby, D. Cox, M. J. Tudor, B. Schiedt, C. Trautmann, A. Bertsch and N. M. White, *J. Micromech. Microeng.*, 2008, **18**, 104015.
40. C. V. Manzano, B. Abad, M. Muñoz Rojo, Y. R. Koh, S. L. Hodson, A. M. Lopez Martinez, X. Xu, A. Shakouri, T. D. Sands, T. Borca-Tasciuc and M. Martin-Gonzalez, *Sci. Rep.*, 2016, **6**, 19129.
41. I. Y. Erdoğan and Ü. Demir, *Electrochim. Acta*, 2011, **56**, 2385.
42. C. Frantz, N. Stein, L. Gravier, S. Granville and C. Boulanger, *J. Electron. Mater.*, 2009, **39**, 2043.
43. A. L. Prieto, M. S. Sander, M. Martín-González, R. Gronsky, T. Sands and A. M. Stacy, *J. Am. Chem. Soc.*, 2001, **123**, 7160.
44. M. S. Sander, R. Gronsky, T. Sands and A. M. Stacy, *Chem. Mater.*, 2003, **15**, 335.
45. M. S. Sander, A. L. Prieto, R. Gronsky, T. Sands and A. M. Stacy, *Adv. Mater.*, 2002, **14**, 665.
46. M. Schlesinger and M. Paunovic, *Modern Electroplating*, Wiley-Interscience, New York, 4th edn, 2000.
47. O. Caballero-Calero, P. Díaz-Chao, B. Abad, C. V. Manzano, M. D. Ynsa, J. J. Romero, M. Muñoz Rojo and M. S. Martín-González, *Electrochim. Acta*, 2014, **123**, 117.
48. M. Norgren and H. Edlund, *Curr. Opin. Colloid Interface Sci.*, 2014, **19**, 409.
49. A. Hazza, D. Pletcher and R. A. Wills, *J. Power Sources*, 2005, **149**, 103.
50. R. A. Wills, Ph.D. thesis, University of Southampton, 2004.
51. L. Qiu, J. Zhou, X. Cheng and R. Ahuja, *J. Phys. Chem. Solids*, 2010, **71**, 1131.
52. J. L. Cui, L. D. Mao, W. Yang, X. B. Xu, D. Y. Chen and W. J. Xiu, *J. Solid State Chem.*, 2007, **180**, 3583.
53. C. Drasar, A. Hovorkova, P. Lostak, S. Ballikaya, C.-P. Li and C. Uher, *J. Electron. Mater.*, 2010, **39**, 1760.
54. S. Wang, W. Xie, H. Li and X. Tang, *J. Phys. D: Appl. Phys.*, 2010, **43**, 335404.
55. A. Zimmer, N. Stein, H. Terryn and C. Boulanger, *J. Phys. Chem. Solids*, 2007, **68**, 1902.
56. S. Michel, S. Diliberto, N. Stein, B. Bolle and C. Boulanger, *J. Solid State Electrochem.*, 2008, **12**, 95.
57. S. Li, H. M. A. Soliman, J. Zhou, M. S. Toprak, M. Muhammed, D. Platzek, P. Ziolkowski and E. Müller, *Chem. Mater.*, 2008, **20**, 4403.
58. L. Bu, W. Wang and H. Wang, *Appl. Surf. Sci.*, 2007, **253**, 3360.

59. D. Bourgault, C. G. Garampon, N. Caillault, L. Carbone and J. Aymami, *Thin Solid Films*, 2008, **516**, 8579.
60. M. Takashiri, T. Shirakawa, K. Miyazaki and H. Tsukamoto, *Sens. Actuators, A*, 2007, **138**, 329.
61. N. Peranio, O. Eibl and J. Nurnus, *J. Appl. Phys.*, 2006, **100**, 114306.
62. J. Zhou, S. Li, H. Soliman, M. Toprak, M. Muhammed, D. Platzek and E. Müller, *J. Alloys Compd.*, 2009, **471**, 278.
63. M.-K. Han, H. Ryu and S.-J. Kim, *J. Electron. Mater.*, 2013, **42**, 2758.
64. F. Wu, H. Song, J. Jia, F. Gao, Y. Zhang and X. Hu, *Phys. Status Solidi A*, 2013, **210**, 1183.
65. J.-H. Yim, S.-H. Baek, H.-Y. Shin, D. B. Hyun and J.-S. Kim, *J. Electron. Mater.*, 2013, **42**, 2178.
66. K. C. Lukas, W. S. Liu, Z. F. Ren and C. P. Opeil, *J. Appl. Phys.*, 2012, **112**, 054509.
67. I. Yutsis, Ph.D. thesis, Tel Aviv University, 2010.
68. A. J. Naylor, E. Koukharenko, I. S. Nandhakumar and N. M. White, *Langmuir*, 2012, **28**, 8296.
69. M. Martín-González, G. J. Snyder, A. L. Prieto, R. Gronsky, T. Sands and A. M. Stacy, *Nano Lett.*, 2003, **3**, 973.
70. H. Köse, M. Biçer, Ç. Tütünoglu, A. O. Aydın and I. Sisman, *Electrochim. Acta*, 2009, **54**, 1680.

CHAPTER 9

Automotive Power Harvesting/ Thermoelectric Applications

STATES CHIWANGA,[a] RICHARD TULEY,[a] KATARZYNA PLACHA,[a] MARK ROBBINS,[a] BOB GILCHRIST[b] AND KEVIN SIMPSON*[a]

[a] European Thermodynamics Ltd, 8 Priory Business Park, Kibworth, Leicestershire LE8 0RX, UK; [b] Jaguar Land Rover Ltd, W/11/8 – Viscount Centre Unit C2, Warwick Science Park, Millburn Hill Rd, Coventry CV4 7HS, UK
*Email: kevin@etdyn.com

9.1 Why are Thermoelectric Devices Suitable for Automotive Applications?

Automotive applications for thermoelectric devices, including thermoelectric generators (TEGs), are growing in importance, driven by increasing legislative and commercial demands from end-users (vehicle owners, operators and drivers) to increase fuel efficiency and reduce emissions of carbon dioxide (CO_2), the main greenhouse gas. Within the European Union (EU), automotive OEMs are being driven to reduce emissions through the implementation of the Euro 6 legislation and the adoption of the voluntary agreement between the European Automobile Manufacturers Association (ACEA) and the European Commission. This provides a continuing drive for reductions in emissions, with average new car passenger car emissions having already fallen from around 160 gCO_2 km^{-1} in 2006 to 132 gCO_2 km^{-1} in 2012.[1] EU regulations will drive this further to 95 gCO_2 km^{-1} in 2020.[1]

RSC Energy and Environment Series No. 17
Thermoelectric Materials and Devices
Edited by Iris Nandhakumar, Neil M. White and Stephen Beeby

Published by the Royal Society of Chemistry, www.rsc.org

At a time when the estimated number of cars within the EU is 231 million[2] with a fuel consumption $\sim$150 billion litres each year, developing more fuel efficient vehicles with low emissions can have a significant impact.

For a typical gasoline fueled internal combustion engine (ICE) vehicle, only around 25% of the generated energy is used for mobility. The rest is lost in the form of waste heat, as well as friction and parasitic loses (Figure 9.1). Many new engines and vehicle technologies are being produced to make vehicle transportation more fuel efficient, and may be divided into one of two categories: energy conservation technologies or energy recovery technologies. Most currently implemented technologies to date, such as powertrain electrification and stop-start technologies, have addressed energy conservation only. While further energy conservation development is ongoing, including displacement on demand, variable compression ratio and parasitic load reduction through variable accessory drives such as oil and coolant pumps, energy recovery technologies are currently being increasingly considered. This is due to the fact that even low efficiency energy recovery can still have a significant impact on the overall efficiency due to the large amount of wasted energy. Energy recovery technologies include regenerative braking, kinetic energy recovery, turbo-generators/turbo-chargers and TEGs.

TEGs are suitable for automotive applications due to their modular nature, allowing compact, low weight systems whose addition has only a very small impact on vehicle performance. Their solid-state nature promises robustness, no added acoustic noise and minimal maintenance. Such a TEG could reduce or eliminate the electric generator's mechanical load on the engine and has been identified as a significant commercial opportunity. The addition of a thermoelectric system could lead to an overall 2% fuel efficiency improvement.[4] Although alternative technologies such as using the Organic Rankine Cycle do offer competitive efficiencies, they have not yet been able to demonstrate that these can be achieved in a low volume, low

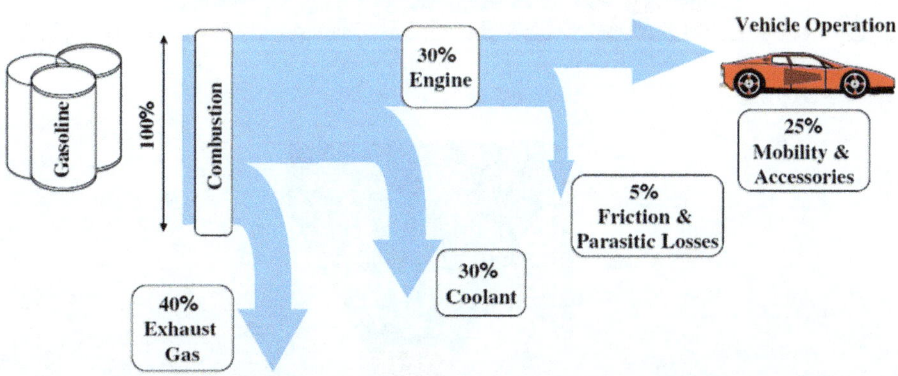

Figure 9.1 Energy diagram for a typical ICE vehicle.
Reproduced from J. Yang, Automotive Applications of Thermoelectric Materials, *J. Electron. Mater.*, **38**, Copyright 2009, with permission of Springer.[3]

weight system with sufficient tolerance to the many transients that can be experienced in automotive applications.

9.2 Automotive TEG Systems

An automotive TEG system typically converts waste heat in the exhaust system to produce electrical power that can be stored and/or utilized to operate subsystem within the vehicle. The TEG consists of several general components (Figure 9.2):

1. An exhaust pipe (hot) heat exchanger to capture and deliver heat from the exhaust gases to the hot side of TEG module. Heat transfer must be maximized while minimizing exhaust pipe back pressure.
2. A TEG module with associated power electronics to provide maximum power output of the TEG system throughout the vehicle drive cycle.
3. A cold heat exchanger to provide continuous cooling on the other part of TEG module by maintaining temperature differential ΔT between the two TEG module sides.
4. A bypass system to direct exhaust gases away from the exhaust pipe heat exchanger for periods of maximum required engine power (minimum back pressure), for when the vehicle energy storage system is at maximum capacity, the exhaust heat exchanger becomes blocked, *e.g.* icing or to avoid overheating the thermoelectric modules as necessary. The bypass system in its simplest form could be a bypass

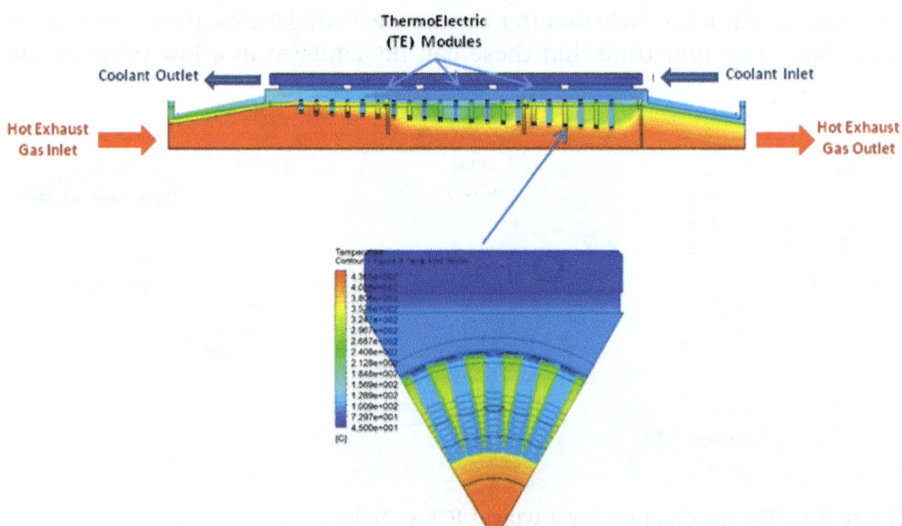

Figure 9.2 Side view of half a thermoelectric generation system showing the heat exchangers design, thermoelectric modules arrangement and cooling channels.

pipe arrangement with a means to divert the exhaust gases when the hot side temperature exceeds the thermoelectric modules maximum operating temperatures.

The main challenges for thermoelectric modules in such a system are electrical performance, cost and robustness. The modules must reach demanding high performance, low cost targets, with for example Toyota aiming for <$1 W^{-1} driving cost reductions and performance improvements.[4] Medium- or high-temperature modules using materials such as Silicides or Skutterudites offer potential lower cost, including stability in cost in high volume due to the security of supply and availability of the thermoelectric base material, as well as being lower toxicity materials compared to traditional bismuth telluride (BiTe). The ability to operate under higher temperatures, increases the power output from the module, as the TEG can operate across most of the speed/load range envelope, enabling the full use of the higher temperature differences which produce the most power. Therefore modules with maximum temperatures 400–500 °C are an important focus for these applications. To realize the potential of these materials, development of modules with excellent thermal and electrical contacts at the interfaces is critical, with a performance stable with time and resistant to oxidation, while ensuring that highly scalable, low cost processing techniques are used from raw materials all the way to completed systems. A key challenge of module integration with the exhaust system is to ensure excellent thermal contact that must accommodate fluctuations in thermo-mechanical strain between heat exchangers and the thermoelectric converters and variations in module sizes.

The robustness of thermoelectric modules can be challenged due to the different thermal expansion coefficient (CTE) between the different components, since stresses generated by CTE mismatch and thermal gradients lead to mechanical failure and thermoelectric modules experience performance reduction. Relatively poor thermal stability and potentially lowered mechanical strength at high temperatures, as well as difficulties in making thermally stable electrical contacts with low contact resistances need to be addressed at both the module and system level to minimize failure. In addition, systems in an automotive environment can experience many rapid mechanical and thermal changes (see Section 9.3.4) that may be partially transmitted onto the modules.

The first attempt at a TEG for automobiles was using the thermoelectric material $FeSi_2$ by the University of Karlsruhe (Germany) in 1988.[5] This system had a zT of 0.2 and produced a few Watts of power at 600 °C. In the USA, several organizations have tried to develop these technologies, such as Hi-Z, who have developed thermoelectric systems based on bulk Bi_2Te_3 for a large diesel truck engine that can produce 1 kW peak power, albeit for a very large 400 kW (500 hp) truck.[6]

A number of the large OEMs have shown interest in applying thermoelectric systems to automotive vehicles to produce electrical energy recovered from the waste heat. In the late 1990s, Nissan Motors, working with Hi-Z published the results of testing its TEG prototype which utilized SiGe thermoelectrics capable of producing 35.6 W peak power for a 3.0 L gasoline engine.[7] Advanced systems have also been developed by BMW and Ford in collaboration with the Department of Energy (DoE) in the US.[8] The prototype system revealed in 2010 has been heated by exhaust gases (before the mid-bed catalyst to improve available heat) and cooled using the engine cooling system and was able to generate up to 250 W peak power of electricity under normal driving conditions, equating to ~50% of the on-board electricity consumption in the BMW Series 5 sedan. They have investigated using bismuth telluride and a number of different high temperature materials in a segmented leg configuration. The use of segmented thermoelectric legs with different materials optimized for different temperatures allows for high efficiency operation over a wider temperature range, but can be more costly to produce. In addition, major challenges exist in achieving good contact resistances between the segments, and avoiding damage to the thermoelectric material due to the different thermal expansion coefficients of the different segments under the range of operating temperatures. As a result, the TEG has been welded solely to the hot side with the cold side attached using thermally conductive grease.

Fiat Thermo-Gen, Fraunhofer, Bosch and Valeo have undertaken work on various thermoelectrics as a part of the EC funded project HeatReCar[9] including BiTe, skutterudite and silicide materials, with a 500 W BiTe system demonstrated. Although this has not yet shown at full scale, HeatReCar concluded that silicides or skutterudites are expected to be the optimal choice in terms of cost (€ W^{-1}) and weight. Renault trucks have investigated silicide materials for exhaust waste heat recovery in the RENOTER project.[10] Recent work by the exhaust supplier Eberspacher[11] has demonstrated the good potential of the high temperature half-Heusler materials in a 1 kW system, although it is not clear if such Hafnium containing compounds can achieve sufficiently low material costs. Jaguar Land Rover and European Thermodynamics have produced a prototype thermoelectric system for automotive applications, with the VIPER project system shown in Figure 9.3, and are continuing development in the Innovate UK funded VIPER 2 project.[12]

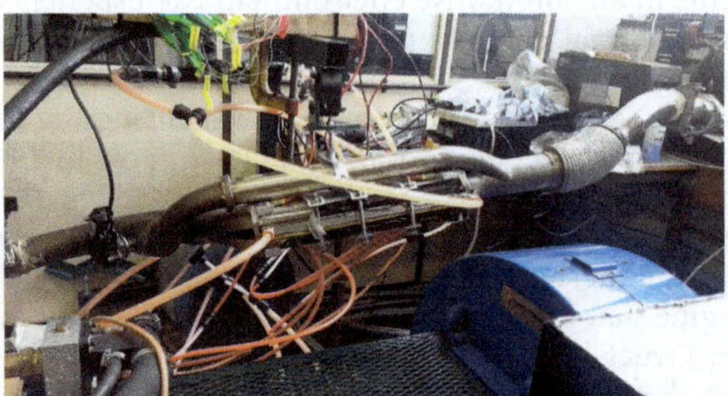

Figure 9.3 TEG system developed by European Thermodynamics and Jaguar Landrover in the VIPER project.

However, each of the revealed technologies is still at an early stage, especially the development of higher temperature modules. The cost to generate the power, and the system size, weight and back pressure of the units are still major concerns that OEMs need to overcome before they are able to bring the technology to the market. In addition the long-term thermal and mechanical behavior and failure modes of the devices is yet to be established.

9.3 Challenges/Trade-offs in Automotive TEGs

The challenges associated with development of a successful automotive TEG system are best discussed in terms of physical and functional categories. The main categories are:

1. conversion of heat energy into electrical energy (thermoelectric modules packaging),
2. heat energy capture and transfer (hot and cold heat exchangers),
3. exhaust gas flow control (valves and sensors),
4. overall mechanical packaging and operating environment,
5. TEG electrical output control measurement and control.

9.3.1 Thermoelectric Modules Packaging

Typical off-the-shelf thermoelectric modules are constructed as shown in Figure 9.4.

The various materials that form layers of the thermoelectric module need to be bonded together in a mechanically robust manner to ensure correct operation throughout the TEG operating regime and product life. For automotive applications this is particularly difficult because of the large environmental and operating temperature range −30 °C to >300 °C. The various material layers have different coefficient of linear thermal expansion (CTE) properties that inherently induce large thermal–mechanical stresses during operation. For example, the ceramic plate (aluminium oxide) has a CTE of 6.8×10^{-6} 1/K compared to copper at 16.6×10^{-6} 1/K and solder

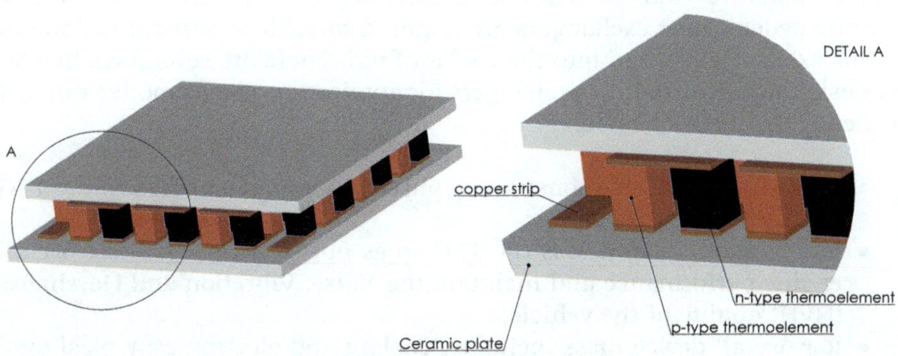

Figure 9.4 Typical thermoelectric module construction.

(95% Pb, 3.5% Sn, 1.5% Ag) has a CTE of 27×10^{-6} 1/K. As well as the required mechanical strength, the bonding material on the hot side of the thermoelectric module is expected to withstand relatively high temperatures (250–450 °C) while maintaining a low electrical resistance.

To generate the target electrical power output when subjected to temperature differential across the hot and cold sides, the thermoelectric modules need to be clamped at a pressure of the order of 1.2 MPa. Hence, the thermoelectric module needs to withstand this level of compressive pressure under all environmental and operating conditions. This is an additional challenge that requires addressing in designing an automotive TEG.

The case against the use of bismuth telluride thermoelectric pellets for the launch of commercial automotive TEG products is well documented.[4,13] Consequently, a lot of effort and investment is currently being applied to develop other commercially viable promising materials such as the half-Heuslers, skutterudites and silicides. One of the challenges for the TEG system mechanical design engineers is the lack of established mechanical and physical properties data for these newly developed thermoelectric materials. Quite often design teams and analysts have to wait for fundamental materials properties such as Young's modulus, thermal conductivity, electrical resistance, density, *etc.*, to be tested and verified by national and internationally established and traceable laboratories. Other challenges include selection of bonding materials between the copper strips and the thermoelectric pellets, as such bonding materials must not suffer from adverse diffusion/migration with time due to the high operating temperatures and the cyclic nature of the TEG operation. Moving away from bismuth telluride to a higher temperature material such as the half-Heuslers, skutterudites or silicides results in the exacerbation of any of these temperature related issues.

9.3.2 TEG Hot and Cold Heat Exchangers

For TEG systems to operate it is necessary to extract heat energy from the exhaust gas flow stream and transfer it to the TEG hot side surface. Similarly, it is also necessary to remove heat from the TEG cold side so as to provide the temperature gradient required to generate a current from the thermoelectric pellets. Heat exchangers are required to achieve heat extraction out of the exhaust gases and into the coolant fluid. There are several challenges to designing effective heat exchangers for application in automotive exhaust systems, the main ones being:

- restricted and highly constrained physical space in which to fit the TEG heat recovery elements;
- low pressure drop across the TEG so as not to adversely affect overall engine performance and maintain the Noise, Vibration and Harshness (NVH) quality of the vehicle;
- low overall device mass including coolant and electronics: typical total TEG mass needs to be less than 12 kg in a 2000 cc petrol engine saloon

vehicle, of which the largest (mass) part is allocated to heat exchangers and clamping mechanism;
- hot heat exchanger material must have a very high melting point to contain and operate exhaust gases temperature exceeding 850 °C for gasoline engines as well as have good thermal conductivity properties;
- provide large hot and cold flat usable surface areas on which to mount the TE modules to maximize TEG power output;
- heat exchangers need to be able to withstand high compressive pressure loads ($\approx$1.2 MPa);
- heat exchangers need to provide 'near-enough' flat surface temperature gradient from their inlet to the outlet;
- they need to have a high energy operating efficiency coefficient and must not foul or block exhaust gas flow throughout the TEG life;
- the impact of heating of the coolant in the cold side heat exchanger on the radiator or additional coolant circuit must be offset against the power generated.

Figure 9.5 shows typical Computational Fluid Dynamics (CFD) predicted temperature drops for a high aspect ratio rectangular section hot air heat exchanger. The hot heat exchanger has thin foil type off set fins that are uniformly distributed throughout the heat exchanger, whilst the cold plate has cylindrical pin fins. The fin type and density has an impact on the temperature distribution.

Figure 9.6 shows typical heat exchanger design configurations used in second generation TEG technology demonstrator prototype applications. The Figure 9.6(a) shows a hexagonal design with internal pin fins that extend into the hot exhaust gas flow stream to extract the heat and transfer it to the six flat surfaces where the thermoelectric modules are mounted. The Figure 9.6(b) shows a high aspect ratio rectangular design where off-set-fin (OSF) technology was used to extract heat from the hot exhaust gas flow stream. For a given overall volume, the high aspect ratio rectangular cross-section design provided a larger flat surface area on which to mount thermoelectric modules. However, it has a higher pressure drop for a given exhaust gas mass flow rate and lower compressive strength than the hexagonal design. This illustrates some of the challenges and trade-offs which require careful balancing associated with design of an optimal heat exchanger for application with automotive TEG systems.

The TEG invariably has to be mounted within the exhaust pipe systems path and layout, typically just downstream of the catalytic converter in order to reduce thermal inertia effects on catalytic converter light-off and quite often there is not much free space available. The available space is usually of irregular shape as shown in the example in Figure 9.7. The free space may even be further restricted, if the design has to be suitable for more than one vehicle, for example for easy scalability and introduction to different vehicles within an automotive manufacturers product line.

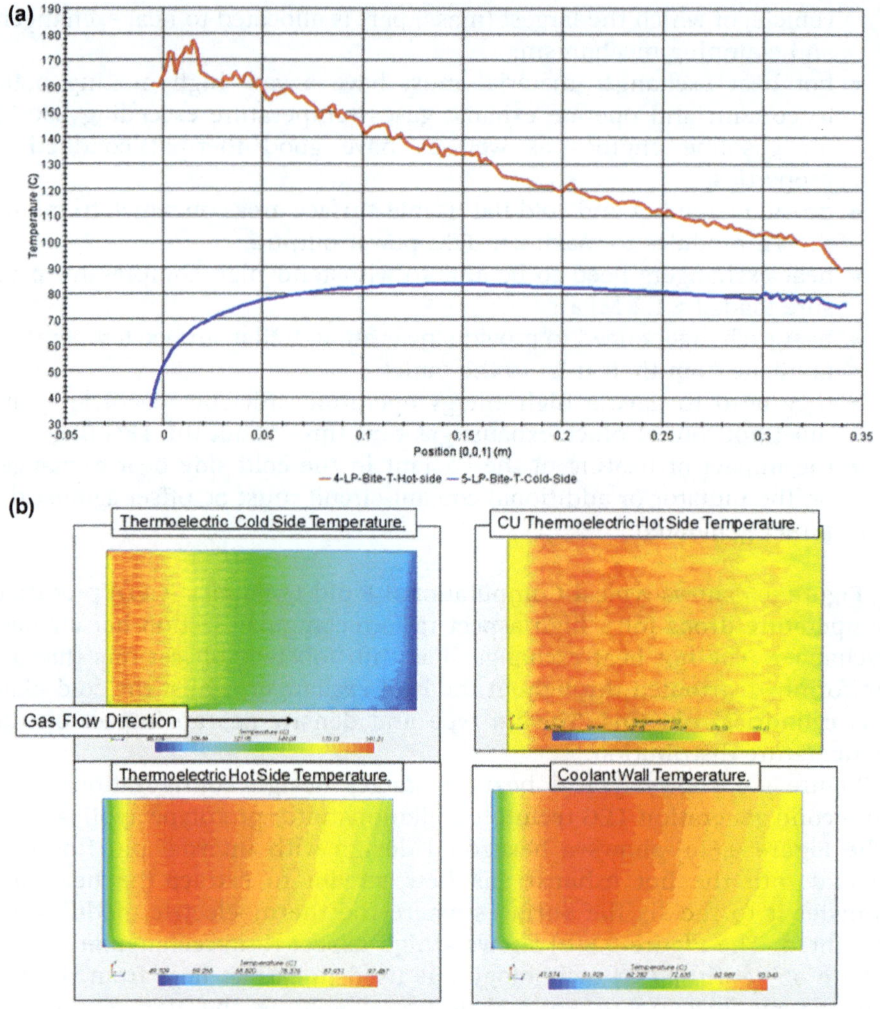

Figure 9.5 (a) CFD predicted thermoelectric temperature gradient of both hot (red) and cold (blue) sides of the thermoelectric modules along the direction of gas flow. (b) Temperature distribution across the hot and cold sides of a flat plate heat exchanger.

From Figure 9.7 it can be seen that the task of designing a TEG system that can retrospectively be fitted within an exhaust system is quite complex and requires that design compromises are brought into effect. In this example the space constraint drove the TEG heat exchanger to be of rectangular shape so as to optimize the available useful space to mount the thermoelectric module, house control valves and fit in bypass pipe-work.

Several design parameters have to be considered which influence the choice of the type of heat exchanger technology. Offset strip fins type,

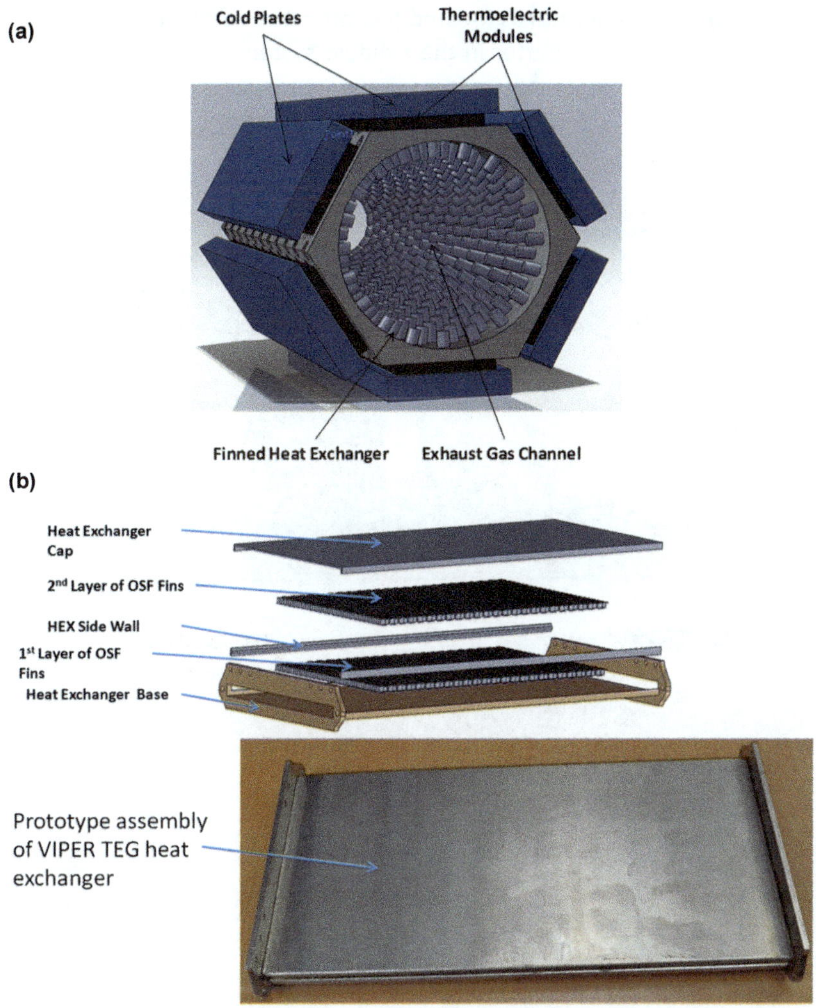

Figure 9.6 Typical heat exchanger designs developed for TEG applications, including a (a) hexagonal and a (b) flat plate design.

shown in Figure 9.8, are widely used in the automotive industry because they result in lightweight high-performance heat exchangers with high heat transfer rate relative to heat exchanger volume. These can therefore satisfy the challenging space claims in this industry. The fins create a greater pressure drop but also allow greater heat transfer coefficient. The fins dimensions, geometry and shapes are designed and selected to minimize exhaust system back pressure that would otherwise be detrimental to overall engine performance when the TEG system is fitted within the vehicle exhaust system.

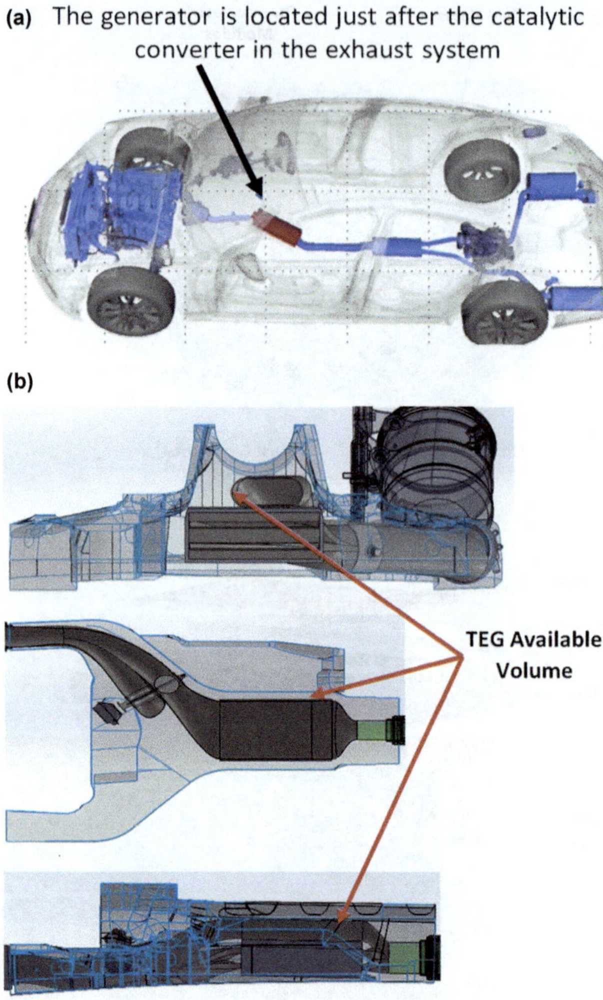

Figure 9.7 (a) Overall TEG system location. (b) Available restricted and irregular volume in which the TEG can be placed in the exhaust system (including ground clearance).

9.3.3 Exhaust Gas Flow Control (Valves and Sensors)

The thermoelectric modules have maximum operating temperature limits that can sometimes be exceeded depending on the regime within which the vehicle is going through on the drive cycle. The two drive cycles normally used to assess vehicle performance and subsystems are: New European Drive Cycle (NEDC) and the Worldwide harmonized Light vehicles Test Cycle (WLTC). In both drive cycles there are acceleration band/periods where the exhaust gas flow and temperature are very high such that the extracted heat energy by the heat exchanger results in one or more thermoelectric modules'

Figure 9.8 Offset Strip Fins used in the exhaust heat exchanger.

maximum temperature being exceeded. To avoid damage to the thermoelectric modules the exhaust gases need to be diverted to bypass the heat exchanger whenever these conditions occur.

On/Off control valves are used to divert the gas flow before it enters the TEG and return it back into the exhaust main stream once past the TEG system. The diverter control valves are either electrically or pneumatically powered and designed to fail safe such that in the event of power source failure the TEG will operate in bypass mode at all times so as to avoid damage to the TEG system. The two images shown in Figure 9.9 show examples of TEG bypass system designs as implemented in prototypes by Jaguar Land Rover and Tenneco.

The bypass control valves are operated by temperature sensors that are placed within the exhaust gas stream upstream of the inlet control valve. The temperature sensors can be thermocouples or thermistors that continually monitor the exhaust gas temperature and upon reaching a set threshold temperature, a signal is activated within the valve control system which causes the control valve to shut the TEG flow path and divert the exhaust gas *via* the bypass route. The TEG system is thus always protected from overheating and related damage. In some cases the control valves are proportionally controlled to achieve greater operating range. Proportionally controlled valves require additional cost and increase the valve complexity compared to on/off valves.

9.3.4 TEG Overall Mechanical Packaging

Overall mechanical packaging refers to TEG design features that make it possible for the assembly to operate as intended in all operating conditions as defined in the TEG specification; *e.g.* applying the correct pressure on thermoelectric modules, sustaining operation under shock and vibration conditions, maintaining Intrinsic Protection (IP rating) at all times, achieve intended product life span, easy product maintenance and serviceability, *etc.*

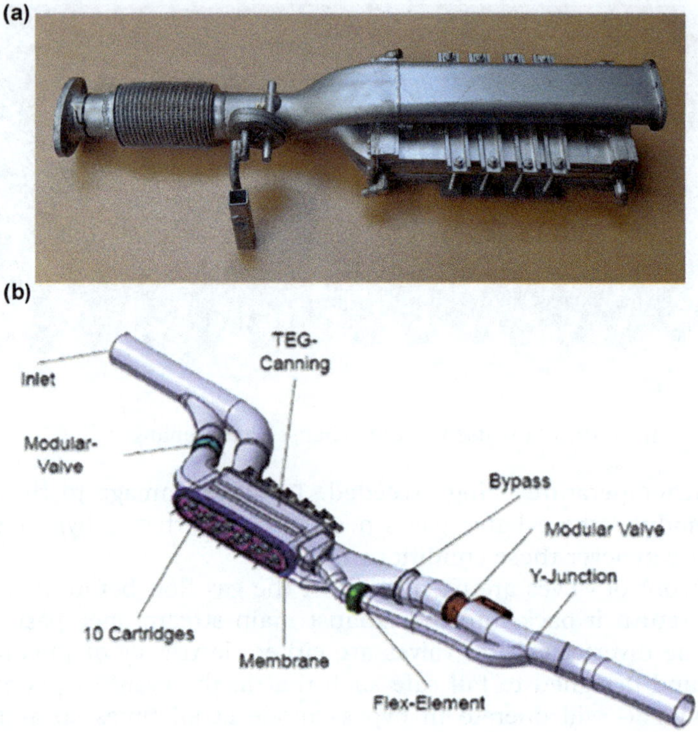

Figure 9.9 TEG Bypass Designs from (a) Jaguar Land Rover and (b) Tenneco (courtesy of Tenneco).

In this chapter only the most challenging aspects of the TEG system mechanical packaging will be discussed.

Applying the correct pressure on thermoelectric modules is important to ensure good thermal contact between the modules and heat exchangers, even when using an interface material. Achieving this is most challenging because the correct magnitude of pressure needs to be applied uniformly across a large surface area, over a very wide temperature range, in a system made out of materials that have a wide range of linear coefficient of expansion. It is therefore required to design a clamping mechanism that satisfies all of these challenges, as well as keeping the overall mass within acceptable limits. The difficulties of achieving this task are illustrated by some of the mechanical failure images shown in Section 9.4.

Automotive exhaust systems are subjected to some of the harshest environmental conditions in the car; like severe water splashes, partial immersion under muddy water, mechanical shocks and knocks from obstacles on rough roads, *etc.* The thermoelectric modules must be mechanically packaged in a design that they are able to survive these environmental and operating conditions to name but a few. Figure 9.10 shows several examples of ways that different automotive teams have tried to tackle these challenges. The wide range of designs illustrate how difficult this task is.

Figure 9.10 Four examples of complete TEG Mechanical Package Designs.

All but one of the four examples of mechanical packaging solutions shown in Figure 9.10 employ a lot of bolts and high strength beams to achieve the required pressure, and the only exception shows a welded together metal structure again emphasizing the need for a high integrity mechanical design. In the latter case, the cylindrical shape offers some good advantages in terms of uniform pressure distribution without any potential stress concentration points, however its surface area is a lot smaller and as such it is likely to produce less overall TEG power output than the other three. Note also that the cylindrical shape does not lend itself well to the space envelope shown in Figure 9.7, for example. This illustrates once again the challenges of automotive TEG systems.

9.3.5 TEG Electrical Output Control Measurement and Control

The electrical power output from the thermoelectric modules that make up the TEG is inherently variable and unstable because of the varying heat transfer coefficients and fluctuating exhaust gas temperatures and flow rates both along its length from the inlet to the outlet and with time. It is not acceptable nor desirable to have highly fluctuating power out from the TEG to feed the vehicle electrical rail as it will cause malfunction or potentially damage electronics hardware components. The fluctuating output power of automotive TEGs is similar to that which occurs in solar photovoltaic power supply systems. It is therefore essential to harness some of the experience from this industry and utilize it in TEG technology. It follows, therefore, that digital switch-mode power converters (SMPC) technology already widely implemented in solar power industry can be utilized in TEGs. The electronic system that carries out management of power output from TE modules is generically known as Electronic Control Unit (ECU). A common technique that has been adapted to stabilize and control output of thermoelectric modules in TEGs is a DC/DC converter functionality incorporating Maximum Power Point Tracking (MPPT) firmware. MPPT controllers come in various forms, one such technique that the authors have used in early stage development is shown in Figure 9.11.

The ECU integrates and ensures that the thermoelectric modules maximum power output is tracked and correlated to give the maximum TEG output power possible. Various implementations of this (MPPT) technique have been proposed or built for TEG systems.

9.4 Failure of Thermoelectric Modules in Automotive Applications

As discussed in Section 9.3.4, high uniform clamping of thermoelectric modules is required, under the range of temperatures and temperature

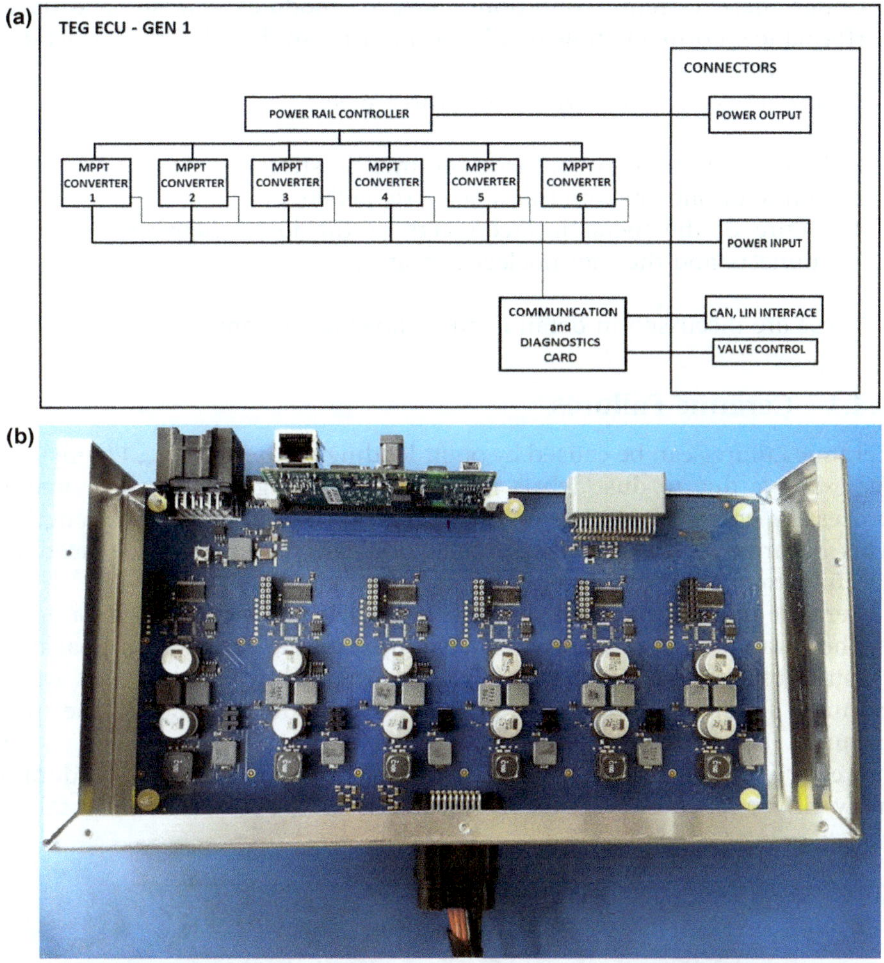

Figure 9.11 Electronic Control unit (a) schematic and (b) hardware, including several maximum power point trackers for different groups of thermoelectric modules.

gradients in operation. Modules can however experience excessive clamping forces if they undergo too high pressures during assembly or too high temperatures or temperature differences (for example uneven temperatures along the heat exchanger, heat spike transients or failure of the bypass).

Commercially available BiTe modules have been examined after exceeding their temperature specification in a TEG system, equivalent to the case of a bypass failure. This case considers an extreme temperature condition for the modules, but by examining and understanding the failure modes, this allows consideration of methods to avoid or reduce failures in long-term lower temperature operation in the field.

Under these extreme conditions, 34% of modules saw some drop in performance, normally indicated by an increase in their internal resistance after the test.

Failure modes can be divided into three types:

1. failure of the hot or cold side ceramic substrates;
2. failure of one or more thermoelectric pellets;
3. failure of the metal interconnects or the joints between the interconnects and the thermoelectric material.

These are discussed in detail in the following sections.

9.4.1 Ceramic Failures

Ceramic failures can be caused by point loading on the ceramic, Figure 9.12, for example due to dust/debris incorporated during the system assembly process or errors during module manufacture (*e.g.* uneven pellet lengths). This occurred in 18% of the failed modules, emphasizing the need for careful assembly of the system for long term reliability.

Ceramic failure can also occur due to uneven loading, or stress build up in the ceramic due to thermal expansion, resulting in cracks propagating through the ceramic from the edge or corners (Figure 9.13). This was a common failure, affecting 64% of the failed modules. However the most common failure was a pellet failure, affecting all the failed modules, so it is not clear if the pellets or the ceramic failed first. The failure is predominantly driven by thermal expansion issues, as any such ceramic failure

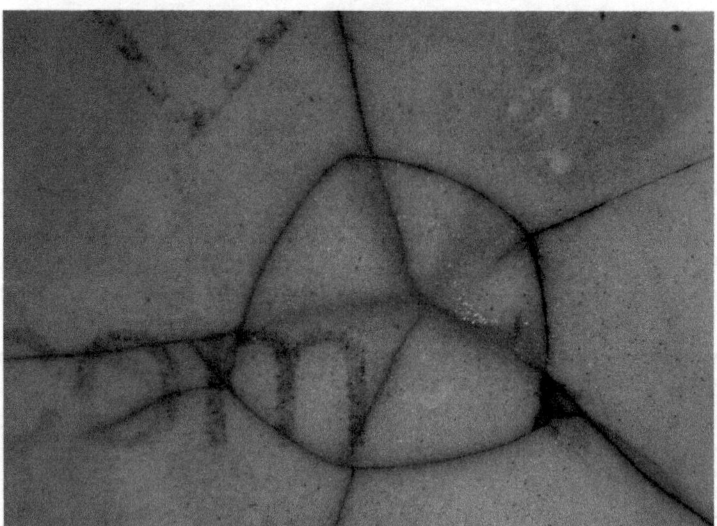

Figure 9.12 Point load failure of ceramic substrate.

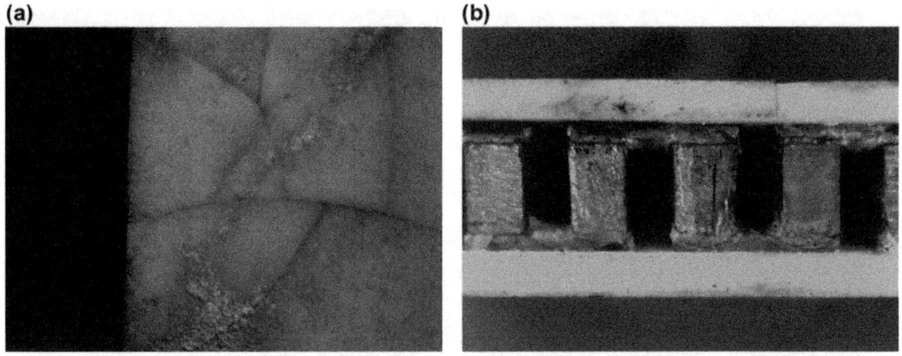

Figure 9.13 Optical images of the edge or corner cracking failure of ceramic substrates shown in (a) plan or in (b) cross section.

occurred in the hot side, and was occasionally additionally accompanied by failure of the cold side ceramic (in 43% of these failures). This form of ceramic failure can even propagate into the metallic interconnects.

9.4.2 Pellet Failures

At least some thermoelectric pellet failure was observed in all failed modules, typically manifested as the chipping of pellets, especially at the corners, and cracking of the pellets, predominantly vertically, with horizontal cracks occurring at the interface (Figure 9.14).

Typical commercial BiTe thermoelectric material is produced by vertical Bridgman or zone melting techniques. This typically leads to a mechanically poor, brittle material, with significant anisotropy in mechanical properties,[14] with many grain boundaries orientated vertically, and so vertical cracking occurs.

9.4.3 Interconnect and Joint Failures

Surprisingly, there was no indication of failure of the metallic diffusion barrier on the ends of the thermoelectric pellets, despite the high temperature, as indicated by the absence of copper in the thermoelectric material (Figure 9.15).

However, some oxidation of the solder joints has occurred, and some solder reflow has possibly occurred due to the high temperature, resulting in drooping of the solder, and in a few cases even wetted across the pellet, shorting it. No failure of the metal interconnect to ceramic bond was observed at these temperatures.

9.4.4 Failure Modelling

This is a complex system to model, due to the range of materials and interfaces involved. A simplified finite element analysis model of a thermoelectric module was constructed in COMSOL Multiphysics to highlight

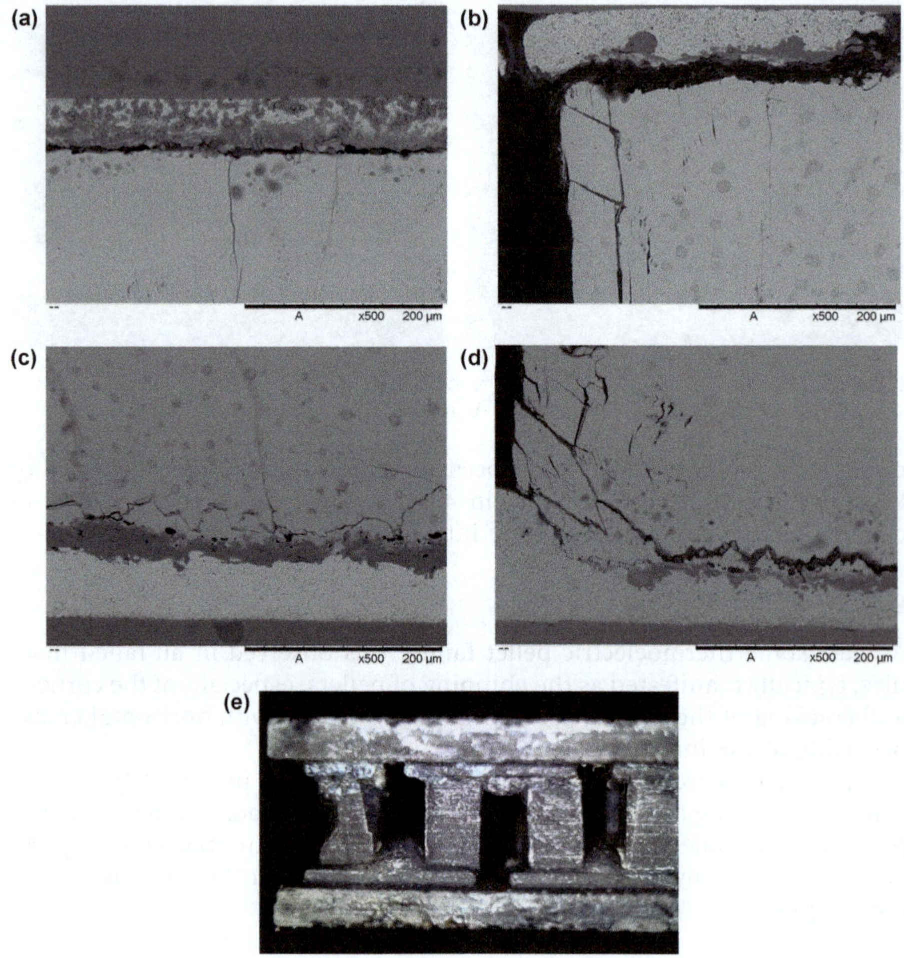

Figure 9.14 Thermoelectric pellet failures: SEM cross sections of a failed module showing (a), (b) vertical cracks through the pellet and (c), (d) Horizontal cracking in the pellet near the interface. (e) Shows an optical image of the module showing an example of chipping of the thermoelectric pellet.

some of the important aspects. A number of significant assumptions were made:

- boundary conditions fixed as 80 °C cold side and 300 °C hot side;
- lateral expansion fixed by the thermal expansion of the surrounding steel system, with no net vertical displacement allowed;
- no pre-stress from clamping or stress induced during module manufacture at elevated temperatures included;
- linear isotropic materials were assumed with no difference in strength in tension or compression.

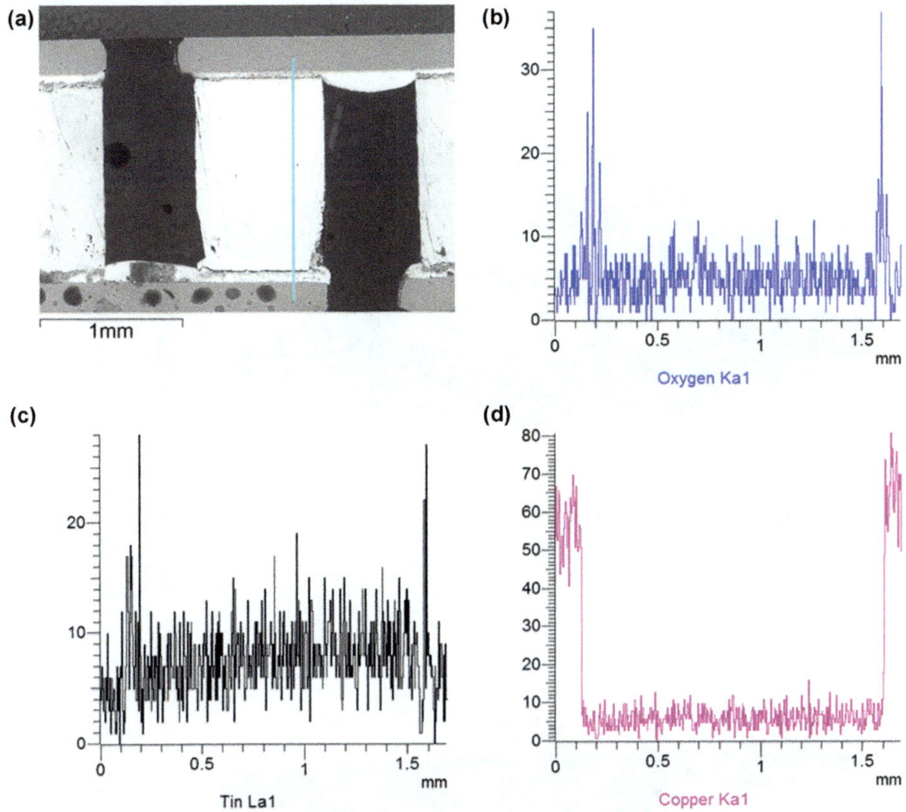

Figure 9.15 (a) SEM image and line scan position. EDX line scan for (b) oxygen, (c) tin and (d) copper illustrate their percentage composition.

These assumptions tend to overestimate the thermal stresses, but can be used to examine the most problematic areas. The Von Mises stress averaged over a 0.02 mm radius sphere divided by the yield stress of the material was calculated to account for the different materials and to avoid results being distorted due to meshing differences around any thermal stress singularities. The results in Figure 9.16 show that the most significant areas of concern are the pellets and then the hot side ceramic, consistent with the experimental results. The model successfully predicts that corners of the pellets are significant stress concentrators, especially at the edges of the module. Improvements in the mechanical properties of the thermoelectric material or methods to relax its thermal expansion would therefore have a significant impact on the robustness of the modules. For a closer match between the model and experiment, further modeling relaxing some of the assumptions listed would be required.

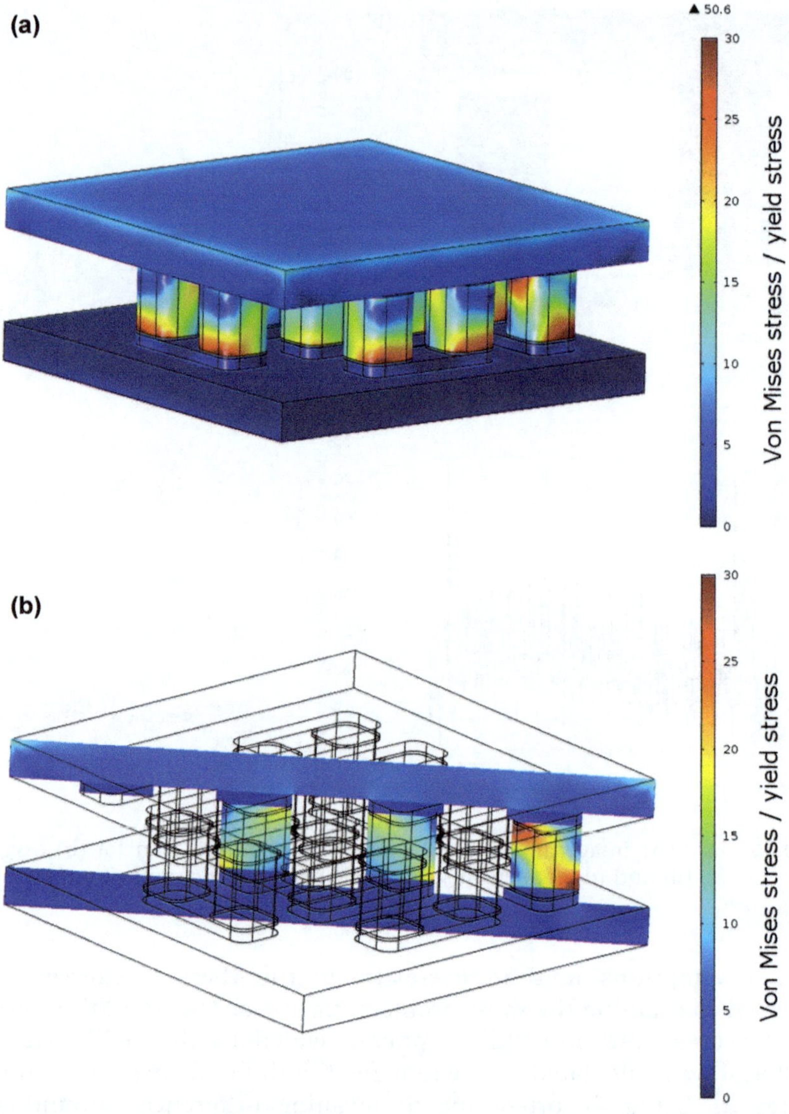

Figure 9.16 (a) The modelled Von Mises stress/yield stress for an example small thermoelectric module. (b) A diagonal slice showing Von Mises stress/yield stress through the module.

9.4.5 Failure Summary

Exceeding maximum temperature specifications on a module applies severe conditions to it. Failure of the module occurs mostly due to thermal expansion mismatches, with significant cracking of pellets and ceramics. This tends to lead to open circuit failures, especially at corners of the module. The maximum temperature limit is also fixed by the solder temperature, while interdiffusion of elements at the contact has not been observed as a problem in these short term high temperature tests. These results show that while thermoelectric devices are robust solid state generators, further improvements in robustness would be advantageous in an automotive environment. Further tests could also be performed under different conditions to test different aspects of the module robustness.

References

1. http://www.theicct.org/sites/default/files/publications/ICCTupdate_EU-95gram_jan2014.pdf (last accessed January 2016).
2. http://ec.europa.eu/eurostat/statistics-explained/index.php/Passenger_cars_in_the_EU (last accessed January 2016).
3. J. Yang and F. R. Stabler, *J. Electron. Mater.*, 2009, **38**(7), 1245.
4. Y. Iriyama, Heat Recovery Technology for fuel efficiency improvement, 4th IAV Thermoelectrics Conference, 2014.
5. U. Birkholz, E. Gross, U. Stohrer and K. Voss, Proceedings of 7th International conference on Thermoelectric Energy Conversion, 1988, **124**.
6. J. C. Bass, N. B. Elsner and F. A. Leavitt, *AIP Conf. Proc.*, 1995, **316**, 295.
7. M. Munekiyo, K. Furuya, M. Kobayashi, T. Izumi and K. Shinohara, Thermoelectric module and generator for gasoline engine vehicles, *Int. Conf. Thermoelectr.*, 1998, 464–498.
8. L. Bell and J. LaGrandeur, Automotive Waste Heat Conversion to Power Program-2010 Vehicle Technologies Program Annual Merit Review, 2010, available at: http://www1.eere.energy.gov/vehiclesandfuels/pdfs/merit_review_2010/solid-state_energy_conv/ace051_lagrandeur_2010_o.pdf (last accessed January 2016).
9. HEATRECAR Final report summary, 2013, available at http://cordis.europa.eu/result/rcn/58791_en.html (last accessed January 2015).
10. L. Aixala, RENOTER Project, 3rd Thermoelectric Applications Workshop, 2012.
11. L. Zhang, M. Cleary, X. Wang *et al.*, *Energy Convers. Manage.*, 2015, **105**, 946.
12. A. Agurto-Goya, S. Chiwanga and P. Shayler, Modelling and optimization of thermoelectric generator systems for automobiles, Vehicle Thermal Management Systems Conference, 2015.
13. A. Miner, Industrialization of Thermoelectrics: Materials, Modules and Manufacturing Scalability, 4th IAV Thermoelectrics Conference, 2014.
14. H. R. Williams, R. M. Ambrosi, K. Chen, U. Friedman, H. Ning, M. J. Reece, M. C. Robbins, K. Simpson and K. Stephenson, *J. Alloys Compd.*, 2015, **626**, 368.

Subject Index